Geophysical Fluid Dynamics for Oceanographers

Joseph J. von Schwind

Naval Postgraduate School
Monterey, California

Geophysical Fluid Dynamics for Oceanographers

Prentice-Hall, Inc., Englewood Cliffs, NJ 07632

Library of Congress Cataloging in Publication Data

Von Schwind, Joseph J.
Geophysical fluid dynamics for oceanographers.

Bibliography: p.
Includes index.
1. Ocean currents. 2. Ocean waves.
3. Ocean-atmosphere interaction. I. Title.
GC239.V66 551.4′7 79-23599
ISBN 0-13-352591-0

Interior design and editorial/production supervision by STEVEN BOBKER
Cover design: Edsal Enterprises
Manufacturing buyers: GORDON OSBOURNE and JOYCE LEVATINO

Printed in the United States of America

10 9 8 7 6 5 4 3 2 1

Prentice-Hall International, Inc., *London*
Prentice-Hall of Australia Pty. Limited, *Sydney*
Prentice-Hall of Canada, Ltd., *Toronto*
Prentice-Hall of India Private Limited, *New Delhi*
Prentice-Hall of Japan, Inc., *Tokyo*
Prentice-Hall of Southeast Asia Pte. Ltd., *Singapore*
Whitehall Books Limited, *Wellington, New Zealand*

To
Charles L. Taylor,
a friend

Contents

TWO **Basic Concepts and Principles in Geophysical Fluid Dynamics** 21

Preface

In teaching introductory geophysical fluid dynamics at the Naval Postgraduate School, I could not find a textbook to suit the needs of graduate students in my oceanography programs.

As is the case in many environmental science departments, the academic background of the students varied greatly. In most instances the students had absolutely no formal education in fluid dynamics beyond undergraduate general physics. Also, for most students, quite a few years had passed since they had completed their undergraduate work. These difficulties probably exist in other similar programs.

My problem was to find a textbook directed to the fluid dynamics of the ocean, suitable for beginning graduate students who were academically capable but "rusty," of sufficient mathematical complexity to be challenging but not overwhelming, and of such detail in exposition as to be convincing. No available single text seemed to meet these requirements and consequently, as many teachers before have done, I began writing and distributing detailed lecture notes. These notes were well enough received so as to encourage me to formalize the effort. This book is the outcome.

The reader will find nothing original in the material covered. I have attempted to assemble and discuss the subject matter in what I believe to be a clear and cohesive manner, capable of being understood to persons with technical backgrounds even though they may have no formal education in theoretical physical oceanography. Further, it has not be my intent to exhaust any single topic, but to make available sufficient material, of a fundamental nature, to serve as an intelligent starting point for independent and more advanced study.

My debt to others is limitless: to my colleagues Robert O. Reid, from whom a part of the material on the hydrothermodynamics of a binary fluid was obtained, Worth D. Nowlin, Jr., Jerry Galt, and Dale F. Leipper; to the many students who have endured my lectures; to the dozen students who have allowed me to work out on the handball court the frustrations resulting from the sedentary job of authorship; to Carol J. Sams for typing the draft of the text; and to my wife Catherine Ann.

It was possible to write this book due to the funding provided by the Naval Sea Systems Command represented by John F. Ropek. This most essential contribution allowed relief from teaching duties and provided the time needed to put it all together.

Joseph J. von Schwind
Monterey, California

ONE

Review of Vectors and Cartesian Tensors

1-1 Introduction

Physical quantities that are encountered in geophysical fluid dynamics can be categorized as scalars, vectors, or second-order tensors. For example, we have as scalars such quantities as time, temperature, volume, and energy; as vectors the quantities of force, velocity, acceleration, and momentum; and as second-order tensors the quantities of shear stress and momentum flux. One of the methods by which we can distinguish among these quantities is as follows:

$$\text{Scalar} = \alpha, s, T$$

$$\text{Vector} = \mathbf{V}, \mathbf{a}, \mathbf{F} \text{ (boldface)}$$

$$\text{Tensor} = \boldsymbol{\mathcal{P}}, \boldsymbol{\mathcal{D}} \text{ (boldface script)}$$

For Cartesian tensors in general (where a scalar can be considered as a tensor of order zero and a vector as a tensor of order one) it is possible to define four different kinds of multiplication: that of the usual arithmetic type and that of three special types: the "dot" product $\cdot$, the "cross" product $\times$, and the "double dot" product :. Typically, we have

$$(b)\,(B) = \text{scalar}$$
$$\mathbf{n} \cdot \mathbf{F} = \text{scalar}$$
$$\boldsymbol{\Omega} \times \mathbf{V} = \text{vector}$$
$$\mathfrak{T} : \mathfrak{D} = \text{scalar}$$

The rules for determining the order of the result of any of these four operations can be summarized as follows:

Multiplication Type	*Order of Result*
None	Σ
$\times$	$\Sigma - 1$
$\cdot$	$\Sigma - 2$
:	$\Sigma - 4$

where Σ is the summation of the order of the terms entering into the operation. Thus, aB is of order $0 + 0 = 0$; $\mathbf{n} \cdot \mathbf{F}$ is of order $1 + 1 - 2 = 0$; $\boldsymbol{\Omega} \times \mathbf{V}$ is of order $1 + 1 - 1 = 1$; and $\mathfrak{T} : \mathfrak{D}$ is of order $2 + 2 - 4 = 0$.

When it is stated that a certain set of quantities is a tensor of order n, it is meant that ways are available to specify its components with respect to any set of axes, and that the components with regard to any two sets of axes are related according to a transformation rule appropriate to tensors of that order. In general, a tensor of any order remains invariant and transforms into itself under a rotation of the coordinate system. The particular rules that apply will not be discussed at this point (see Section 1-17).

1-2 Indicial Notation

Let us introduce the unit vectors $\mathbf{1}_1$, $\mathbf{1}_2$, and $\mathbf{1}_3$, referred to the rectangular Cartesian axes shown in Figure 1-1. Using classical vector notation, any vector, such as $\mathbf{V}$ situated at an arbitrary point in space P, can be written as

$$\mathbf{V} = V_1\mathbf{1}_1 + V_2\mathbf{1}_2 + V_3\mathbf{1}_3 \tag{1-1}$$

where V_1, V_2, and V_3 are the rectangular projections of $\mathbf{V}$ in the x_1, x_2, and x_3 directions, respectively. The Cartesian components V_1, V_2, and V_3 may be represented in *indicial* (*subscript*) notation by the symbol V_i ($i = 1, 2, 3$), where the subscript i is understood to take on the values 1, 2, and 3 in that order. Therefore, the symbol V_i represents the set of three Cartesian components V_1, V_2, and V_3 in the same order. In index notation V_i is the symbol used to designate the vector $\mathbf{V}$ in ordinary three-dimensional space. It completely characterizes and therefore represents the vector $\mathbf{V}$. When particularly

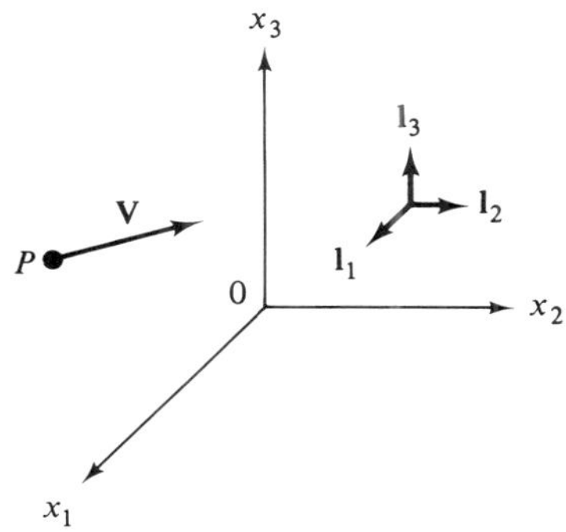

Fig. 1-1. Rectangular Cartesian axes in indicial notation.

appropriate we will still have recourse to the standard notation of **i**, **j**, and **k** as unit vectors referred to the rectangular Cartesian axes x, y, and z as in Figure 1-2. In this case, entirely equivalent to (1-1), we can write

$$\mathbf{V} = V_x\mathbf{i} + V_y\mathbf{j} + V_z\mathbf{k} \tag{1-2}$$

where again V_x, V_y, and V_z are the rectangular projections of $\mathbf{V}$ in the x, y, and z directions.

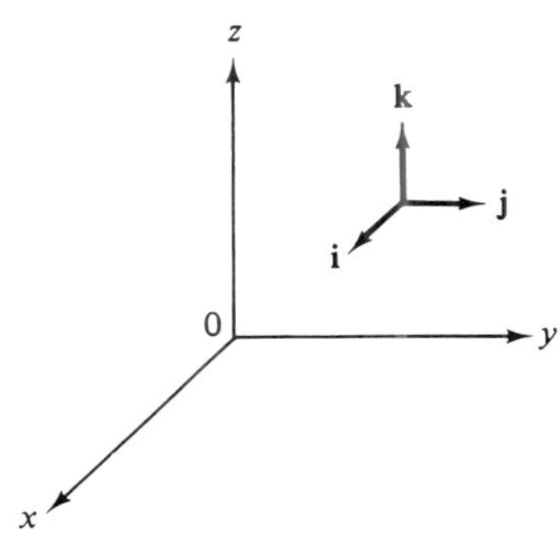

Fig. 1-2. Rectangular Cartesian axes in conventional notation.

With the foregoing in mind, we now introduce a shorthand notation for representing the terms of an equation and the equations in which they appear. These rules, which are valid for tensors of all orders, are the range convention and the summation convention.

Range Convention

Whenever a small Latin subscript occurs *unrepeated* in a term, it is understood to take on the values 1, 2, 3 (unless stated otherwise), the number of dimensions of physical space. We then have a representation for a set of numbers, terms, or equations. For example,

$$F_i = ma_i \qquad (i = 1, 2, 3)$$

implies the single vector equation

$$F_1\mathbf{1}_1 + F_2\mathbf{1}_2 + F_3\mathbf{1}_3 = m(a_1\mathbf{1}_1 + a_2\mathbf{1}_2 + a_3\mathbf{1}_3)$$

or

$$\mathbf{F} = m\mathbf{a}$$

or

$$F_x\mathbf{i} + F_y\mathbf{j} + F_z\mathbf{k} = m(a_x\mathbf{i} + a_y\mathbf{j} + a_z\mathbf{k})$$

or the three equivalent scalar equations.

It is then possible to define the order of a tensor as being equal to the number of range indices appearing. Thus,

$$\begin{aligned} B &= \text{zero-order tensor (a scalar, 1 component)} \\ F_i &= \text{first-order tensor (a vector, 3 components)} \\ \tau_{ij} &= \text{second-order tensor (9 components)} \\ \epsilon_{ijk} &= \text{third-order tensor (27 components)} \end{aligned}$$

The second-order tensor τ_{ij} having 9 components can be written out as the array

$$\tau_{ij} = \begin{bmatrix} \tau_{11} & \tau_{12} & \tau_{13} \\ \tau_{21} & \tau_{22} & \tau_{23} \\ \tau_{31} & \tau_{32} & \tau_{33} \end{bmatrix}$$

where i stands for the number of the row and j stands for the number of the column. The square brackets are used here to distinguish between the 3×3 tensor and the 3×3 determinant of the tensor, which will be encountered later.

Summation Convention

Whenever a small Latin subscript occurs *twice* in a term, it is understood to represent a summation over the range 1, 2, 3 (unless stated otherwise). Thus,

$$\tau_{ii} \equiv \sum_{i=1}^{3} \tau_{ii} \equiv \tau_{11} + \tau_{22} + \tau_{33}$$

It is to be noted also that in this example the number of unrepeated subscripts is zero and application of the range convention tells us that τ_{ii} is a scalar.

1-3 The Kronecker Delta and the Alternating Unit Tensor

Two special tensors that appear repeatedly, because of the fact that many equations can be expressed more compactly through their use, are the *Kronecker delta* or *unity matrix* and the *alternating unit tensor* or *permutation symbol.*

The Kronecker Delta

The Kronecker delta is a second-order tensor symbolized by δ_{ij} and defined as

$$\delta_{ij} = \begin{cases} 1 & \text{if } i = j \\ 0 & \text{if } i \neq j \end{cases} \tag{1-3}$$

or

$$\delta_{ij} = \begin{bmatrix} 1 & 0 & 0 \\ 0 & 1 & 0 \\ 0 & 0 & 1 \end{bmatrix}$$

The Alternating Unit Tensor

The alternating unit tensor is of third order, symbolized by ϵ_{ijk}, and defined as

$$\epsilon_{ijk} = \begin{cases} +1 & \text{if } ijk = 123, 231, 312 \text{ (all different indices and even permutations of 123)} \\ -1 & \text{if } ijk = 321, 132, 213 \text{ (all different indices and odd permutations of 123)} \\ 0 & \text{if any two indices are alike} \end{cases} \tag{1-4}$$

1-4 Addition and Subtraction of Vectors and Tensors

The addition or subtraction of two Cartesian tensors of the same order is defined by

$$V_{ij\ldots} \pm U_{ij\ldots} = W_{ij\ldots} \tag{1-5}$$

where $V_{ij\ldots}$, $U_{ij\ldots}$, and $W_{ij\ldots}$ are all of the same order. This definition implies that the addition or subtraction is to be carried out for each pair of corresponding elements of $V_{ij\ldots}$ and $U_{ij\ldots}$. It is to be noted that the indices of each term of (1-5) are the same, thus satisfying the general requirement that all equations must be homogeneous in range indices.

Equation (1-5) clearly holds for vectors,

$$V_i \pm U_i = W_i$$

or

$$\mathbf{V} \pm \mathbf{U} = \mathbf{W}$$

the geometric interpretation of which is illustrated via the familiar parallelogram construction. For $\mathbf{V} + \mathbf{U} = \mathbf{W}$ see Figure 1-3, and for $\mathbf{V} - \mathbf{U} = \mathbf{W}$ see Figure 1-4.

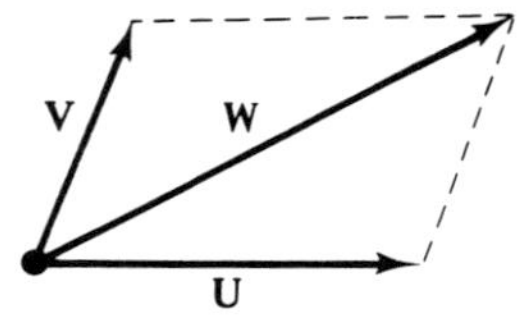

Fig. 1-3. Parallelogram summing of two vectors.

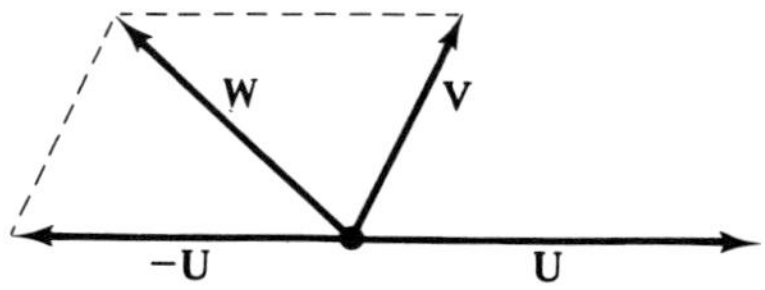

Fig. 1-4. Parallelogram subtraction of two vectors.

1-5 The Dot Product of Two Vectors

The dot (or scalar) product of two vectors **V** and **U**, denoted by **V** • **U**, is a scalar defined by

$$\mathbf{V} \cdot \mathbf{U} = |\mathbf{V}||\mathbf{U}| \cos \theta = VU \cos \theta \tag{1-6}$$

where $|\mathbf{V}|$, $|\mathbf{U}|$, V, and U are the magnitudes of the vectors **V** and **U** defined as $(V_1^2 + V_2^2 + V_3^2)^{1/2}$ and $(U_1^2 + U_2^2 + U_3^2)^{1/2}$, respectively, and θ is the angle (less than 180°) between **V** and **U**. From vector analysis we recall that

$$\mathbf{V} \cdot \mathbf{U} = V_1U_1 + V_2U_2 + V_3U_3$$

or in index notation

$$\mathbf{V} \cdot \mathbf{U} = V_iU_i$$

To find the component of a vector V_i in a given direction, it is only necessary to take the dot product of the vector with a unit vector n_i in the given direction. Thus, the component of V_i in the direction of n_i is V_in_i, where n_i satisfies the condition

$$n_in_i = 1$$

Finally, recall that the dot product is commutative,

$$\mathbf{V} \cdot \mathbf{U} = \mathbf{U} \cdot \mathbf{V}$$

is not associative,

$$(\mathbf{V} \cdot \mathbf{U})\mathbf{W} \neq \mathbf{V}(\mathbf{U} \cdot \mathbf{W})$$

but is distributive,

$$\mathbf{V} \cdot (\mathbf{U} + \mathbf{W}) = (\mathbf{V} \cdot \mathbf{U}) + (\mathbf{V} \cdot \mathbf{W})$$

1-6 The Cross Product of Two Vectors

The cross (or vector) product of two vectors **V** and **U**, denoted by **V** × **U**, is a vector **W** that is normal to the plane defined by **V** and **U** and points in the direction of advance of a right-handed screw when rotated from **V** to **U** by the shortest route. The magnitude of **W** is given by

$$|\mathbf{W}| = |\mathbf{V}||\mathbf{U}| \sin \theta$$

where θ is the smaller of the two angles between $\mathbf{V}$ and $\mathbf{U}$. From vector analysis it is known that the Cartesian components of $\mathbf{W}$ are related to the Cartesian components of $\mathbf{V}$ and $\mathbf{U}$ as follows:

$$\begin{aligned} W_1 &= V_2U_3 - V_3U_2 \\ W_2 &= V_3U_1 - V_1U_3 \\ W_3 &= V_1U_2 - V_2U_1 \end{aligned} \tag{1-7}$$

and that

$$\mathbf{V} \times \mathbf{U} = \begin{vmatrix} \mathbf{1}_1 & \mathbf{1}_2 & \mathbf{1}_3 \\ V_1 & V_2 & V_3 \\ U_1 & U_2 & U_3 \end{vmatrix}$$

Using the permutation symbol ϵ_{ijk}, (1-7) is written in index notation as

$$W_i = \epsilon_{ijk} V_j U_k \tag{1-8}$$

The cross product can be geometrically interpreted in that the magnitude of $\mathbf{W}$ is just the area of the parallelogram defined by $\mathbf{V}$ and $\mathbf{U}$ (Figure 1-5).

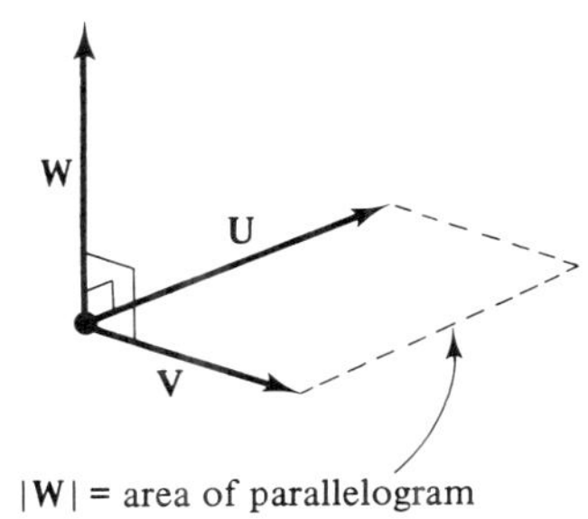

Fig. 1-5. Geometric interpretation of the cross product.

Concerning the laws of algebra, the cross product is not commutative,

$$\mathbf{V} \times \mathbf{U} = -\mathbf{U} \times \mathbf{V}$$

is not associative,

$$\mathbf{U} \times (\mathbf{V} \times \mathbf{W}) \neq (\mathbf{U} \times \mathbf{V}) \times \mathbf{W}$$

but is distributive,

$$(\mathbf{V} + \mathbf{U}) \times \mathbf{W} = \mathbf{V} \times \mathbf{W} + \mathbf{U} \times \mathbf{W}$$

1-7 The Gradient of a Scalar Field

The gradient of a scalar function of position $\phi(x_1, x_2, x_3)$ in a region R is expressed as

$$\text{grad}\,\phi \equiv \nabla\phi \equiv \frac{\partial\phi}{\partial x_1}\mathbf{1}_1 + \frac{\partial\phi}{\partial x_2}\mathbf{1}_2 + \frac{\partial\phi}{\partial x_3}\mathbf{1}_3 \tag{1-9}$$

where the *del operator* (or *nabla*) is defined by

$$\nabla \equiv \mathbf{1}_1 \frac{\partial}{\partial x_1} + \mathbf{1}_2 \frac{\partial}{\partial x_2} + \mathbf{1}_3 \frac{\partial}{\partial x_3}$$

The del operator is a vector operator; it has components as has a vector and it cannot stand alone; it must operate on a scalar, a vector, or a tensor. From (1-9) it is apparent that the gradient operation on a scalar can be expressed in index notation as

$$\nabla\phi \equiv \frac{\partial \phi}{\partial x_i} \qquad (1\text{-}10)$$

Consider a given equiscalar surface

$$\phi(x_1, x_2, x_3) = \text{constant}$$

By assigning different values to ϕ, a systematic set of surfaces are defined which describe the space distribution of the variable ϕ. As illustrated in Figure 1-6, the geometrical spacing of these surfaces, when drawn for equal

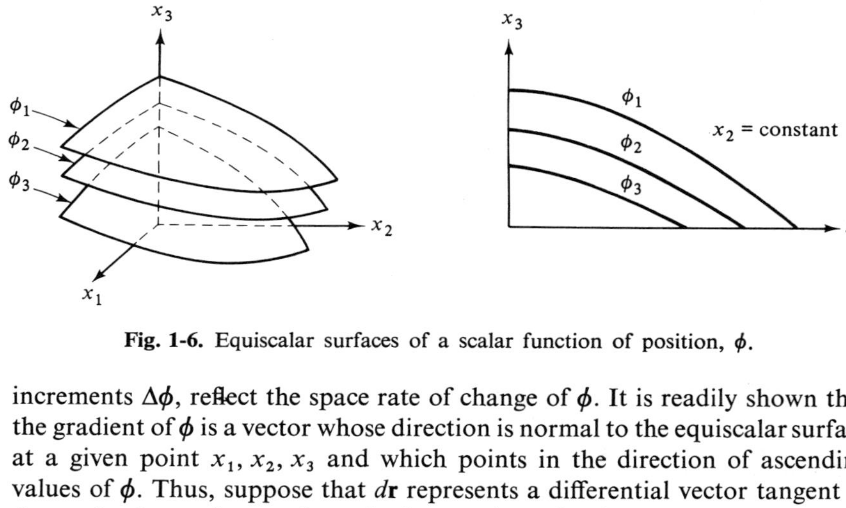

Fig. 1-6. Equiscalar surfaces of a scalar function of position, ϕ.

increments $\Delta\phi$, reflect the space rate of change of ϕ. It is readily shown that the gradient of ϕ is a vector whose direction is normal to the equiscalar surface at a given point x_1, x_2, x_3 and which points in the direction of ascending values of ϕ. Thus, suppose that $d\mathbf{r}$ represents a differential vector tangent to the equiscalar surface at the point in question; that is,

$$d\mathbf{r} = \mathbf{1}_1 dx_1 + \mathbf{1}_2 dx_2 + \mathbf{1}_3 dx_3$$

where dx_1, dx_2, and dx_3 are related through the equation of the surface

$$\phi(x_1, x_2, x_3) = \text{constant}$$

Differentiating the preceding expression, we obtain

$$d\phi = \frac{\partial \phi}{\partial x_1} dx_1 + \frac{\partial \phi}{\partial x_2} dx_2 + \frac{\partial \phi}{\partial x_3} dx_3 = 0$$

which can be written as

$$d\mathbf{r} \cdot \nabla\phi = 0$$

This states that $\nabla\phi$ is perpendicular to $d\mathbf{r}$ and therefore normal to the equiscalar surface at the point in question.

The magnitude of $\nabla\phi$ represents the maximum space rate of change of ϕ at the point x_1, x_2, x_3. If n is the distance measured normal to the equiscalar surface, then

$$|\nabla\phi| = \frac{d\phi}{dn} = \left[\left(\frac{\partial\phi}{\partial x_1}\right)^2 + \left(\frac{\partial\phi}{\partial x_2}\right)^2 + \left(\frac{\partial\phi}{\partial x_3}\right)^2\right]^{1/2}$$

1-8 The Divergence of a Vector Field

In classical vector notation the divergence of a vector function of position in a region R is defined as the scalar function of position,

$$\operatorname{div} \mathbf{V} \equiv \nabla \cdot \mathbf{V} \equiv \frac{\partial V_1}{\partial x_1} + \frac{\partial V_2}{\partial x_2} + \frac{\partial V_3}{\partial x_3} \tag{1-11}$$

or equivalently in index notation as

$$\operatorname{div} \mathbf{V} = \frac{\partial V_i}{\partial x_i} \tag{1-12}$$

At this point of our discussion the divergence is purely a mathematical construct devoid of physical meaning. In a later section we will examine the divergence of the fluid velocity field; when so applied the operation takes on definite physical significance.

1-9 The Curl of a Vector Function of Position

The curl of a vector function of position is a vector function of position and is defined as

$$\begin{aligned}\operatorname{curl} \mathbf{V} \equiv \nabla \times \mathbf{V} &\equiv \begin{vmatrix} \mathbf{1}_1 & \mathbf{1}_2 & \mathbf{1}_3 \\ \partial/\partial x_1 & \partial/\partial x_2 & \partial/\partial x_3 \\ V_1 & V_2 & V_3 \end{vmatrix} \\ &= \left(\frac{\partial V_3}{\partial x_2} - \frac{\partial V_2}{\partial x_3}\right)\mathbf{1}_1 + \left(\frac{\partial V_1}{\partial x_3} - \frac{\partial V_3}{\partial x_1}\right)\mathbf{1}_2 + \left(\frac{\partial V_2}{\partial x_1} - \frac{\partial V_1}{\partial x_2}\right)\mathbf{1}_3\end{aligned} \tag{1-13}$$

Since the curl of $\mathbf{V}$ can be formally considered as the cross product of ∇ and $\mathbf{V}$, the operation can be expressed in terms of the permutation symbol

$$\nabla \times \mathbf{V} = \epsilon_{ijk} \frac{\partial V_k}{\partial x_j} \tag{1-14}$$

Again, as with the divergence, the definition of (1-13) at this point is merely mathematical; later we will examine the curl of the fluid velocity field and then attach a physical meaning to the operation.

1-10 The Dyadic Product

If we adjoin two vectors **V** and **U** to form the combination **VU**, with no multiplication sign of any sort, we have a *dyad*, another name for a second-order tensor. The *dyadic product* **VU** is then defined as

$$\mathbf{VU} \equiv \begin{bmatrix} V_1U_1 & V_1U_2 & V_1U_3 \\ V_2U_1 & V_2U_2 & V_2U_3 \\ V_3U_1 & V_3U_2 & V_3U_3 \end{bmatrix} \tag{1-15}$$

In general, the dyadic product **VU** is different from the dyadic product **UV**. Applying (1-15) to the dyadic **∇V**, we obtain

$$\nabla\mathbf{V} = \begin{bmatrix} \partial V_1/\partial x_1 & \partial V_2/\partial x_1 & \partial V_3/\partial x_1 \\ \partial V_1/\partial x_2 & \partial V_2/\partial x_2 & \partial V_3/\partial x_2 \\ \partial V_1/\partial x_3 & \partial V_2/\partial x_3 & \partial V_3/\partial x_3 \end{bmatrix} \tag{1-16}$$

a second-order tensor which in index notation is written as $\partial V_i/\partial x_j$.

1-11 The Double Dot Product of Two Tensors

Two tensors may be multiplied according to the double dot operation to produce the double dot or scalar product:

$$\begin{aligned} \mathcal{P} : \mathcal{D} &= P_{ij}D_{ij} \qquad\qquad (1\text{-}17) \\ &= (P_{11}D_{11} + P_{12}D_{12} + P_{13}D_{13} + P_{21}D_{21} + P_{22}D_{22} + P_{23}D_{23} \\ &\quad + P_{31}D_{31} + P_{32}D_{32} + P_{33}D_{33}) \end{aligned}$$

the result clearly being a scalar.

1-12 The Single Dot Product of Two Tensors

The single dot product or tensor product of two tensors results from the multiplication of two tensors according to the single dot operation of

$$\begin{aligned} \mathcal{P} \cdot \mathcal{D} &= P_{ij}D_{jk} \qquad\qquad (1\text{-}18) \\ &= P_{i1}D_{1k} + P_{i2}D_{2k} + P_{i3}D_{3k} \end{aligned}$$

the result being a second-order tensor.

1-13 The Dot Product of a Tensor and a Vector

A tensor may be dotted into a vector, the result being a vector according to

$$
\begin{aligned}
\mathcal{P} \cdot \mathbf{V} &= P_{ij}V_j \qquad (1\text{-}19)\\
&= (P_{11}V_1 + P_{12}V_2 + P_{13}V_3)\mathbf{1}_1 + (P_{21}V_1 + P_{22}V_2 + P_{23}V_3)\mathbf{1}_2\\
&\quad + (P_{31}V_1 + P_{32}V_2 + P_{33}V_3)\mathbf{1}_3
\end{aligned}
$$

Similarly,

$$
\begin{aligned}
\mathbf{V} \cdot \mathcal{P} &= V_jP_{ji} \qquad (1\text{-}20)\\
&= (P_{11}V_1 + P_{21}V_2 + P_{31}V_3)\mathbf{1}_1 + (P_{12}V_1 + P_{22}V_2 + P_{32}V_3)\mathbf{1}_2\\
&\quad + (P_{13}V_1 + P_{23}V_2 + P_{33}V_3)\mathbf{1}_3
\end{aligned}
$$

Clearly,

$$\mathcal{P} \cdot \mathbf{V} \neq \mathbf{V} \cdot \mathcal{P}$$

unless the tensor $\mathcal{P}$ is symmetric, by which we mean that $P_{ij} = P_{ji}$.

1-14 The Laplacian of a Scalar Field

If we take the divergence of the gradient of the scalar function ϕ, we obtain

$$
\begin{aligned}
\nabla \cdot \nabla\phi &= \left(\frac{\partial}{\partial x_1}\mathbf{1}_1 + \frac{\partial}{\partial x_2}\mathbf{1}_2 + \frac{\partial}{\partial x_3}\mathbf{1}_3\right) \cdot \left(\frac{\partial\phi}{\partial x_1}\mathbf{1}_1 + \frac{\partial\phi}{\partial x_2}\mathbf{1}_2 + \frac{\partial\phi}{\partial x_3}\mathbf{1}_3\right)\\
&= \frac{\partial^2\phi}{\partial x_1^2} + \frac{\partial^2\phi}{\partial x_2^2} + \frac{\partial^2\phi}{\partial x_3^2} \qquad (1\text{-}21)\\
&= \frac{\partial^2\phi}{\partial x_i^2}\\
&= \nabla^2\phi
\end{aligned}
$$

where the symbol ∇^2, called the *Laplacian operator*, has been defined as

$$
\begin{aligned}
\nabla^2 &\equiv \nabla \cdot \nabla\\
&= \frac{\partial^2}{\partial x_1^2} + \frac{\partial^2}{\partial x_2^2} + \frac{\partial^2}{\partial x_3^2}\\
&= \frac{\partial^2}{\partial x_i^2}
\end{aligned}
$$

1-15 Matrices and Their Transpose

We have seen that the second order-tensor $T_{ij}(\mathfrak{T})$ can be conveniently displayed in a 3×3 array,

$$T_{ij} = \begin{bmatrix} T_{11} & T_{12} & T_{13}\\ T_{21} & T_{22} & T_{23}\\ T_{31} & T_{32} & T_{33} \end{bmatrix} \qquad (1\text{-}22)$$

Thus, the nine scalar quantites associated with T_{ij} can be written in what is called a *square matrix array*, T, and it is occasionally convenient to treat it as such. The tensor T_{ji} is called the *transpose* of T_{ij}; or in matrix notation, T^T is the transpose of T. The transpose is formed by interchanging rows and columns or by interchanging the two subscripts on each element. Typically,

$$T_{ji} = \begin{bmatrix} T_{11} & T_{21} & T_{31} \\ T_{12} & T_{22} & T_{32} \\ T_{13} & T_{23} & T_{33} \end{bmatrix} \tag{1-23}$$

In a similar manner, a transpose can also be associated with the column matrix V, which is simply a method of displaying the three Cartesian components of the vector V_i to give the row matrix V^T. If

$$\mathsf{V} = \begin{bmatrix} V_1 \\ V_2 \\ V_3 \end{bmatrix} \tag{1-24}$$

then

$$\mathsf{V}^T = [V_1 \quad V_2 \quad V_3] \tag{1-25}$$

1-16 Matrix Multiplication

Application of the rules concerning index notation cited in Section 1-2 permits us to write the expression

$$U_i = T_{ij}V_j \tag{1-26}$$

as

$$U_i = T_{i1}V_1 + T_{i2}V_2 + T_{i3}V_3$$

and letting i take on the values 1, 2, and 3, we see that the components of U_i are

$$\begin{aligned} U_1 &= T_{11}V_1 + T_{12}V_2 + T_{13}V_3 \\ U_2 &= T_{21}V_1 + T_{22}V_2 + T_{23}V_3 \\ U_3 &= T_{31}V_1 + T_{32}V_2 + T_{33}V_3 \end{aligned}$$

The indicial equation (1-26) could also have been written as a matrix equation,

$$\mathsf{U} = \mathsf{TV} \tag{1-27}$$

or

$$\begin{bmatrix} U_1 \\ U_2 \\ U_3 \end{bmatrix} = \begin{bmatrix} T_{11} & T_{12} & T_{13} \\ T_{21} & T_{22} & T_{23} \\ T_{31} & T_{32} & T_{33} \end{bmatrix} \begin{bmatrix} V_1 \\ V_2 \\ V_3 \end{bmatrix} = \begin{bmatrix} T_{11}V_1 + T_{12}V_2 + T_{13}V_3 \\ T_{21}V_1 + T_{22}V_2 + T_{23}V_3 \\ T_{31}V_1 + T_{32}V_2 + T_{33}V_3 \end{bmatrix} \tag{1-28}$$

This illustrates the rule for multiplying a 3×3 matrix T into a 3×1 column matrix V: each element in the first row of T is multiplied by the corresponding element in the first column of V and the results summed to obtain the first row/first column element of the resulting matrix U; the remaining elements of U are obtained in similar fashion. It is seen then that the number of columns in T (three in this example) must be the same as the number of rows in V (three) and that the result will be a matrix having the number of rows in T (three) and the number of columns in V (one). For the particular case of (1-28), the product is the column matrix U representing the vector U_i.

Consider now the product

$$U_j = V_i T_{ij} \tag{1-29}$$

which after summing over i yields

$$U_j = V_1 T_{1j} + V_2 T_{2j} + V_3 T_{3j} \tag{1-30}$$

a vector equation having the three components

$$\begin{aligned} U_1 &= V_1 T_{11} + V_2 T_{21} + V_3 T_{31} \\ U_2 &= V_1 T_{12} + V_2 T_{22} + V_3 T_{32} \\ U_3 &= V_1 T_{13} + V_2 T_{23} + V_3 T_{33} \end{aligned} \tag{1-31}$$

we now might be tempted to write a matrix equation in the form

$$\mathsf{U} = \mathsf{VT}$$

However, it is immediately apparent that we have violated the rule for matrix multiplication, in that the column vector V does not have the same number of columns as the matrix T has rows. To rectify this, we can make V a row matrix through its transpose V^T and then apply the rule of rows into columns, to obtain

$$\begin{aligned} \mathsf{V}^T\mathsf{T} &= [V_1 \quad V_2 \quad V_3] \begin{bmatrix} T_{11} & T_{12} & T_{13} \\ T_{21} & T_{22} & T_{23} \\ T_{31} & T_{32} & T_{33} \end{bmatrix} \\ &= [(V_1 T_{11} + V_2 T_{21} + V_3 T_{31})\,(V_1 T_{12} + V_2 T_{22} + V_3 T_{32}) \\ &\quad (V_1 T_{13} + V_2 T_{23} + V_3 T_{33})] \\ &= [U_1\ U_2\ U_3] \\ &= \mathsf{U}^T, \end{aligned} \tag{1-32}$$

and thus obtain the same result as in (1-31).

From (1-26) and (1-27) we have seen that $U_i = T_{ij} V_j$ can be written as U = TV, where U and V are column matrices, while from (1-29) and (1-32) we have seen that $U_j = V_i T_{ij}$ can be written as $\mathsf{U}^T = \mathsf{V}^T\mathsf{T}$, where U^T and V^T are row matrices. This result can be formalized by asserting that, if the rules of matrix multiplication are to be used, the repeated subscripts of the indicial equivalent must be located adjacent to each other. Thus, if we want to write

$U_j = V_i T_{ij}$ as a matrix equation involving column matrices, we know that V must follow T (rows into columns), that in indicial notation the order of arrangement is immaterial ($U_j = V_i T_{ij} = T_{ij} V_i$), and that $U \neq TV$ since the repeated subscripts i are not adjacent. To make them adjacent we can take the transpose of T (i.e., by changing T_{ij} to T_{ji}). If, as in (1-32), $U^T = V^T T$, where U^T and V^T are row matrices, then we obtain the same components of U^T by writing $U = T^T V$, where U and V are now column matrices.

1-17 Transformation of Vectors and Tensors Between Cartesian Coordinate Systems

Consider the two different Cartesian coordinate systems (x_1, x_2, x_3) and (x'_1, x'_2, x'_3), as shown in Figure 1-7. Let α_{ij} be the cosine of the angle between the axes x'_i and x_j, that is,

$$\alpha_{ij} = \cos(x'_i, x_j) = \begin{bmatrix} \alpha_{11} & \alpha_{12} & \alpha_{13} \\ \alpha_{21} & \alpha_{22} & \alpha_{23} \\ \alpha_{31} & \alpha_{32} & \alpha_{33} \end{bmatrix} \tag{1-33}$$

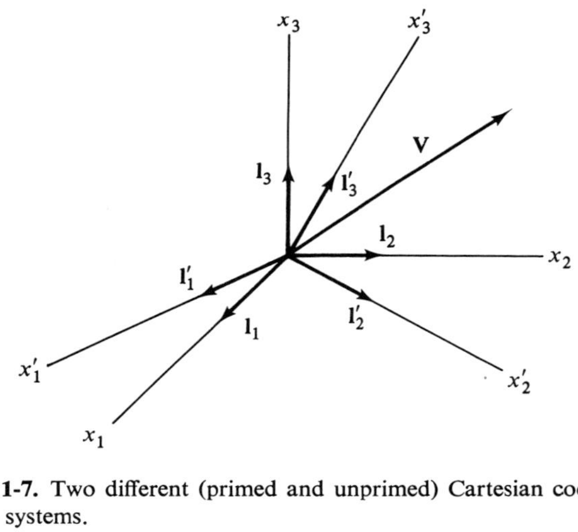

Fig. 1-7. Two different (primed and unprimed) Cartesian coordinate systems.

For example, α_{23} is the cosine of the angle between the x'_2 axis and the x_3 axis. Now the vector **V** has the Cartesian components V_i in an unprimed system and V'_i in the primed system. It is easy to relate these two sets of Cartesian components, V_i and V'_i, since the projection of **V** onto the x'_1 axis (giving V'_1) is equal to the sum of the projections of the V_i components on the same axis. Thus,

$$V'_1 = \alpha_{11} V_1 + \alpha_{12} V_2 + \alpha_{13} V_3$$

Similarly,

$$\begin{aligned} V_2' &= \alpha_{21}V_1 + \alpha_{22}V_2 + \alpha_{23}V_3 \\ V_3' &= \alpha_{31}V_1 + \alpha_{32}V_2 + \alpha_{33}V_3 \end{aligned} \tag{1-34}$$

all of which can be written as the matrix equation

$$\begin{bmatrix} V_1' \\ V_2' \\ V_3' \end{bmatrix} = \begin{bmatrix} \alpha_{11} & \alpha_{12} & \alpha_{13} \\ \alpha_{21} & \alpha_{22} & \alpha_{23} \\ \alpha_{31} & \alpha_{32} & \alpha_{33} \end{bmatrix} \begin{bmatrix} V_1 \\ V_2 \\ V_3 \end{bmatrix} \tag{1-35}$$

or

$$\mathsf{V}' = \mathsf{AV} \tag{1-36}$$

where A is the matrix α_{ij} and is called the *rotation matrix*. In indicial notation (1-35) can be written as

$$V_i' = \alpha_{ij}V_j \tag{1-37}$$

Interchanging the roles of the primed and unprimed axes but otherwise repeating the same procedure, the unprimed components V_i can be written in terms of the primed components V_i' as

$$\begin{aligned} V_1 &= \alpha_{11}V_1' + \alpha_{21}V_2' + \alpha_{31}V_3' \\ V_2 &= \alpha_{12}V_1' + \alpha_{22}V_2' + \alpha_{32}V_3' \\ V_3 &= \alpha_{13}V_1' + \alpha_{23}V_2' + \alpha_{33}V_3' \end{aligned} \tag{1-38}$$

or

$$\mathsf{V} = \mathsf{A}^T\mathsf{V}' \tag{1-39}$$

or

$$V_i = \alpha_{ji}V_j' \tag{1-40}$$

In terms of the row matrices V^T and $(\mathsf{V}')^T$, the transformation equation of (1-39) can be written as

$$\mathsf{V}^T = (\mathsf{V}')^T\mathsf{A} \tag{1-41}$$

We also observe that

$$\mathsf{AA}^T = \mathsf{I} \tag{1-42}$$

and

$$\mathsf{A}^T\mathsf{A} = \mathsf{I}$$

where I is the Kronecker delta δ_{ij}. A matrix having this property is said to be *orthogonal*.

From (1-40) we see that two vectors V_i and U_j transform according to

$$\begin{aligned} V_i &= \alpha_{mi}V_m' \\ U_j &= \alpha_{nj}U_n' \end{aligned}$$

Suppose that we let T_{ij} and T'_{mn} equal the dyadic products $V_i U_j$ and $V'_m U'_n$, respectively. Then

$$T_{ij} = V_i U_j = \alpha_{mi} \alpha_{nj} T'_{mn} \tag{1-43}$$

which is frequently taken as the definition of a second-order tensor; that is, a second-order tensor is defined as a quantity whose nine components transform according to (1-43). The inverse of this expression is found by multiplying through by $\alpha_{pi}\alpha_{qj}$. Thus,

$$\alpha_{pi}\alpha_{qj} T_{ij} = \alpha_{pi}\alpha_{mi}\alpha_{qj}\alpha_{nj} T'_{mn}$$

which from using (1-42) becomes

$$\alpha_{pi}\alpha_{qj} T_{ij} = \delta_{pm}\delta_{qn} T'_{mn} = T'_{pq}$$

or, after renaming the indexes,

$$T'_{ij} = \alpha_{im}\alpha_{jn} T_{mn} \tag{1-44}$$

In matrix notation the transformations of (1-43) and (1-44) can be written as

$$\mathsf{T} = \mathsf{A}^T \mathsf{T}' \mathsf{A} \tag{1-45}$$

and

$$\mathsf{T}' = \mathsf{A} \mathsf{T} \mathsf{A}^T \tag{1-46}$$

respectively.

1-18 Principal Coordinates

Consider the vector equation

$$T_{ij} V_j = \lambda V_i \tag{1-47}$$

where T_{ij} is a symmetrical tensor, V_i is some vector, and λ is some scalar. If a vector V_i other than $V_i = 0$ satisfies this equation for any value of λ, it is called an *eigenvector* of T_{ij} and the associated value of λ is called an *eigenvalue*. The matrix form of (1-47),

$$\begin{bmatrix} T_{11} & T_{12} & T_{13} \\ T_{21} & T_{22} & T_{23} \\ T_{31} & T_{32} & T_{33} \end{bmatrix} \begin{bmatrix} V_1 \\ V_2 \\ V_3 \end{bmatrix} = \lambda \begin{bmatrix} V_1 \\ V_2 \\ V_3 \end{bmatrix} \tag{1-48}$$

represents the three homogeneous algebraic equations

$$\begin{aligned} (T_{11} - \lambda)V_1 + T_{12}V_2 + T_{13}V_3 &= 0 \\ T_{21}V_1 + (T_{22} - \lambda)V_2 + T_{23}V_3 &= 0 \\ T_{31}V_1 + T_{32}V_2 + (T_{33} - \lambda)V_3 &= 0 \end{aligned} \tag{1-49}$$

The condition that there are solutions other than the trivial one $V_i = 0$ is that the determinant of the coefficients be equal to zero:

$$\det (T_{ij} - \lambda\delta_{ij}) = 0$$

or

$$\begin{vmatrix} T_{11} - \lambda & T_{12} & T_{13} \\ T_{21} & T_{22} - \lambda & T_{23} \\ T_{31} & T_{32} & T_{33} - \lambda \end{vmatrix} = 0 \tag{1-50}$$

Evaluating (1-50), we have that

$$-\lambda^3 + I_1\lambda^2 - I_2\lambda + I_3 = 0 \tag{1-51}$$

where

$$\begin{aligned} I_1 &= T_{11} + T_{22} + T_{33} \\ I_2 &= T_{11}T_{22} + T_{22}T_{33} + T_{11}T_{33} - T_{12}T_{21} \\ &\quad - T_{23}T_{32} - T_{13}T_{31} \\ I_3 &= T_{11}T_{22}T_{33} + T_{12}T_{23}T_{31} + T_{13}T_{21}T_{32} \\ &\quad - T_{13}T_{22}T_{31} - T_{12}T_{21}T_{33} - T_{11}T_{23}T_{32} \end{aligned}$$

If we regard the elements of the tensor T_{ij} as known, (1-51) is a cubic in λ and, it can be shown, as such has three real roots as long as T_{ij} is symmetric. This cubic, called the *characteristic equation*, has the three roots (*eigenvalues*) $\lambda^{(1)}$, $\lambda^{(2)}$, and $\lambda^{(3)}$.

Choosing one of these eigenvalues, say $\lambda^{(1)}$, and substituting into (1-40), we can solve any two of these equations (only two of the three are independent) for the ratios V_1/V_3 and V_2/V_3, or the ratios of the direction cosines of V_i, namely α_1/α_3 and α_2/α_3, since

$$\begin{aligned} V_1 &= \alpha_1\sqrt{V_1^2 + V_2^2 + V_3^2} \\ V_2 &= \alpha_2\sqrt{V_1^2 + V_2^2 + V_3^2} \\ V_3 &= \alpha_3\sqrt{V_1^2 + V_2^2 + V_3^2} \end{aligned}$$

Combining these ratios with the requirement that

$$\alpha_1^2 + \alpha_2^2 + \alpha_3^2 = 1 \tag{1-52}$$

permits us to solve completely for the unit eigenvector $e_i^{(1)}$, where

$$\pm e_i^{(1)} = \alpha_1\mathbf{1}_1 + \alpha_2\mathbf{1}_2 + \alpha_3\mathbf{1}_3 \tag{1-53}$$

By repeating the operation with $\lambda^{(2)}$ and $\lambda^{(3)}$, we may determine the other two unit eigenvectors, $\pm e_i^{(2)}$ and $\pm e_i^{(3)}$.

Now if the three eigenvalues are all distinct, that is, $\lambda^{(1)} \neq \lambda^{(2)} \neq \lambda^{(3)}$, then the three eigenvectors are mutually perpendicular. To show this we can write from (1-47) that

$$T_{ij}e_j^{(1)} = \lambda^{(1)}e_i^{(1)}$$

and

$$T_{ij}e_j^{(2)} = \lambda^{(2)}e_i^{(2)}$$

Taking the scalar product of these equations by $e_i^{(2)}$ and $e_i^{(1)}$ and subtracting, we obtain

$$T_{ij}e_j^{(1)}e_i^{(2)} - T_{ij}e_j^{(2)}e_i^{(1)} = \lambda^{(1)}e_i^{(1)}e_i^{(2)} - \lambda^{(2)}e_i^{(2)}e_i^{(1)}$$

and then rewriting after interchanging the dummy suffixes i and j in the second term on the left to yield

$$(T_{ij} - T_{ji})e_j^{(1)}e_i^{(2)} = (\lambda^{(1)} - \lambda^{(2)})e_i^{(1)}e_i^{(2)} = 0$$

since T_{ij} is symmetric. So it follows that if $\lambda^{(1)} \neq \lambda^{(2)}$, $e_i^{(1)}$ is perpendicular to $e_i^{(2)}$. This result may be extended to show that $e_i^{(1)}$, $e_i^{(2)}$, and $e_i^{(3)}$ are all mutually perpendicular provided that no two values of $\lambda^{(i)}$ are equal.

Suppose now that we construct a new Cartesian coordinate system as in Figure 1-8, in which the three unit eigenvectors are the three unit vectors.

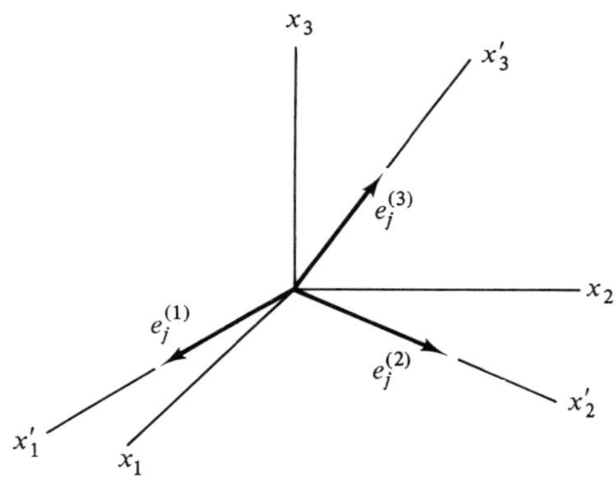

Fig. 1-8. Three unit eigenvectors of the unprimed Cartesian coordinates form unit vectors of primed Cartesian coordinates.

The rotation matrix A will consist of the elements $\alpha_{ij} = \cos(x_i', x_j)$ as given in (1-33). Since $e_1^{(1)}$, $e_2^{(1)}$, and $e_3^{(1)}$ are the components of a unit vector in the x_1' direction, $e_1^{(2)}$, $e_2^{(2)}$, and $e_3^{(3)}$ the components of a unit vector in the x_2' direction, and $e_1^{(3)}$, $e_2^{(3)}$, and $e_3^{(3)}$ the components of a unit vector in the x_3' direction, the rotation matrix written in terms of the components of the eigenvectors is

$$\mathsf{A} = \alpha_{ij} = \begin{bmatrix} e_1^{(1)} & e_2^{(1)} & e_3^{(1)} \\ e_1^{(2)} & e_2^{(2)} & e_3^{(2)} \\ e_1^{(3)} & e_2^{(3} & e_3^{(3)} \end{bmatrix} \tag{1-54}$$

A question we might ask at this point is: How does a matrix T, with elements T_{ij} in the (x_1, x_2, x_3) coordinate system transform when referred to the (x_1', x_2', x_3') coordinate system defined by the eigenvectors of T? To answer this, consider (1-46),

$$\mathsf{T}' = \mathsf{ATA}^T$$

which when written out in expanded form is

$$\mathsf{T}' = \begin{bmatrix} e_1^{(1)} & e_2^{(1)} & e_3^{(1)} \\ e_1^{(2)} & e_2^{(2)} & e_3^{(2)} \\ e_1^{(3)} & e_2^{(3)} & e_3^{(3)} \end{bmatrix} \begin{bmatrix} T_{11} & T_{12} & T_{13} \\ T_{21} & T_{22} & T_{23} \\ T_{31} & T_{32} & T_{33} \end{bmatrix} \begin{bmatrix} e_1^{(1)} & e_1^{(2)} & e_1^{(3)} \\ e_2^{(1)} & e_2^{(2)} & e_2^{(3)} \\ e_3^{(1)} & e_3^{(2)} & e_3^{(3)} \end{bmatrix} \tag{1-55}$$

For the eigenvalues the defining equation is, from (1-47),

$$T_{ij} e_j^{(n)} = \lambda^{(n)} e_i^{(n)} \tag{1-56}$$

where $n = 1$, 2, or 3. Writing this out in turn in expanded form gives

$$\begin{bmatrix} T_{11} & T_{12} & T_{13} \\ T_{21} & T_{22} & T_{23} \\ T_{31} & T_{32} & T_{33} \end{bmatrix} \begin{bmatrix} e_1^{(n)} \\ e_2^{(n)} \\ e_3^{(n)} \end{bmatrix} = \begin{bmatrix} \lambda^{(n)} e_1^{(n)} \\ \lambda^{(n)} e_2^{(n)} \\ \lambda^{(n)} e_3^{(n)} \end{bmatrix} \tag{1-57}$$

But the left-hand side of (1-57) is equivalent to the product of the last two matrices on the right-hand side of (1-55). Combining, then, we have that

$$\mathsf{T}' = \begin{bmatrix} e_1^{(1)} & e_2^{(1)} & e_3^{(1)} \\ e_1^{(2)} & e_2^{(2)} & e_3^{(2)} \\ e_1^{(3)} & e_2^{(3)} & e_3^{(3)} \end{bmatrix} \begin{bmatrix} \lambda^{(1)} e_1^{(1)} & \lambda^{(2)} e_1^{(2)} & \lambda^{(3)} e_1^{(3)} \\ \lambda^{(1)} e_2^{(1)} & \lambda^{(2)} e_2^{(2)} & \lambda^{(3)} e_2^{(3)} \\ \lambda^{(1)} e_3^{(1)} & \lambda^{(2)} e_3^{(2)} & \lambda^{(3)} e_3^{(3)} \end{bmatrix}$$

Multiplying this out and recalling that since the three vectors, $e_i^{(1)}$, $e_i^{(2)}$, and $e_i^{(3)}$ are unit vectors and are mutually perpendicular that

$$e_i^{(1)} e_i^{(2)} = e_i^{(2)} e_i^{(3)} = e_i^{(1)} e_i^{(3)} = 0$$

and

$$e_i^{(1)} e_i^{(1)} = e_i^{(2)} e_i^{(2)} = e_i^{(3)} e_i^{(3)} = 1$$

we obtain

$$\mathsf{T}' = \begin{bmatrix} \lambda^{(1)} & 0 & 0 \\ 0 & \lambda^{(2)} & 0 \\ 0 & 0 & \lambda^{(3)} \end{bmatrix} \tag{1-58}$$

Now, the set of coordinate axes formed by the three eigenvectors is called the *principal coordinate axes* of the symmetric tensor T.

Previously, we assumed that the three eigenvalues were all distinct. Suppose now that two of them, say $\lambda^{(1)}$ and $\lambda^{(2)}$, are equal while $\lambda^{(3)}$ is distinct. For this case it is possible to show that any two perpendicular vectors in the plane of $e_i^{(1)}$ and $e_i^{(2)}$, which are both perpendicular to $e_i^{(3)}$, combine with

$e_i^{(3)}$ to yield principal axes. Finally, given the case that all three eigenvalues are equal, then every mutually perpendicular set of axes constitutes a *principal set*.

1-19 Two Integral Theorems for Vectors and Tensors

Several integral theorems, which are summarized below, are extremely useful in developing the equations of fluid dynamics.

The *divergence theorem* states that if v is a closed region in space surrounded by the surface A, then

$$\int_v \nabla \cdot \mathbf{V}\, dv = \int_A \mathbf{n} \cdot \mathbf{V}\, dA \tag{1-59}$$

where $\mathbf{n}$ is the outward directed unit normal vector. This result merely relates the integral of the divergence of a vector $\mathbf{V}$ throughout the volume v to the integral of the normal component of $\mathbf{V}$ over the closed surface A bounding the volume. The right side of (1-59) may be written alternatively as

$$\int_A \mathbf{V} \cdot \mathbf{n}\, dA \qquad \text{or} \qquad \int_A V_n\, dA \qquad \text{or} \qquad \int_A \mathbf{V} \cdot d\mathbf{A}$$

Two theorems for scalars and tensors closely associated with the divergence theorem are

$$\int_v \nabla s\, dv = \int_A \mathbf{n} s\, dA \tag{1-60}$$

and

$$\int_v \nabla \cdot \boldsymbol{\mathcal{P}}\, dv = \int_A \mathbf{n} \cdot \boldsymbol{\mathcal{P}}\, dA \tag{1-61}$$

The last relation also happens to be true for the dyadic product $\mathbf{VU}$.

If A is a surface bounded by the closed curve C, then

$$\int_A \mathbf{n} \cdot (\nabla \times \mathbf{V}) dA = \oint_C \mathbf{V} \cdot \mathbf{t}\, dC \tag{1-62}$$

in which $\mathbf{t}$ denotes a unit vector tangent in the direction of integration along C, the dot product $\mathbf{V} \cdot \mathbf{t}$ is the tangential component of $\mathbf{V}$ along C, and $\mathbf{n}$ is the unit normal vector to A in the direction of advance of a right-hand screw when it is twisted in the direction of integration along C. Equation (1-62), known as the *Stokes theorem*, asserts that the integral of the normal component of the curl of a vector $\mathbf{V}$ over any portion of a surface A is equal to the line integral of the tangential component of $\mathbf{V}$ around the curve C bounding A. A similar relation holds for tensors.

TWO

Basic Concepts and Principles in Geophysical Fluid Dynamics

2-1 Notation

As discussed in Section 1-2, a vector can be written in the form of (1-1) as

$$\mathbf{V} = V_1\mathbf{1}_1 + V_2\mathbf{1}_2 + V_3\mathbf{1}_3$$

(see Figure 1-1), or as in (1-2),

$$\mathbf{V} = V_x\mathbf{i} + V_y\mathbf{j} + V_z\mathbf{k}$$

(see Figure 1-2). If $\mathbf{V}$ is explicitly assigned the meaning of a fluid velocity, we may also write

$$\mathbf{V} = u\mathbf{i} + v\mathbf{j} + w\mathbf{k}$$

where u, v, and w are the x, y, and z components of $\mathbf{V}$. In that which follows we will use, without further discussion, whichever of the foregoing representations appears to be most appropriate. The reader should become mentally adept at operating with any of these equivalent forms.

2-2 The Material Derivative

In general, a given thermal or kinematic property of the ocean (or atmosphere) depends upon both position and time. Consider the scalar quantity $s = s(x, y, z, t)$. The total time rate of change of s is then given by

$$\frac{\partial s}{\partial t} + \frac{\partial s}{\partial x}\frac{dx}{dt} + \frac{\partial s}{\partial y}\frac{dy}{dt} + \frac{\partial s}{\partial z}\frac{dz}{dt} \tag{2-1}$$

where the differential displacements dx, dy, and dz are specified for the differential time dt. Now letting

$$u \equiv \frac{dx}{dt}, \qquad v \equiv \frac{dy}{dt}, \qquad w \equiv \frac{dz}{dt}$$

be the components of the fluid velocity $\mathbf{V}$ in the x, y, and z directions at the point x, y, z, the total derivative (2-1) has a meaning of such importance as to attach a special symbol to the operation. We thus define the operator D/Dt such that

$$\frac{D}{Dt} \equiv \frac{\partial}{\partial t} + \mathbf{V} \cdot \nabla \tag{2-2}$$

or when operating on the scalar s,

$$\frac{Ds}{Dt} = \frac{\partial s}{\partial t} + \mathbf{V} \cdot \nabla s \tag{2-3}$$

Equation (2-3) is referred to as the *material derivative* of s (alternatively as the *comoving*, *substantial*, *convective*, or *particle derivative*). The first term on the right is the *local time-rate-of-change term*; it represents the change of s with time at a fixed point x, y, z and will vanish if the field of s is not time dependent. The second term on the right, called the *advectional* or *field-rate-of-change term*, is associated with the motion of the fluid in a field where a gradient of s exists. In general, this term will vanish if (1) $\mathbf{V} = 0$, (2) $\nabla s = 0$ (i.e., the field of s is homogeneous and at most is only time dependent), or (3) $\mathbf{V}$ is directed along equiscalar surfaces of s. We can attach a physical significance to the material derivative as the individual time-rate-of-change of the property s following a particular parcel of fluid.

2-3 Lagrangian versus Eulerian Representation

A field of motion $\mathbf{V}$ can be described in two ways:

1. The Eulerian system seeks the description of the velocity at each point in space-time through a direct specification $\mathbf{V}$. This implies that

$$u = u(x, y, z, t)$$
$$v = v(x, y, z, t)$$
$$w = w(x, y, z, t)$$

or

$$\mathbf{V} = \mathbf{V}(x, y, z, t) \tag{2-4}$$

is known. The variables x, y, z, and t are called *Eulerian variables.*

2. The Lagrangian system seeks the description of the particle position as time passes through a specification of the particle path (trajectory). This representation is familiar from particle mechanics. Suppose that at $t = 0$ the coordinates of a particular fluid particle are $x = a$, $y = b$, and $z = c$. At any later time t the flow is specified if we can give the coordinates x, y, and z of the fluid particle as functions of the Lagrangian variables a, b, c, and t. That is,

$$x = x(a, b, c, t)$$
$$y = y(a, b, c, t)$$
$$z = z(a, b, c, t)$$

or, alternatively,

$$x_i = x_i(a_i, t) \qquad (i = 1, 2, 3) \tag{2-5}$$

This representation is illustrated in Figure 2-1.

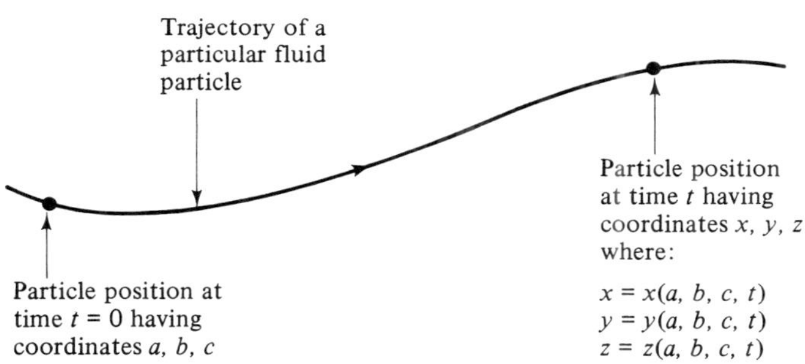

Fig. 2-1. Lagrangian specification of fluid flow.

Frequently, a connection is sought between the Lagrangian displacement field $x_i(a_i, t)$ and the Eulerian velocity field $V_i(x_i, t)$, where we take the Lagrangian reference time as $t = 0$ such that

$$a_i = x_i(a_i, 0)$$

First we note that the Lagrangian velocity field is given by

$$V_i(a_i, t) = \frac{\partial x_i(a_i, t)}{\partial t} \tag{2-6}$$

Clearly, it is required then that

$$[V_i(a_i, t)]_{\mathrm{La}} = [V_i(x_i(a_i, t), t)]_{\mathrm{Eu}} \tag{2-7}$$

where the subscripts La and Eu refer to Lagrangian and Eulerian, respectively, and where x_i is considered as a function of a_i, t. Then:

1. If the Lagrangian representation $x_i = x_i(a_i, t)$ is given, the Lagrangian velocity is defined by (2-6), and to obtain the Eulerian velocity we must invert the three expressions $x_i = x_i(a_i, t)$ to obtain $a_i = a_i(x_i, t)$ and substitute the result into the left side of (2-7).

2. If the Eulerian representation $V_i = V_i(x_i, t)$ is given, the Lagrangian velocity field cannot be obtained directly from (2-7) since we have no way to substitute for $x_i(a_i, t)$ in (2-7). However, combining (2-6) and (2-7), we have that

$$V_i(x_i, t) = \frac{\partial x_i(a_i, t)}{\partial t} \tag{2-8}$$

which is a set of three simultaneous partial differential equations in the Langrangian form. These three equations, given a reference time and a set of initial conditions, can be integrated.

Consider the following examples:

1. Given the two-dimensional Eulerian flow field

$$\begin{aligned} u &= u_0\left(1 - \frac{y^2}{h^2}\right) \\ v &= 0 \end{aligned} \tag{2-9}$$

which for u_0 and h being constants, is illustrated in Figure 2-2.

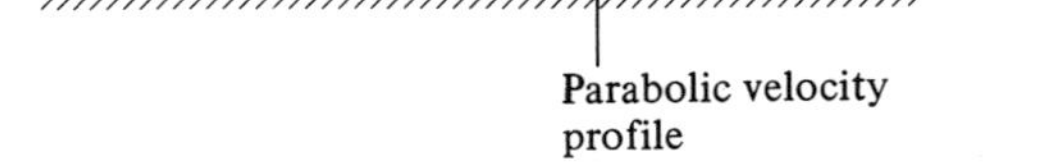

Fig. 2-2. Flow field for (2-9).

From (2-8),

$$\frac{\partial x(a, b, t)}{\partial t} = u_0\left(1 - \frac{y^2}{h^2}\right)$$

and

$$\frac{\partial y(a, b, t)}{\partial t} = 0$$

Integrating yields

$$x = u_0 \int \left(1 - \frac{y^2}{h^2}\right) dt + f(a, b)$$

$$y = g(a, b)$$

where f and g are functions of a and b to be determined. Rewriting, we have that

$$x = u_0\left(1 - \frac{y^2}{h^2}\right)t + f(a, b)$$

$$y = g(a, b)$$

Given the initial conditions that at $t = 0$, $x = a$ and $y = b$, we find that $f(a, b) = a$ and $g(a, b) = b$. Therefore, the Lagrangian equivalent of (2-9) is

$$x(a, b, t) = u_0\left(1 - \frac{b^2}{h^2}\right)t + a \tag{2-10}$$

$$y(a, b, t) = b$$

2. Given the Lagrangian equations of (2-10) we take the time derivatives:

$$\frac{\partial x(a, b, t)}{\partial t} = u_0\left(1 - \frac{b^2}{h^2}\right)$$

$$\frac{\partial y(a, b, t)}{\partial t} = 0$$

and substitute to eliminate the Lagrangian variable b in the first of these to obtain the Eulerian equations

$$u = u_0\left(1 - \frac{y^2}{h^2}\right)$$

$$v = 0$$

In many cases it is not quite so simple to eliminate Lagrangian variables from the Eulerian flow equations; it might not even be possible to do so.

Concerning the relative usefulness of these two representations, it is clear that they are interchangable; the Eulerian to Lagrangian transformation requiring integration, the Lagrangian to Eulerian transformation requiring differentiation. In the majority of cases the Eulerian formulation is simpler, this being particularly true if viscous effects are taken into account. It is also the more desirable in any case that can be reduced to a steady-state condition through the introduction of a moving coordinate system. An example of this is that of the flow about an object moving through a fluid at a constant velocity. By attaching the coordinate system to the object we reduce the independent variables from four, $V_i(x_i, t)$, to three $V_i(x_i)$. In general, this is not possible in the Lagrangian system.

The Lagrangian formulation is helpful in problems involving diffusion and mixing of mass attached properties. In this case we can "tag" the particles in terms of the characteristic quantities whose transport we wish to study.

2-4 Representation of a Fluid in Motion

We now introduce some methods for representing fluid flow, not all of which are of equal utility.

Streamline. A *streamline* is defined as a line which is everywhere tangent to the fluid velocity vector at a given instant of time. Thus, a set of streamlines merely indicates the direction of the flow at a given time; that is, they give the synoptic flow pattern. No number is associated with them and their spacing is quite arbitrary. Utilizing the Eulerian form of the velocity field representation, $\mathbf{V} = \mathbf{V}(x, y, z, t)$, a streamline is defined from

$$d\mathbf{r} \times \mathbf{V} \equiv 0 \tag{2-11}$$

where

$$d\mathbf{r} = (dx)\mathbf{i} + (dy)\mathbf{j} + (dz)\mathbf{k}$$

is a differential line element along a streamline (Figure 2-3). Writing out (2-11) in detail, we have that

$$\begin{vmatrix} \mathbf{i} & \mathbf{j} & \mathbf{k} \\ dx & dy & dz \\ u & v & w \end{vmatrix} = 0$$

yielding the two equations

$$\frac{dx}{u} = \frac{dy}{v} = \frac{dz}{w} \tag{2-12}$$

These equations allow evaluation of the shape of streamlines for non-time-dependent flow or at a given instant for time-dependent flow when the time dependent terms in the equations for u, v, and w equal a constant. For the non-time-dependent case, streamlines and trajectories (for the definition of

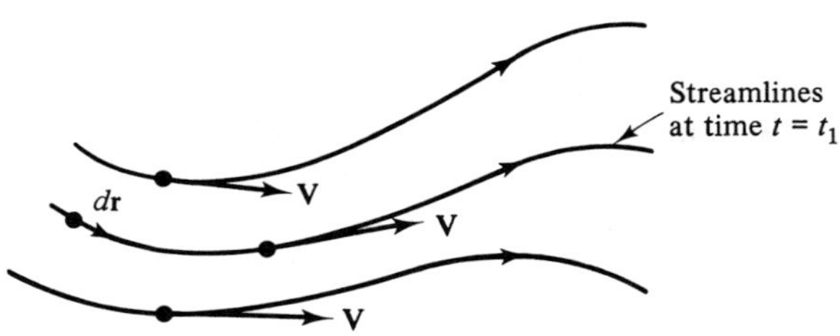

Fig. 2-3. Streamlines, fluid velocity, and the differential line element.

a trajectory, see below) are identical. By fixing the time in the time-dependent case we actually evaluate the trajectory equations, which for any particular instant are identical to the streamline equations. Take, for example, the flow field as given by

$$u = u_0$$

$$v = v_0 \cos (kx - \sigma t)$$

where u_0, v_0, k, and σ are constant. We have then, from (2-12), for $t = t_1$

$$\frac{dx}{u_0} = \frac{dy}{v_0 \cos (kx - \sigma t_1)}$$

or

$$v_0 \cos (kx - \sigma t_1)\, dx = u_0\, dy$$

and, upon integrating,

$$y = \frac{v_0}{u_0 k} \sin (kx - \sigma t_1) + C$$

where the constant C specifies any particular streamline at time t_1.

Trajectory. The *trajectory* or *pathline* of a fluid particle is merely its Lagrangian displacement history and as such is represented by the Lagrangian form of the displacement field equations, $x_i = x_i(a_i, t)$. It is a line that is traced out in time by a given fluid particle.

Streakline. A line connecting all the particles that have passed a given fixed Eulerian coordinate position is called a *streakline*, a plume of smoke from a chimney being an example.

Fluid Line. A *fluid line* is a line connecting fluid particles at a time t_1 which was identified at an earlier time t_0 (Figure 2-4).

Examples of some of these concepts are illustrated in (1) Figure 2-5 for a flow pattern of straight, parallel streamlines that rotate at a constant rate, and (2) Figure 2-6 for inviscid flow past an infinite circular cylinder.

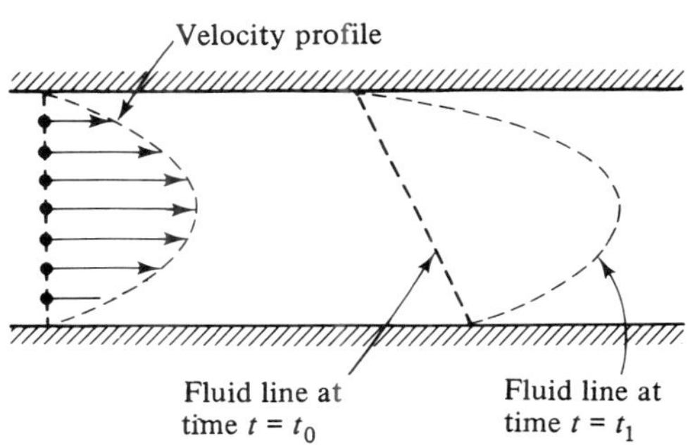

Fig. 2-4. Fluid line representation.

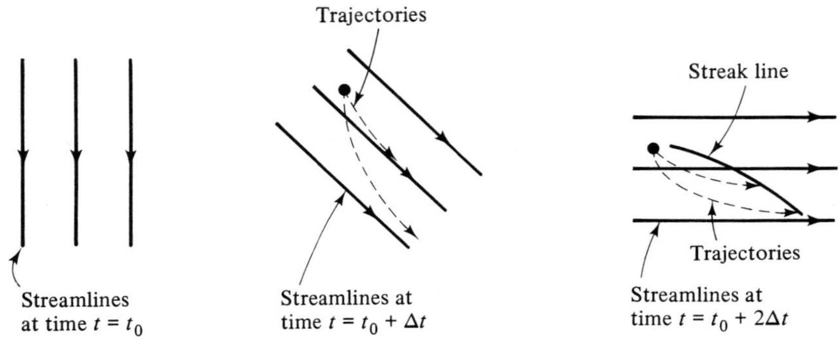

Fig. 2-5. Streamlines, trajectories, and streakline for a rotating flow.

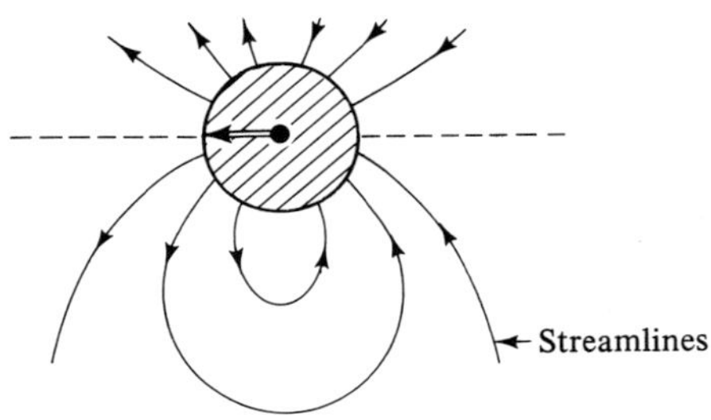

(a) Short time exposure with dust in the fluid

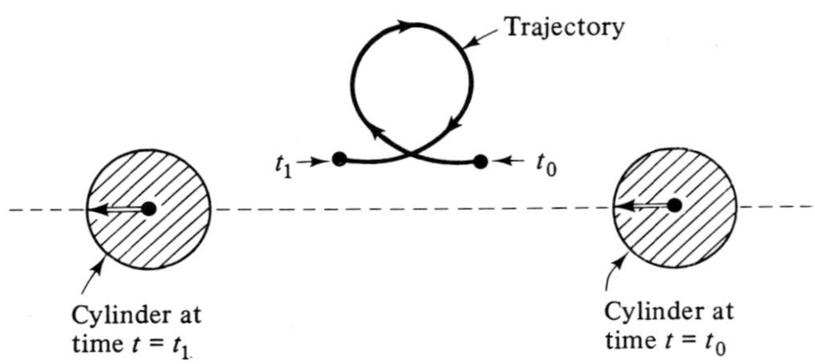

(b) Long time exposure with a fluid particle tagged

Fig. 2-6. Streamlines, trajectories, streaklines, and fluid lines for flow around an infinite cylinder: (a) Short time exposure with dust in the fluid; (b) long time exposure with a fluid particle tagged; (c) photograph of dye line; (d) a fluid line, paint a line of dye of any shape before cylinder approaches, photograph same dyed line at any later time; (e) cylinder moving at a constant velocity with coordinate system fixed in the cylinder.

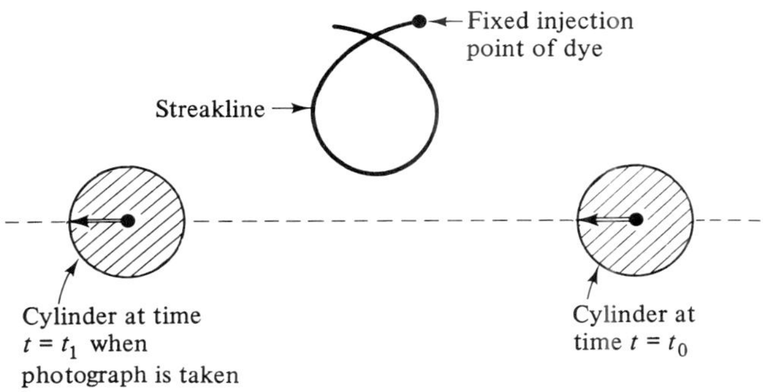

(c) Photograph of dye line

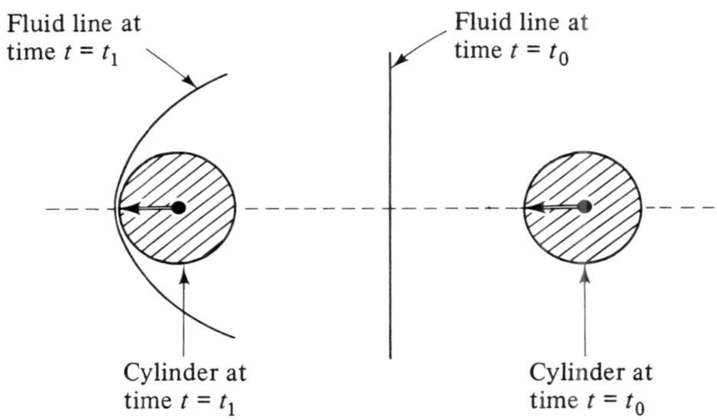

(d) A fluid line, paint a line of dye of any shape before cylinder approaches, photograph same dyed line at any later time.

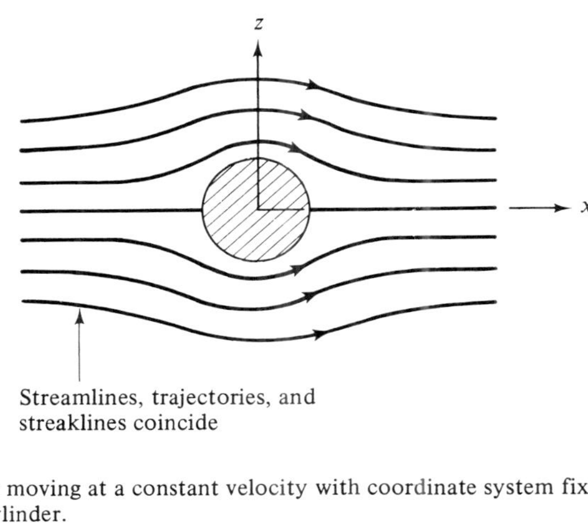

(e) Cylinder moving at a constant velocity with coordinate system fixed in the cylinder.

Fig 2-6. (Continued)

2-5 Kinematic Modes of a Fluid

A parcel of fluid can undergo, in general, three distinct types of kinematic behavior: translation, rotation (or spin), and deformation. The first of these, *translation*, is essentially the gross motion of the fluid parcel. Rigorously, translation is defined as the motion that the center of mass of a fluid particle undergoes. Superimposed on translation we have the possiblity of the relative motions of *rotation* and *deformation*. In general, the three phenomena are independent of each other; each can occur in the absence of the other (Figure 2-7). For example: (1) Consider the motion of the suspended seats

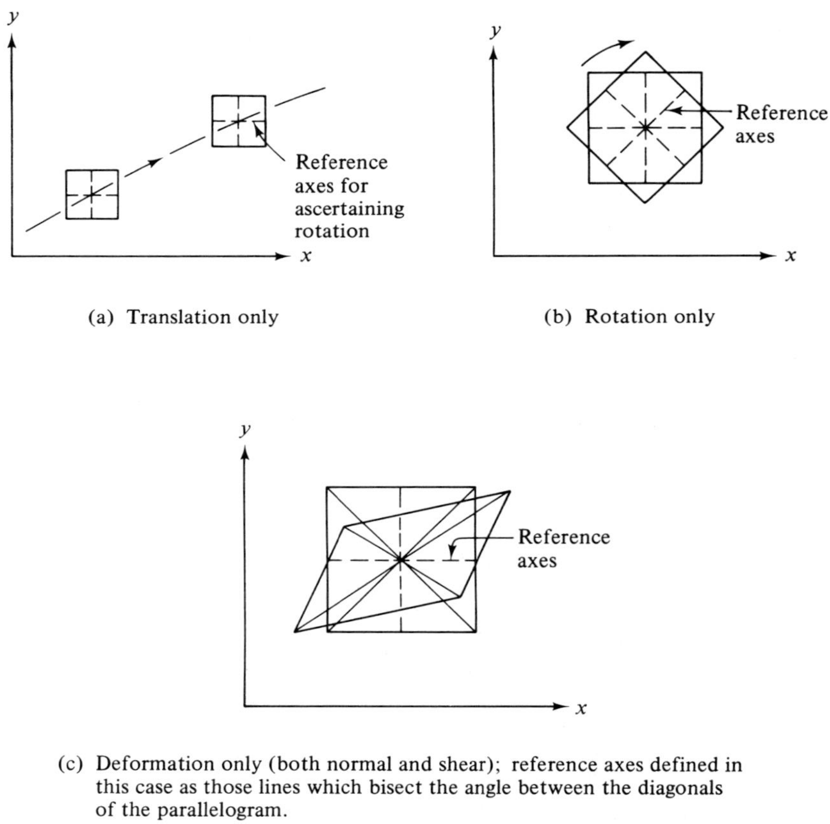

Fig. 2-7. Small two dimensional fluid parcel undergoing: (a) Translation only; (b) rotation only; (c) deformation only (both normal and shear); reference axes defined in this case as those lines which bisect the angle between the diagonals of the parallelogram.

of a ferris wheel. The seats have no spin and are not subject to deformation as they move along a curved path in space. (2) Particles of dust on a phonographic turntable are not deformed, but undergo both rotation and translation.

Practically, it is difficult to find examples of fluid motion without deformation since it is just this property that distinguishes a fluid from a solid; a fluid can be deformed without limit. In most fluid motions the three phenomena of translation, rotation, and deformation occur simultaneously.

Deformation can be of two distinct types: normal (dilatational or linear) deformation and shear deformation. In *normal deformation* the volume of the fluid parcel may or may not change, but all motion of the boundaries of the parcel is normal to the surface of the fluid parcel. If the volume changes, the fluid is said to have undergone *bulk deformation* (i.e., divergence), while if the volume is constant, the fluid is said to have undergone *pure stretching*. In the case of *shear deformation* the volume is constant and the motion of the boundaries of the parcel is not normal to the fluid parcel surface. These concepts are illustrated in Figure 2-8.

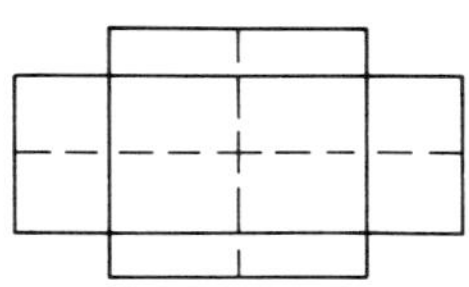

(a) Normal deformation (either bulk deformation or pure stretching).

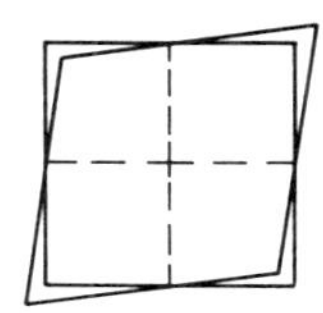

(b) Shear deformation

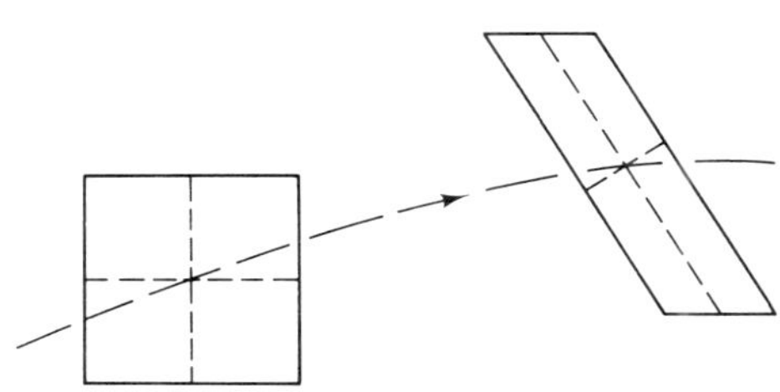

(c) Translation, rotation, normal, and shear deformation

Fig. 2-8. Deformation and the general mode: (a) Normal deformation (either bulk deformation or pure stretching); (b) shear deformation; (c) translation, rotation, normal, and shear deformation.

2-6 Deformation and Vorticity Tensors

We have seen that deformation is always accompanied by a change in the distance between particles making up a parcel. In normal deformation the spacing between particles located on opposite faces of an elemental parallelopiped (or parallelogram in two dimensions) changes, while for shear deformation the spacing between particles located at opposite corners of the parallelopiped change.

When dealing with a solid undergoing deformation one is concerned with the amount of deformation (or strain). With a fluid, however, there is no limit to the amount of deformation which results from the application of an external deforming force; as pointed out in the previous section, this is a distinguishing property of a fluid. Thus, unlike a solid, a fluid seeks no equilibrium of strain due to the action of a given stress. For a fluid the *rate of deformation* (*rate of strain*) is the important factor. This rate can be evaluated in terms of the rate of change of distance between adjacent fluid particles and is characterized by a tensor, which is determined by the velocity field.

Consider two particles A and B, located very near each other, whose positions relative to some fixed reference O is given by the vectors x_i' and y_i, respectively. The difference in positions A and B is denoted by the differential vector dx_i such that

$$y_i \equiv x_i' + dx_i \tag{2-13}$$

(Figure 2-9). The distance between the particles, denoted as L, is the magnitude of dx_i,

$$\begin{aligned} L^2 &\equiv (y_i - x_i')^2 = dx_i\, dx_i \\ &= dx_1^2 + dx_2^2 + dx_3^2 \end{aligned} \tag{2-14}$$

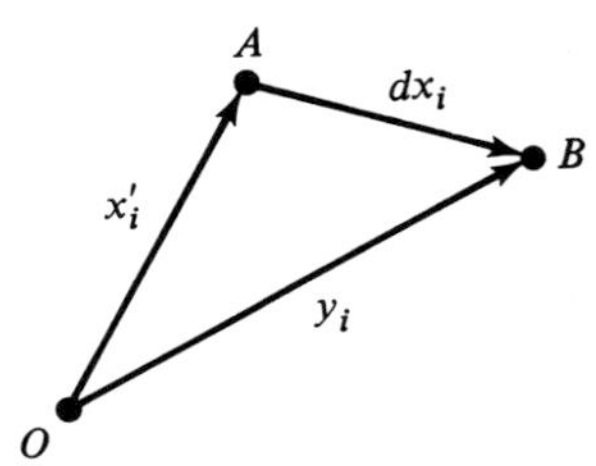

Fig. 2-9. Position diagram.

The time rate of change of L^2 is then

$$\frac{dL^2}{dt} = 2(y_i - x_i')\left(\frac{dy_i}{dt} - \frac{dx_i'}{dt}\right) \tag{2-15}$$

However, dx_i'/dt and dy_i/dt are merely the velocities of particles A and B. If we now let V_i be the velocity of A; that is,

$$V_i \equiv \frac{dx_i'}{dt} \tag{2-16}$$

then by Taylor's theorem the velocity of the adjacent particle B is given in terms of V_i as

$$\frac{dy_i}{dt} = V_i + \frac{\partial V_i}{\partial x_j} dx_j \tag{2-17}$$

Substituting into (2-15) yields

$$\frac{dL^2}{dt} = 2\frac{\partial V_i}{\partial x_j} dx_i\, dx_j \tag{2-17}$$

or the fractional rate of change of L is

$$\frac{1}{L}\frac{dL}{dt} = \frac{dx_i\, dx_j}{L^2}\frac{\partial V_i}{\partial x_j} \tag{2-18}$$

The quantity $\partial V_i/\partial x_j$, a tensor representing the gradient of the velocity at point A, is called the *velocity gradient tensor* and has the nine components

$$\begin{bmatrix} \dfrac{\partial V_1}{\partial x_1} & \dfrac{\partial V_1}{\partial x_2} & \dfrac{\partial V_1}{\partial x_3} \\ \dfrac{\partial V_2}{\partial x_1} & \dfrac{\partial V_2}{\partial x_2} & \dfrac{\partial V_2}{\partial x_3} \\ \dfrac{\partial V_3}{\partial x_1} & \dfrac{\partial V_3}{\partial x_2} & \dfrac{\partial V_3}{\partial x_3} \end{bmatrix}$$

Now the velocity gradient tensor can be written as the sum of two tensors, one being symmetric and one being antisymmetric:

$$\frac{\partial V_i}{\partial x_j} = \frac{1}{2}\left(\frac{\partial V_i}{\partial x_j} + \frac{\partial V_j}{\partial x_i}\right) + \frac{1}{2}\left(\frac{\partial V_i}{\partial x_j} - \frac{\partial V_j}{\partial x_i}\right) \tag{2-19}$$

whose components are

Symmetric tensor $\Longrightarrow$

$$\begin{bmatrix} \dfrac{\partial V_1}{\partial x_1} & \dfrac{1}{2}\left(\dfrac{\partial V_1}{\partial x_2} + \dfrac{\partial V_2}{\partial x_1}\right) & \dfrac{1}{2}\left(\dfrac{\partial V_1}{\partial x_3} + \dfrac{\partial V_3}{\partial x_1}\right) \\ \dfrac{1}{2}\left(\dfrac{\partial V_2}{\partial x_1} + \dfrac{\partial V_1}{\partial x_2}\right) & \dfrac{\partial V_2}{\partial x_2} & \dfrac{1}{2}\left(\dfrac{\partial V_2}{\partial x_3} + \dfrac{\partial V_3}{\partial x_2}\right) \\ \dfrac{1}{2}\left(\dfrac{\partial V_3}{\partial x_1} + \dfrac{\partial V_1}{\partial x_3}\right) & \dfrac{1}{2}\left(\dfrac{\partial V_3}{\partial x_2} + \dfrac{\partial V_2}{\partial x_3}\right) & \dfrac{\partial V_3}{\partial x_3} \end{bmatrix} \tag{2-20}$$

and

Antisymmetric tensor $\Longrightarrow$

$$\begin{bmatrix} 0 & \dfrac{1}{2}\left(\dfrac{\partial V_1}{\partial x_2} - \dfrac{\partial V_2}{\partial x_1}\right) & \dfrac{1}{2}\left(\dfrac{\partial V_1}{\partial x_3} - \dfrac{\partial V_3}{\partial x_1}\right) \\ \dfrac{1}{2}\left(\dfrac{\partial V_2}{\partial x_1} - \dfrac{\partial V_1}{\partial x_2}\right) & 0 & \dfrac{1}{2}\left(\dfrac{\partial V_2}{\partial x_3} - \dfrac{\partial V_3}{\partial x_2}\right) \\ \dfrac{1}{2}\left(\dfrac{\partial V_3}{\partial x_1} - \dfrac{\partial V_1}{\partial x_3}\right) & \dfrac{1}{2}\left(\dfrac{\partial V_3}{\partial x_2} - \dfrac{\partial V_2}{\partial x_3}\right) & 0 \end{bmatrix} \tag{2-21}$$

The symmetric tensor of (2-20) is called the *deformation tensor* (or rate of deformation tensor, strain rate tensor, rate of strain tensor, velocity strain tensor) and will be denoted by the symbol D_{ij},

$$D_{ij} \equiv \frac{1}{2}\left(\frac{\partial V_i}{\partial x_j} + \frac{\partial V_j}{\partial x_i}\right) \tag{2-22}$$

The antisymmetric tensor of (2-21), called the *vorticity tensor*, will be denoted by the symbol ξ_{ij},

$$\xi_{ij} = \frac{1}{2}\left(\frac{\partial V_i}{\partial x_j} - \frac{\partial V_j}{\partial x_i}\right) \tag{2-23}$$

From (2-20) and (2-21) it is clear that D_{ij} requires only six different quantities for its specification while ξ_{ij} requires only three. In the latter case, the three quantities are simply one-half of the components of the vector that we will come to know as the fluid vorticity. The entire nine components, then, specify the velocity gradient tensor,

$$\frac{\partial V_i}{\partial x_j} = D_{ij} + \xi_{ij} \tag{2-24}$$

Returning to (2-17), substitution of (2-24) yields

$$\frac{dL^2}{dt} = 2(D_{ij} + \xi_{ij})\, dx_i\, dx_j \tag{2-25}$$

However,

$$\xi_{ij} dx_i\, dx_j = 0$$

this being a characteristic property of any antisymmetric tensor, which can be readily verified by writing out the expression in full. We have, then, that

$$\frac{dL^2}{dt} = 2D_{ij}\, dx_i\, dx_j \tag{2-26}$$

confirming the fact that the rate of change of the spacing of adjacent particles of a fluid is governed by the deformation tensor alone—that the vorticity tensor plays no part in such a change.

In the following sections we will examine the physical significance of the deformation and vorticity tensors in detail; in preview however, we can state that:

1. The terms of the vorticity tensor measure the rigid-body rotation of a fluid particle.
2. The three major diagonal terms of the deformation tensor,

$$\frac{\partial V_1}{\partial x_1} \qquad \frac{\partial V_2}{\partial x_2} \qquad \frac{\partial V_3}{\partial x_3}$$

determine the rate of normal deformation.

3. The remaining three terms of the deformation tensor,

$$\frac{1}{2}\left(\frac{\partial V_1}{\partial x_2}+\frac{\partial V_2}{\partial x_1}\right) \qquad \frac{1}{2}\left(\frac{\partial V_1}{\partial x_3}+\frac{\partial V_3}{\partial x_1}\right) \qquad \frac{1}{2}\left(\frac{\partial V_2}{\partial x_3}+\frac{\partial V_3}{\partial x_2}\right)$$

determine the rate of shear deformation of a fluid element.

2-7 Normal Deformation

In this section, as well as the two that follow, the examples for the sake of simplicity will be that of two-dimensional flow. At the end of each section the results will be generalized for the case of three-dimensional motion.

Consider an infinitesimal fluid parcel, as shown in Figure 2-10, where the fluid velocity components at point x_1, x_2 are V_1 and V_2 in the x_1 and x_2 directions, respectively. Further, let us assume that V_1 is not a function of x_2 and V_2 is not a function of x_1, so that the orientation in space of the sides AB, BC, CD, and AD are invariant. From Taylor's theorem the velocity of AB is given by $V_1 - (\partial V_1/\partial x_1)(dx_1/2)$, of CD by $V_1 + (\partial V_1/\partial x_1)(dx_1/2)$, of AD by $V_2 - (\partial V_2/\partial x_2)(dx_2/2)$, and of BC by $V_2 + (\partial V_2/\partial x_2)(dx_2/2)$, to the first order. After a time interval dt, BC becomes $B'C'$ and increases in length such that

$$B'C' = dx_1 + \frac{\partial V_1}{\partial x_1} dx_1\, dt$$

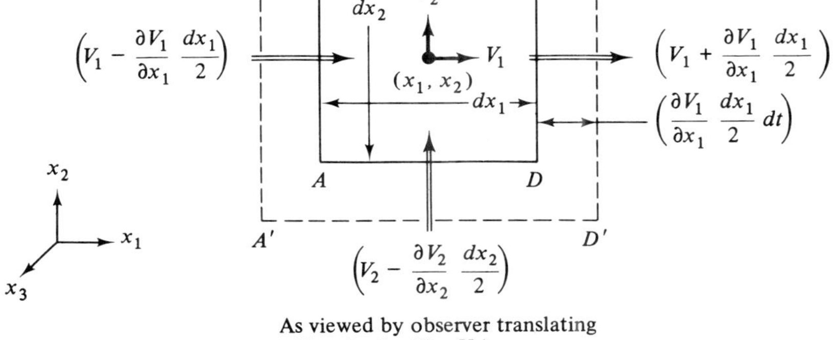

As viewed by observer translating with velocity (V_1, V_2).

Fig. 2-10. Normal deformation of an infinitesimal fluid element.

Similarly, CD becomes $C'D'$ and increases in length to

$$C'D' = dx_2 + \frac{\partial V_2}{\partial x_2}\, dx_2\, dt$$

The fractional rate of change of volume (taking a unit distance in the x_3 direction) is

$$\frac{1}{\upsilon}\frac{d\upsilon}{dt} = \frac{(A'B')(C'D') - (AB)(CD)}{(AB)(CD)\, dt}$$

$$= \frac{\left(dx_2 + \frac{\partial V_2}{\partial x_2}\, dx_2\, dt\right)\left(dx_1 + \frac{\partial V_1}{\partial x_1}\, dx_1\, dt\right) - dx_1\, dx_2}{dx_1\, dx_2\, dt}$$

which upon neglecting higher-order terms reduces to

$$\frac{1}{\upsilon}\frac{d\upsilon}{dt} = \left(\frac{\partial V_1}{\partial x_1} + \frac{\partial V_2}{\partial x_2}\right) \tag{2-27}$$

where υ is the volume.

In the general case of three-dimensional flow, (2-27) becomes

$$\frac{1}{\upsilon}\frac{d\upsilon}{dt} = \left(\frac{\partial V_1}{\partial x_1} + \frac{\partial V_2}{\partial x_2} + \frac{\partial V_3}{\partial x_3}\right) \tag{2-28}$$

the right side of which is merely the sum of the terms on the major diagonal of the deformation tensor D_{ij} as defined by (2-22). It represents the *normal deformation rate* (*dilatational* or *linear deformation rate*) of the fluid element as well as giving physical meaning to the divergence of the fluid velocity, since (2-28) can be written as

$$\frac{1}{\upsilon}\frac{d\upsilon}{dt} = \nabla \cdot \mathbf{V} \tag{2-29}$$

This asserts that for a compressible fluid the divergence of the velocity gives the fractional rate of change of volume, a positive value representing expansion, a negative value representing compression.

2-8 Shear Deformation

Again let us consider an infinitesimal fluid parcel where the fluid velocity components at the point x_1, x_2 are V_1 and V_2 (Figure 2-11). Now, however, we assume that V_1 is not a function of x_1 and V_2 is not a function of x_2 and that the gradient of V_1 in the x_2 direction is equal to the gradient of V_2 in the x_1 direction. The first requirement eliminates any normal deformation and the second eliminates any rotation. From Taylor's theorem, the component of the velocity of side AD in the x_1 direction is $V_1 - (\partial V_1/\partial x_2)(dx_2/2)$ and that of side BC in the x_1 direction is $V_1 + (\partial V_1/\partial x_2)(dx_2/2)$. Thus, CC'

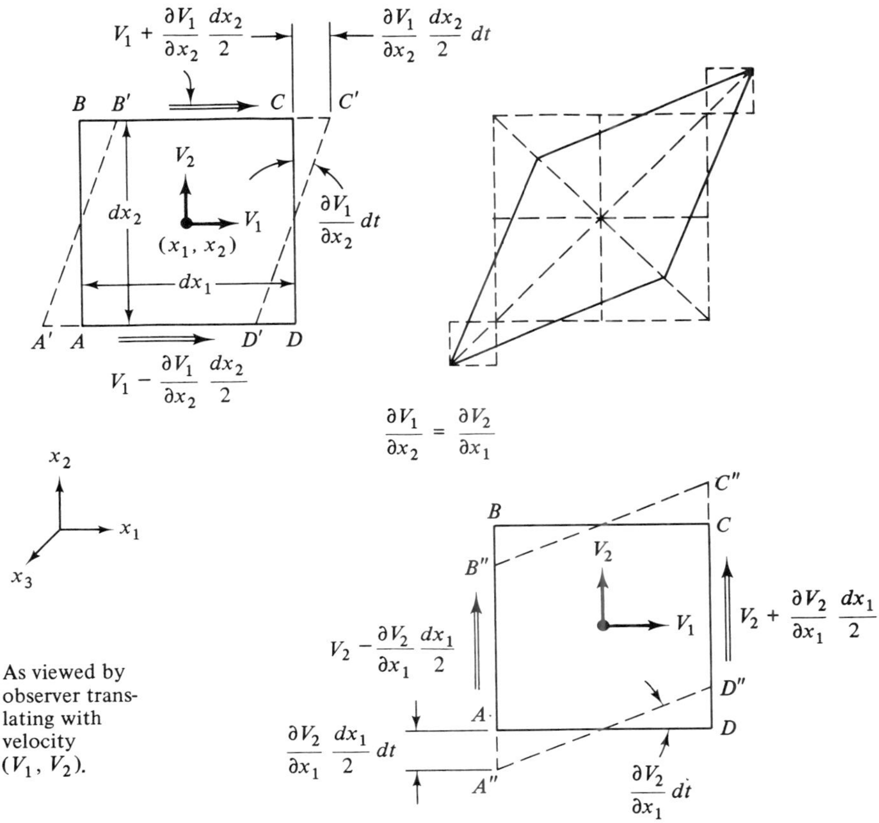

Fig. 2-11. Shear deformation of an infinitesimal fluid element.

(or AA', BB', or DD') after a time interval dt is given by $(\partial V_1/\partial x_2)(dx_2/2)\,dt$ and the angular speed of CD (or AB) is

$$\frac{\left(\dfrac{\partial V_1}{\partial x_2}\dfrac{dx_2}{2}\,dt\right)}{\dfrac{dx_2}{2}\,dt} = \frac{\partial V_1}{\partial x_2}$$

In a similar way, CC'' is given by $(\partial V_2/\partial x_1)(dx_1/2)\,dt$ and the angular speed of BC (or AD) is $\partial V_2/\partial x_1$. The rate of shear deformation is taken as the sum of these two angular speeds, $\partial V_1/\partial x_2 + \partial V_2/\partial x_1$.

For the general case of three-dimensional flow the rate of shear deformation is specified by the three terms $\partial V_1/\partial x_2 + \partial V_2/\partial x_1$, $\partial V_1/\partial x_3 + \partial V_3/\partial x_1$, and $\partial V_2/\partial x_3 + \partial V_3/\partial x_2$, which are twice the off-diagonal terms of the deformation tensor of (2-20) or (2-22).

2-9 Rotation and Vorticity

In the discussion of shear deformation we made the assumptions that for two-dimensional flow the fluid velocity component V_1 was not a function of x_1, that V_2 was not a function of x_2, and that the gradient of V_1 in the x_2 direction was equal to the gradient of V_2 in the x_1 direction. These conditions resulted in the fluid element undergoing only shear deformation. The upper right diagram of Figure 2-11 illustrates the fact that the bisectors of the angles made by the diagonals of the fluid particle remain parallel to their original positions; thus, no rotation has occurred.

Consider now the case identical to that just discussed with respect to shear deformation with the exception that the space derivatives of V_1 and V_2 are the negative of each other, that is,

$$-\frac{\partial V_1}{\partial x_2} = \frac{\partial V_2}{\partial x_1}$$

Taking $\partial V_1/\partial x_2$ as positive, sides AB, BC, CD, and AD all rotate clockwise through an angle $(\partial V_2/\partial x_1)\,dt$, or $(\partial V_1/\partial x_2)\,dt$, in a time interval dt as seen in Figure 2-12. Here we then have no shear deformation (nor normal deformation), but only rotation about an axis through the center of the element and

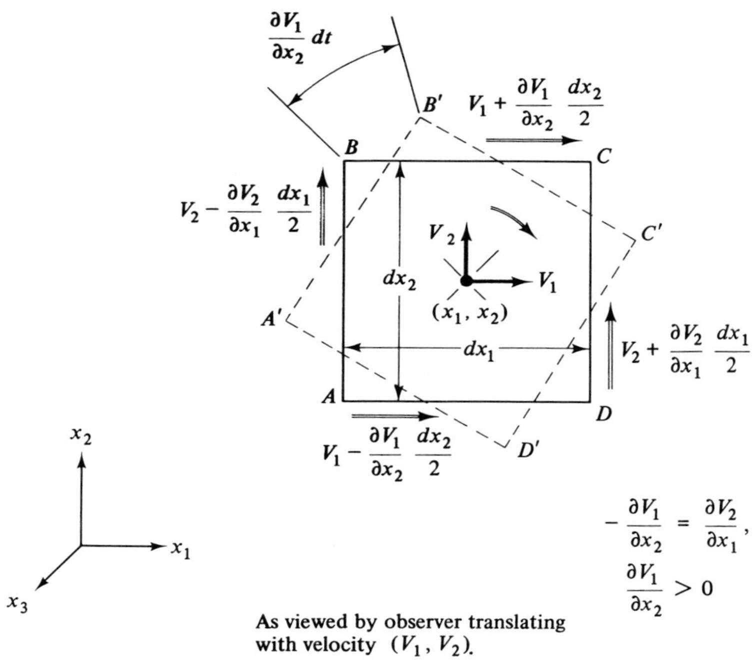

Fig. 2-12. Rotation of an infinitesimal fluid element.

parallel to the x_3 axis with an angular speed of $\partial V_1/\partial x_2$ or $\partial V_2/\partial x_1$. Both of these represent a rate of rotation in the same sense in that by convention we take counterclockwise rotation as positive and clockwise rotation as negative (Figure 2-13).

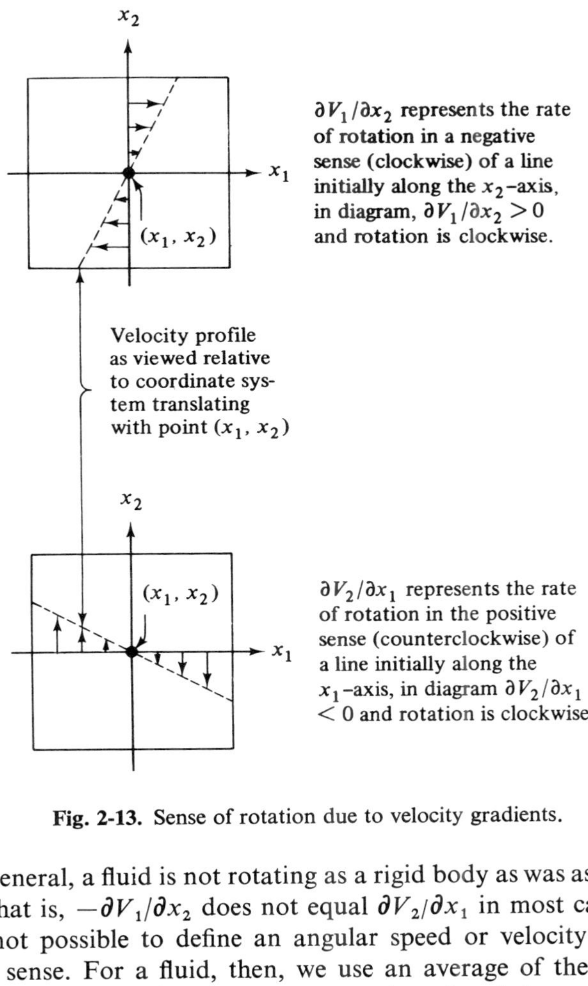

Fig. 2-13. Sense of rotation due to velocity gradients.

In general, a fluid is not rotating as a rigid body as was assumed in our example, that is, $-\partial V_1/\partial x_2$ does not equal $\partial V_2/\partial x_1$ in most cases, and it is therefore not possible to define an angular speed or velocity in the usual rigid-body sense. For a fluid, then, we use an average of the two angular speeds, $1/2(\partial V_2/\partial x_1 - \partial V_1/\partial x_2)$, the negative sign being required since $\partial V_2/\partial x_1 > 0$ results in positive rotation while $\partial V_1/\partial x_2 > 0$ produces negative rotation. We can now define the *vorticity* of the fluid about the x_3 axis as ζ_3 such that

$$\zeta_3 \equiv \frac{\partial V_2}{\partial x_1} - \frac{\partial V_1}{\partial x_2}$$

For the case of three-dimensional flow, we would expect to have rotation (or vorticity) about all three of the Cartesian axes and the preceding relation can be expanded to the three equations:

$$\zeta_1 = \frac{\partial V_3}{\partial x_2} - \frac{\partial V_2}{\partial x_3}$$
$$\zeta_2 = \frac{\partial V_1}{\partial x_3} - \frac{\partial V_3}{\partial x_1} \tag{2-30}$$
$$\zeta_3 = \frac{\partial V_2}{\partial x_1} - \frac{\partial V_1}{\partial x_2}$$

It is apparent that these can be represented as the single vector equation

$$\boldsymbol{\zeta} = \zeta_1 \mathbf{1}_1 + \zeta_2 \mathbf{1}_2 + \zeta_3 \mathbf{1}_3 \tag{2-31}$$

where $\boldsymbol{\zeta}$ is the vorticity vector and $\mathbf{1}_1$, $\mathbf{1}_2$, and $\mathbf{1}_3$ are unit vectors in the x_1, x_2, and x_3 directions, respectively. Clearly, (2-31) is identical to (1-13); the vorticity is merely the curl of the fluid velocity vector

$$\zeta = \nabla \times \mathbf{V} \tag{2-32}$$

or, in indicial notation,

$$\zeta_i = \epsilon_{ijk} \frac{\partial V_k}{\partial x_j} \tag{2-33}$$

Finally, it is seen that ζ_1, ζ_2, and ζ_3 are, from (2-23), just twice the value of the three terms required to specify the vorticity tensor ξ_{ij}.

2-10 Summary of Relations Concerning Two-Dimensional Rotation and Shear

For the two-dimensional flow situation the differential requirements on the velocity components are, in summary, as follows:

1. Shear deformation without rotation:

$$\frac{\partial V_1}{\partial x_2} - \frac{\partial V_2}{\partial x_1} = 0$$

and

$$\frac{\partial V_1}{\partial x_2} + \frac{\partial V_2}{\partial x_1} \neq 0$$

2. Rotation without shear deformation;

$$\frac{\partial V_1}{\partial x_2} - \frac{\partial V_2}{\partial x_1} \neq 0$$

and

$$\frac{\partial V_1}{\partial x_2} + \frac{\partial V_2}{\partial x_1} = 0$$

3. Rotation and shear deformation;

$$\frac{\partial V_1}{\partial x_2} - \frac{\partial V_2}{\partial x_1} \neq 0$$

and

$$\frac{\partial V_1}{\partial x_2} + \frac{\partial V_2}{\partial x_1} \neq 0$$

2-11 Relative Motion Analysis

In this section we will examine and make an analysis of the character of the motion of a fluid in the neighborhood of a point in the fluid. To proceed, assume that the velocity of the fluid at a point P with position vector x_i at time t is $V_i(x_i, t)$, and that simultaneously the velocity at an adjacent point Q with position vector $(x_i + r_i)$ is $(V_i + \delta V_i)$; Figure 2-14. Then, to the first order in r_i, at time t

$$\delta V_i = r_1 \frac{\partial V_i}{\partial x_1} + r_2 \frac{\partial V_i}{\partial x_2} + r_3 \frac{\partial V_i}{\partial x_3}$$

or

$$\delta V_i = r_j \frac{\partial V_i}{\partial x_j} \tag{2-34}$$

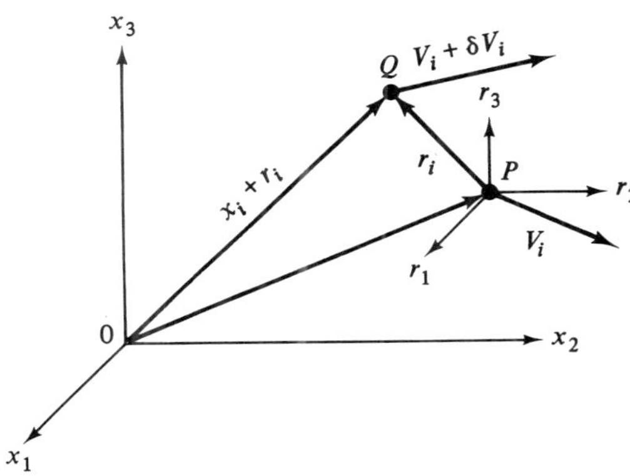

Fig. 2-14. Relative motion of adjacent fluid particles.

We have already seen that the tensor $\partial V_i/\partial x_j$, the velocity gradient tensor, can be divided into two tensors, one of which is symmetrical and one of which is antisymmetrical. Therefore, we can write that

$$\delta V_i = \delta V_i^{(s)} + \delta V_i^{(a)} \tag{2-35}$$

where

$$\delta V_i^{(s)} \equiv r_j D_{ij} \tag{2-36}$$

and

$$\delta V_i^{(a)} \equiv r_j \xi_{ij} \tag{2-37}$$

D_{ij} and ξ_{ij} being defined by (2-22) and (2-23), respectively, as the deformation tensor and the vorticity tensor.

The first term on the right of (2-35), $\delta V_i^{(s)}$, can be written as the negative gradient of a scalar function ϕ. (Since we are dealing only with the symmetric portion of the motion, that portion being irrotational, ϕ is actually what we will later call the *velocity potential function*, the negative gradient of which gives the velocity.) Thus,

$$\delta V_i^{(s)} = r_j D_{ij} = -\frac{\partial \phi}{\partial r_i} \tag{2-38}$$

where

$$\phi = -\tfrac{1}{2} r_k r_m D_{km} \tag{2-39}$$

To verify (2-39) we can expand first on k to obtain

$$\phi = -(\tfrac{1}{2} r_1 r_m D_{1m} + \tfrac{1}{2} r_2 r_m D_{2m} + \tfrac{1}{2} r_3 r_m D_{3m})$$

and then on m,

$$\begin{aligned} \phi = -(&\tfrac{1}{2} r_1^2 D_{11} + \tfrac{1}{2} r_1 r_2 D_{12} + \tfrac{1}{2} r_1 r_3 D_{13} \\ &+ \tfrac{1}{2} r_2 r_1 D_{21} + \tfrac{1}{2} r_2^2 D_{22} + \tfrac{1}{2} r_2 r_3 D_{23} \\ &+ \tfrac{1}{2} r_3 r_1 D_{31} + \tfrac{1}{2} r_3 r_2 D_{32} + \tfrac{1}{2} r_3^2 D_{33}) \end{aligned}$$

the negative gradient of which satisfies (2-38). $\phi(r_i) =$ constant defines a family of surfaces which are similar *quadrics* (i.e., family of ellipsoids, hyperboloids, or some special form of these); and from (2-38), $\delta V_i^{(s)}$ is normal to the quadric passing through the point determined by r_i.

Additional insight can be gained as to the character of $\delta V_i^{(s)}$ by considering the following. In terms of matrix notation (2-39) can be written as

$$\phi = -\tfrac{1}{2} \mathsf{r}^T \mathsf{D} \mathsf{r} \tag{2-40}$$

where r is a 3×1 matrix, D is a 3×3 matrix, and r^T is the transpose of r; thus,

$$\phi = -\tfrac{1}{2}[r_1 r_2 r_3] \begin{bmatrix} D_{11} & D_{12} & D_{13} \\ D_{21} & D_{22} & D_{23} \\ D_{31} & D_{32} & D_{33} \end{bmatrix} \begin{bmatrix} r_1 \\ r_2 \\ r_3 \end{bmatrix} \tag{2-41}$$

The transformation of the vector r_i from one Cartesian coordinate system to another can be written, as seen from (1-36), as the matrix equation

$$\tilde{\mathsf{r}} - \mathsf{A}\mathsf{r} \tag{2-42}$$

or

$$\begin{bmatrix} \tilde{r}_1 \\ \tilde{r}_2 \\ \tilde{r}_3 \end{bmatrix} = \begin{bmatrix} \alpha_{11} & \alpha_{12} & \alpha_{13} \\ \alpha_{21} & \alpha_{22} & \alpha_{23} \\ \alpha_{31} & \alpha_{32} & \alpha_{33} \end{bmatrix} \begin{bmatrix} r_1 \\ r_2 \\ r_3 \end{bmatrix}$$

where the matrix A is the rotation matrix, which we have seen has as its members the direction cosines between the coordinate axes of the two systems. Recall, for example, that α_{32} is the cosine of the angle between the $\tilde{r}_3$ axis and the r_2 axis. Now multiplying both sides of (2-42) by the transpose of A, A^T, we get

$$\mathsf{A}^T\tilde{\mathsf{r}} = \mathsf{A}^T\mathsf{A}\mathsf{r}$$

which since A is orthogonal can be written, from (1-42), as

$$\mathsf{A}^T\tilde{\mathsf{r}} = \mathsf{r} \tag{2-43}$$

and again taking the transpose to obtain

$$\mathsf{r}^T = (\mathsf{A}^T\tilde{\mathsf{r}})^T = \tilde{\mathsf{r}}^T\mathsf{A} \tag{2-44}$$

Returning to (2-40), substitution of (2-43) and (2-44) results in

$$\phi = -\tfrac{1}{2}\tilde{\mathsf{r}}^T\mathsf{A}\mathsf{D}\mathsf{A}^T\tilde{\mathsf{r}} \tag{2-45}$$

Since D is a symmetric matrix, we have from (1-46) and (1-58) that

$$\mathsf{A}\mathsf{D}\mathsf{A}^T = \begin{bmatrix} \lambda^{(1)} & 0 & 0 \\ 0 & \lambda^{(2)} & 0 \\ 0 & 0 & \lambda^{(3)} \end{bmatrix} \equiv \tilde{\mathsf{D}} \tag{2-46}$$

where $\tilde{\mathsf{D}}$ is a diagonal matrix with the three eigenvalues $\lambda^{(1)}$, $\lambda^{(2)}$, and $\lambda^{(3)}$ on the diagonal. Using this, the velocity potential function ϕ has the particularly simple form

$$\phi = -\tfrac{1}{2}\tilde{\mathsf{r}}^T\tilde{\mathsf{D}}\tilde{\mathsf{r}} \tag{2-47}$$

or

$$\phi = -\tfrac{1}{2}[\tilde{r}_1\tilde{r}_2\tilde{r}_3] \begin{bmatrix} \lambda^{(1)} & 0 & 0 \\ 0 & \lambda^{(2)} & 0 \\ 0 & 0 & \lambda^{(3)} \end{bmatrix} \begin{bmatrix} \tilde{r}_1 \\ \tilde{r}_2 \\ \tilde{r}_3 \end{bmatrix} \tag{2-48}$$

which upon expansion yields

$$\phi = -\tfrac{1}{2}(\lambda^{(1)}\tilde{r}_1^2 + \lambda^{(2)}\tilde{r}_2^2 + \lambda^{(3)}\tilde{r}_3^2) \tag{2-49}$$

where $\tilde{r}_1$, $\tilde{r}_2$, and $\tilde{r}_3$ are the components of the vector r_i referred to the tilde coordinate system. We recognize (2-49) as that of a family of quadrics whose axes of reference coincide with the principal axes of the tensor D_{ij}. Furthermore, it is readily verified that

$$\tilde{D}_{ii} = \alpha_{ki}\alpha_{mi}D_{km} = \delta_{km}D_{km} = D_{kk} \tag{2-50}$$

or

$$\lambda^{(1)} + \lambda^{(2)} + \lambda^{(3)} = \frac{\partial V_k}{\partial x_k} \tag{2-51}$$

which states that the trace of a symmetric tensor is invariant under any coordinate transformation, the trace of D_{ij} being the divergence of the velocity.

From (2-38) we see that in the principal coordinate system the contribution $\delta V_i^{(s)}$ to the relative velocity has the three components

$$\delta V_i^{(s)} = (\lambda_1 \tilde{r}_1, \lambda_2 \tilde{r}_2, \lambda_3 \tilde{r}_3) \tag{2-52}$$

This means that any material line element near the position x_i that is parallel to the r_1 axis (thus all points on the line element have the same r_2 and r_3 coordinates) continues to maintain its orientation but is stretched at a rate $\tilde{D}_{11} = \lambda^{(1)}$. Similarly, all material line elements parallel to the $\tilde{r}_2$ and $\tilde{r}_3$ axes are being stretched at rates $\lambda^{(2)}$ and $\lambda^{(3)}$, respectively, without rotation (Figure 2-15). Any material line element not parallel to one of the $\tilde{r}_1$, $\tilde{r}_2$, or $\tilde{r}_3$ axes will, in general, experience both stretching and rotation. This stretching and rotation, however, will be only to the extent required so as to maintain continuity with the pure stretching motion undergone by the line elements parallel to any one of the principal axes.

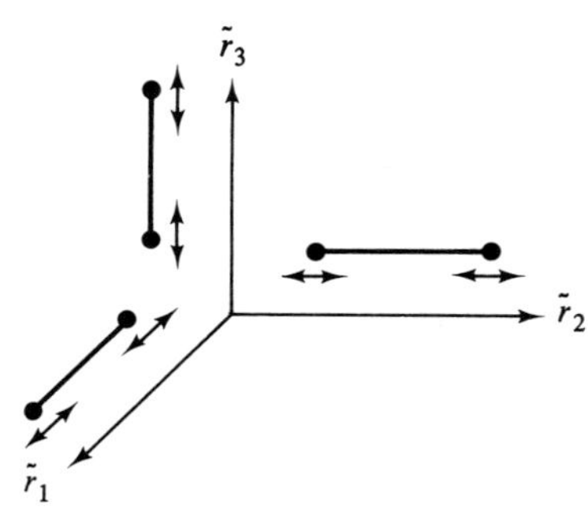

Fig. 2-15. Symmetrical portion of relative motion of material line elements parallel to the principal axes.

The contribution of $\delta V_i^{(s)}$ to the relative motion represents pure straining motion. The deformation tensor D_{ij} is completely determined by the direction of its principal axes and by the three principal rates of strain $\lambda^{(1)}$, $\lambda^{(2)}$, and $\lambda^{(3)}$. Another way of looking at the symmetric portion of the relative velocity field is that a material surface element near point P (Figure 2-14) which initially is spherical is converted into an ellipsoid with principal diameters which do not rotate and having rates of extension proportional to $\lambda^{(1)}$, $\lambda^{(2)}$, and $\lambda^{(3)}$. If the fluid is incompressible, the volume of the ellipsoid is constant ($\partial V_i/\partial x_i = 0$) and $D_{ii} = 0$. On the other hand, if the fluid is compressible, it is convenient at times to consider the pure straining motion as the superposition of an isotropic expansion, for which the rate of extension of all line elements parallel to the principal axes is $\frac{1}{3}D_{ii}$, and a straining motion without a change in volume as follows:

$$D_{ij} = \tfrac{1}{3}\,\delta_{ij}D_{kk} + (D_{ij} - \tfrac{1}{3}\,\delta_{ij}D_{kk})$$
$$= \frac{1}{3}\begin{bmatrix} D_{ii} & 0 & 0 \\ 0 & D_{ii} & 0 \\ 0 & 0 & D_{ii} \end{bmatrix} + \begin{bmatrix} D_{11} - \frac{1}{3}D_{ii} & D_{12} & D_{13} \\ D_{21} & D_{22} - \frac{1}{3}D_{ii} & D_{23} \\ D_{31} & D_{32} & D_{33} - \frac{1}{3}D_{ii} \end{bmatrix} \qquad (2\text{-}53)$$

where the first term on the right represents the isotropic expansion and the second the nondivergent straining motion. The velocity potential ϕ, from (2-39), is then written

$$\phi = -\tfrac{1}{2}r_i r_j(D_{ij} - \tfrac{1}{3}\delta_{ij}D_{kk}) - \tfrac{1}{6}r_i r_j \delta_{ij} D_{kk} \qquad (2\text{-}54)$$

Returning to (2-37), we next examine that portion of the relative velocity field represented by the antisymmetric tensor ξ_{ij}. Since ξ_{ij} is antisymmetric, it has only three independent components and can be written as

$$\xi_{ij} = \frac{1}{2}\begin{bmatrix} 0 & -\zeta_3 & \zeta_2 \\ \zeta_3 & 0 & -\zeta_1 \\ -\zeta_2 & \zeta_1 & 0 \end{bmatrix}$$

or

$$\xi_{ij} = -\tfrac{1}{2}\epsilon_{ijk}\zeta_k \qquad (2\text{-}55)$$

where ζ_k is the vorticity vector with components ζ_1, ζ_2, and ζ_3 and the factor of $-\frac{1}{2}$ is included so as to make the definition of ξ_{ij} as given by (2-33) compatible with the definition of the vorticity as being the curl of the velocity in (2-32). The contribution then of the antisymmetrical part of the relative velocity $\partial V_i^{(a)}$ to the relative velocity δV_i is, from (2-37) and (2-55),

$$\delta V_i^{(a)} = r_j \xi_{ij} = -\tfrac{1}{2}\epsilon_{ijk} r_j \zeta_k \qquad (2\text{-}56)$$

or, in conventional vector notation,

$$\delta \mathbf{V}^{(a)} = -\tfrac{1}{2}(\mathbf{r} \times \boldsymbol{\zeta}) = \tfrac{1}{2}(\boldsymbol{\zeta} \times \mathbf{r}) \qquad (2\text{-}57)$$

from which we see that $\delta \mathbf{V}^{(a)}$ can be viewed as the velocity produced at the position $\mathbf{r}$ relative to a point about which there exists a rigid-body rotation of angular velocity $\frac{1}{2}\boldsymbol{\zeta}$.

To summarize, it has been shown that to the first order in the linear dimensions of a small region surrounding the position x_i, the velocity field can be considered to consist of the superposition of:

1. A uniform translation with the velocity $V_i(x_i)$.
2. A pure straining motion characterized by the deformation tensor D_{ij}, which in turn may be resolved into an isotropic expansion and a straining motion without a change of volume.
3. A rigid-body rotation with an angular velocity of $\frac{1}{2}\zeta_i$.

With this in mind, the fluid velocity at the point $(x_i + r_i)$ may be written to the first-order as

$$V_i(x_i + r_i) = V_i(x_i) + \frac{\partial}{\partial r_i}(\tfrac{1}{2}r_j r_k D_{jk} + \tfrac{1}{2}\epsilon_{ijk}\zeta_j r_k \qquad (2\text{-}58)$$

where D_{ij} and ζ_j are evaluated at point x_i.

To firm up the concepts just covered concerning relative motion, we will analyze two simple examples in some detail.

First, consider the case of simple shearing motion in which plane layers of fluid slide over each other. The relative velocity δV_i then is taken everywhere in the same direction and is assumed to vary linearly in magnitude with respect to the distance perpendicular to δV_i. Let coordinate axes, as in Figure 2-16, be chosen such that the velocity gradient $\partial V_i/\partial x_j$ is nonzero only for $i = 1, j = 2$; that is, from (2-34),

$$\delta V_i = \left(r_2 \frac{\partial V_1}{\partial x_2}, 0, 0\right) \qquad (2\text{-}59)$$

Also, from (2-39),

$$\phi = -\tfrac{1}{2}r_1 r_2 \frac{\partial V_1}{\partial x_2} \qquad (2\text{-}60)$$

and from (2-33),

$$\zeta_i = \left(0, 0, -\frac{\partial V_1}{\partial x_2}\right) \qquad (2\text{-}61)$$

To examine the various symmetric and antisymmetric features of the flow and to see how they fall out of the analysis, we can start with the

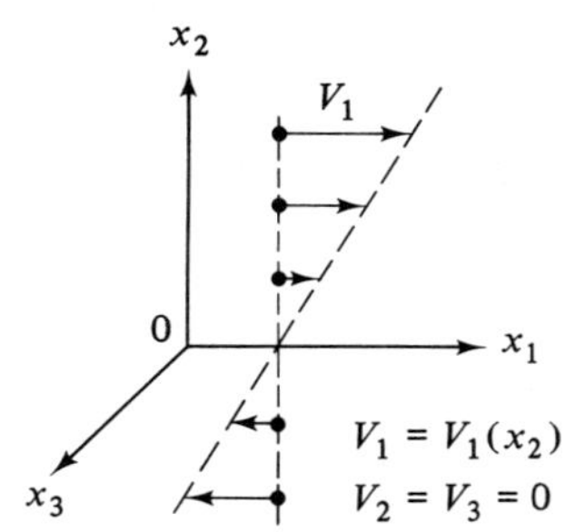

Fig. 2-16. Simple shearing motion.

deformation tensor written as

$$D_{ij} = \frac{1}{2}\left(\frac{\partial V_i}{\partial x_j} + \frac{\partial V_j}{\partial x_i}\right) = \begin{bmatrix} 0 & \frac{1}{2}S & 0 \\ \frac{1}{2}S & 0 & 0 \\ 0 & 0 & 0 \end{bmatrix}$$

where

$$S \equiv \frac{\partial V_1}{\partial x_2}$$

From this and (2-41), ϕ is given by

$$\phi = -\tfrac{1}{2}[r_1 r_2 r_3] \begin{bmatrix} 0 & \frac{1}{2}S & 0 \\ \frac{1}{2}S & 0 & 0 \\ 0 & 0 & 0 \end{bmatrix} \begin{bmatrix} r_1 \\ r_2 \\ r_3 \end{bmatrix}$$

$$= -\tfrac{1}{2}[r_1 r_2 r_3] \begin{bmatrix} \frac{1}{2}r_2 S \\ \frac{1}{2}r_1 S \\ 0 \end{bmatrix} = -\tfrac{1}{2}r_1 r_2 S$$

From (2-38) we can obtain the symmetric component of the flow directly from the negative of the gradient of ϕ, and in this particular case it is not so complicated that we cannot see what it means. Thus,

$$\delta V_i^{(s)} = -\frac{\partial \phi}{\partial r_i} = (\tfrac{1}{2}r_2 S, \tfrac{1}{2}r_1 S, 0) \tag{2-62}$$

a plot of which is shown in Figure 2-17.

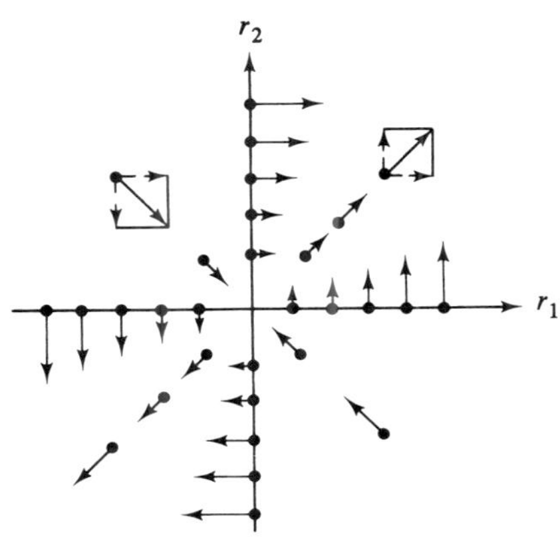

Fig. 2-17. Symmetric part of relative flow for simple shear.

Since we are trying to understand how the mathematics relates to the flow characteristics, let us proceed to transform the given coordinates into a principal-axes system. The eigenvalues of D, from (2-50), are found by solving the determinant

$$\begin{vmatrix} D_{11} - \lambda & D_{12} & D_{13} \\ D_{21} & D_{22} - \lambda & D_{23} \\ D_{31} & D_{32} & D_{33} - \lambda \end{vmatrix} = 0$$

which results in the characteristic equation

$$\lambda^3 - \tfrac{1}{4}S^2\lambda = 0$$

whose roots are

$$\lambda = -\tfrac{1}{2}S, \quad 0, \quad \tfrac{1}{2}S$$

Utilizing (1-56), the eigenvectors are given by

$$D_{ij}e_j^{(n)} = \lambda^{(n)}e_i^{(n)} \qquad n = 1, 2, 3$$

Applying this to the eigenvalue of $\lambda^{(1)} = -\frac{1}{2}S$, we have that

$$\begin{bmatrix} 0 & \frac{1}{2}S & 0 \\ \frac{1}{2}S & 0 & 0 \\ 0 & 0 & 0 \end{bmatrix} \begin{bmatrix} e_1^{(1)} \\ e_2^{(1)} \\ e_3^{(1)} \end{bmatrix} = -\tfrac{1}{2}S \begin{bmatrix} e_1^{(1)} \\ e_2^{(1)} \\ e_3^{(1)} \end{bmatrix}$$

or

$$\begin{aligned} \tfrac{1}{2}Se_2^{(1)} &= -\tfrac{1}{2}Se_1^{(1)} \\ \tfrac{1}{2}Se_1^{(1)} &= -\tfrac{1}{2}Se_2^{(1)} \\ 0 &= -\tfrac{1}{2}Se_3^{(1)} \end{aligned}$$

In addition, we require the eigenvector $e_i^{(1)}$ to be of unit magnitude, imposing the condition

$$e_i^{(1)^2} = 1$$

Solving this set of equations, we obtain

$$e_1^{(1)} = -\frac{1}{\sqrt{2}} \qquad e_2^{(1)} = \frac{1}{\sqrt{2}} \qquad e_3^{(1)} = 0$$

which is the first eigenvector, and as such, is the first row $(\alpha_{11}, \alpha_{12}, \alpha_{13})$ of the rotation matrix of (1-33) and (1-54), giving the direction of the principal axis designated by the bracked superscript (1) with respect to the original axes designated by the subscripts 1, 2, and 3. Continuing, for the second eigenvector,

$$\begin{bmatrix} 0 & \frac{1}{2}S & 0 \\ \frac{1}{2}S & 0 & 0 \\ 0 & 0 & 0 \end{bmatrix} \begin{bmatrix} e_1^{(2)} \\ e_2^{(2)} \\ e_3^{(2)} \end{bmatrix} = 0 \begin{bmatrix} e_1^{(2)} \\ e_2^{(2)} \\ e_3^{(2)} \end{bmatrix}$$

or

$$e_2^{(2)} = 0$$
$$e_1^{(2)} = 0$$

and

$$e_i^{(2)^2} = 1$$

which gives, upon solving,

$$e_1^{(2)} = 0, \qquad e_2^{(2)} = 0, \qquad e_3^{(2)} = 1$$

the second row of the rotation matrix; the direction of the principal axis designated by the superscript (2) with respect to the original axes designated by the subscripts 1, 2, and 3. For the last eigenvector,

$$\begin{bmatrix} 0 & \frac{1}{2}S & 0 \\ \frac{1}{2}S & 0 & 0 \\ 0 & 0 & 0 \end{bmatrix} \begin{bmatrix} e_1^{(3)} \\ e_2^{(3)} \\ e_3^{(3)} \end{bmatrix} = \frac{1}{2}S \begin{bmatrix} e_1^{(3)} \\ e_2^{(3)} \\ e_3^{(3)} \end{bmatrix}$$

yielding

$$\tfrac{1}{2}Se_2^{(3)} = \tfrac{1}{2}Se_1^{(3)}$$
$$\tfrac{1}{2}Se_1^{(3)} = \tfrac{1}{2}Se_2^{(3)}$$
$$0 = \tfrac{1}{2}Se_3^{(3)}$$

which, along with the condition for unit magnitude, gives the solution

$$e_1^{(3)} = \frac{1}{\sqrt{2}} \qquad e_2^{(3)} = \frac{1}{\sqrt{2}} \qquad e_3^{(3)} = 0$$

Putting all of this together, we have the complete rotation matrix

$$\mathsf{A} = \alpha_{ij} = \begin{bmatrix} -\dfrac{1}{\sqrt{2}} & \dfrac{1}{\sqrt{2}} & 0 \\ 0 & 0 & 1 \\ \dfrac{1}{\sqrt{2}} & \dfrac{1}{\sqrt{2}} & 0 \end{bmatrix}$$

As a check on our work up to this point we can apply (1-46) in the form

$$\tilde{\mathsf{D}} = \mathsf{ADA}^T$$

where $\tilde{\mathsf{D}}$ is given by

$$\tilde{\mathsf{D}} = \begin{bmatrix} \lambda^{(1)} & 0 & 0 \\ 0 & \lambda^{(2)} & 0 \\ 0 & 0 & \lambda^{(3)} \end{bmatrix}$$

Substituting then for this particular flow,

$$\mathsf{ADA}^T = \begin{bmatrix} -\frac{1}{\sqrt{2}} & \frac{1}{\sqrt{2}} & 0 \\ 0 & 0 & 1 \\ \frac{1}{\sqrt{2}} & \frac{1}{\sqrt{2}} & 0 \end{bmatrix} \begin{bmatrix} 0 & \frac{1}{2}S & 0 \\ \frac{1}{2}S & 0 & 0 \\ 0 & 0 & 0 \end{bmatrix} \begin{bmatrix} -\frac{1}{\sqrt{2}} & 0 & \frac{1}{\sqrt{2}} \\ \frac{1}{\sqrt{2}} & 0 & \frac{1}{\sqrt{2}} \\ 0 & 1 & 0 \end{bmatrix}$$

$$= \begin{bmatrix} \frac{1}{2\sqrt{2}}S & -\frac{1}{2\sqrt{2}}S & 0 \\ 0 & 0 & 0 \\ \frac{1}{2\sqrt{2}}S & \frac{1}{2\sqrt{2}}S & 0 \end{bmatrix} \begin{bmatrix} -\frac{1}{\sqrt{2}} & 0 & \frac{1}{\sqrt{2}} \\ \frac{1}{\sqrt{2}} & 0 & \frac{1}{\sqrt{2}} \\ 0 & 1 & 0 \end{bmatrix}$$

$$= \begin{bmatrix} -\frac{1}{2}S & 0 & 0 \\ 0 & 0 & 0 \\ 0 & 0 & \frac{1}{2}S \end{bmatrix}$$

$$= \begin{bmatrix} \lambda^{(1)} & 0 & 0 \\ 0 & \lambda^{(2)} & 0 \\ 0 & 0 & \lambda^{(3)} \end{bmatrix} = \tilde{\mathsf{D}} \qquad \text{Q.E.D.}$$

To determine the orientation of the principal coordinates, we write (2-42) as

$$\tilde{r}_i = \alpha_{ij} r_j$$

which after substituting for α_{ij} gives

$$\tilde{r}_1 = -\frac{1}{\sqrt{2}} r_1 + \frac{1}{\sqrt{2}} r_2$$

$$\tilde{r}_2 = r_3$$

$$\tilde{r}_3 = \frac{1}{\sqrt{2}} r_1 + \frac{1}{\sqrt{2}} r_2$$

and in the tilde coordinate system the velocity potential ϕ is, from (2-47),

$$\phi = -\tfrac{1}{2}\tilde{\mathsf{r}}^T \tilde{\mathsf{D}} \tilde{\mathsf{r}}$$

$$= -\tfrac{1}{2}[r_1 r_2 r_3] \begin{bmatrix} -\frac{1}{2}S & 0 & 0 \\ 0 & 0 & 0 \\ 0 & 0 & \frac{1}{2}S \end{bmatrix} \begin{bmatrix} \tilde{r}_1 \\ \tilde{r}_2 \\ \tilde{r}_3 \end{bmatrix}$$

$$= \tfrac{1}{4}S\tilde{r}_1^2 - \tfrac{1}{4}S\tilde{r}_3^2$$

which, as shown in Figure 2-18, plots as a family of hyperbolas in the $\tilde{r}_1$, $\tilde{r}_3$ plane. Differentiating the ϕ expression, one obtains the negative of the symmetrical velocity components in the $\tilde{r}_1$ and $\tilde{r}_3$ directions,

$$\frac{\partial \phi}{\partial \tilde{r}_1} = \tfrac{1}{2} S \tilde{r}_1$$

$$\frac{\partial \phi}{\partial \tilde{r}_3} = -\tfrac{1}{2} S \tilde{r}_3$$

respectively. These results show that we have a contraction along the $\tilde{r}_1$ axis, expansion along the $\tilde{r}_3$ axis, and that these axes are rotated 45° clockwise around the r_3 axis from the r_1 and r_2 axes, respectively, as shown in Figure 2-19. This, then, completes the analysis of the symmetric portion of the relative flow.

Going on to the antisymmetrical part of the relative motion, we have from (2-55),

$$\xi_{ij} = \frac{1}{2}\left(\frac{\partial V_i}{\partial x_j} - \frac{\partial V_j}{\partial x_i}\right) = \frac{1}{2}\begin{bmatrix} 0 & S & 0 \\ -S & 0 & 0 \\ 0 & 0 & 0 \end{bmatrix}$$

which we have seen gives for the components of the vorticity

$$\zeta_i = (0, 0, -S)$$

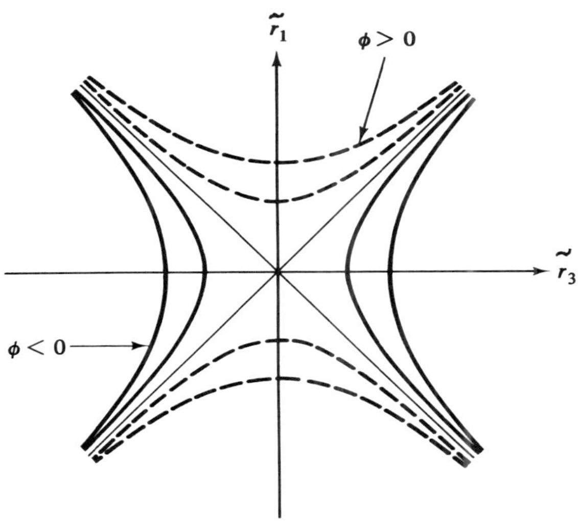

Fig. 2-18. Isolines of the velocity potential ϕ with reference to the principal coordinates for simple shear flow and positive velocity gradient.

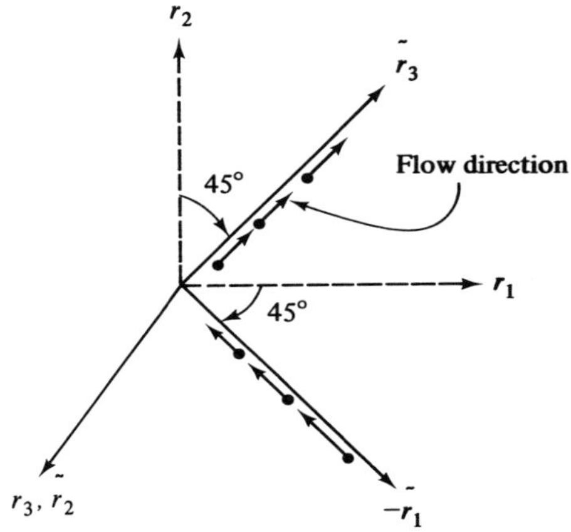

Fig. 2-19. Orientation of principal axes and direction of symmetrical portion of motion for simple shear.

The antisymmetrical part of the motion is thus

$$\delta V_i^{(a)} = \tfrac{1}{2}\epsilon_{ijk}\zeta_j r_k = (\tfrac{1}{2}Sr_2, -\tfrac{1}{2}Sr_1, 0)$$

and is illustrated in Figure 2-20.

As a final check on our results, we can reconstruct the flow field from its symmetrical and antisymmetrical parts as in Figure 2-21, to obtain the original field of simple shearing flow.

As a second example, consider the fluid flow represented by the deformation tensor

$$D_{ij} = \begin{bmatrix} 0 & 1 & 1 \\ 1 & 0 & 1 \\ 1 & 1 & 0 \end{bmatrix}$$

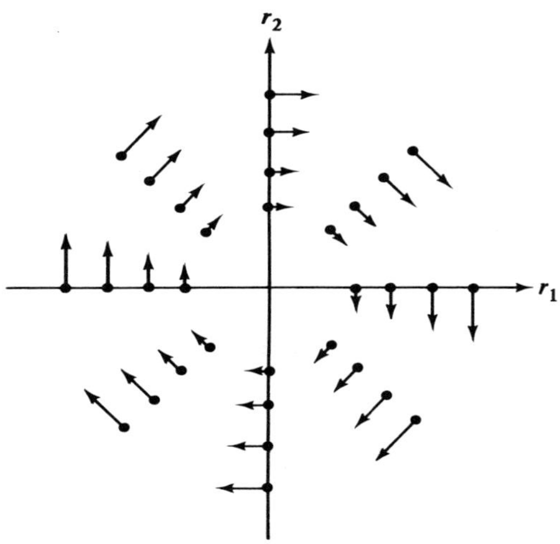

Fig. 2-20. Antisymmetrical portion of motion for the case of simple shear flow.

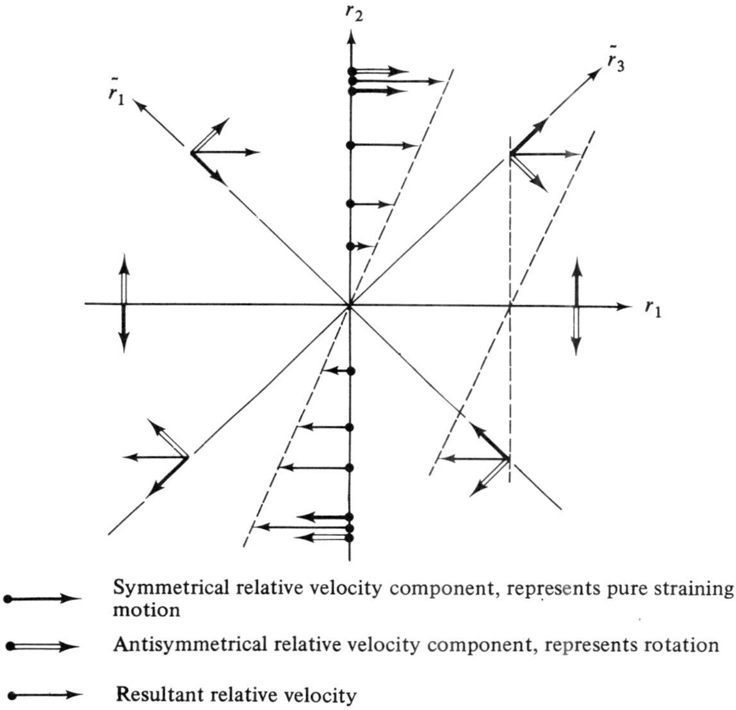

Fig. 2-21. Symmetrical and antisymmetrical relative velocity components representing pure straining motion and rotation for a simple shear flow.

which is nondivergent since $D_{ii} = 0$. The eigenvalues of the matrix are given by

$$\begin{vmatrix} -\lambda & 1 & 1 \\ 1 & -\lambda & 1 \\ 1 & 1 & -\lambda \end{vmatrix} = 0$$

with the resulting characteristic equation

$$\lambda^3 - 3\lambda - 2 = 0$$

whose roots are $\lambda = -1, -1$, and 2.

This example is a rather special case in that a multiple root (-1) exists. As we will see, this implies that there is only one unique principal axis of strain, which is also an axis of symmetry for strain in the plane normal to that axis. This result is illustrated from a determination of the eigenvectors;

for $\lambda^{(1)} = -1$ (or equivalently for $\lambda^{(2)} = -1$),

$$\begin{bmatrix} 0 & 1 & 1 \\ 1 & 0 & 1 \\ 1 & 1 & 0 \end{bmatrix} \begin{bmatrix} e_1^{(1)} \\ e_2^{(1)} \\ e_3^{(1)} \end{bmatrix} = - \begin{bmatrix} e_1^{(1)} \\ e_2^{(1)} \\ e_3^{(1)} \end{bmatrix}$$

yielding the three equations

$$\begin{aligned} e_2^{(1)} + e_3^{(1)} &= -e_1^{(1)} \\ e_1^{(1)} + e_3^{(1)} &= -e_2^{(1)} \\ e_1^{(1)} + e_2^{(1)} &= -e_3^{(1)} \end{aligned}$$

which reduce to the single relation

$$e_1^{(1)} + e_2^{(1)} + e_3^{(1)} = 0$$

and when combined with the condition for a unit vector,

$$e_1^{(1)^2} + e_2^{(2)^2} + e_3^{(1)^2} = 1$$

points out the indeterminacy of the direction of the eigenvector obtained from the root(s), $\lambda^{(1)} = \lambda^{(2)} = -1$.

However, for $\lambda^{(3)} = 2$,

$$\begin{bmatrix} 0 & 1 & 1 \\ 1 & 0 & 1 \\ 1 & 1 & 0 \end{bmatrix} \begin{bmatrix} e_1^{(3)} \\ e_2^{(3)} \\ e_3^{(3)} \end{bmatrix} = 2 \begin{bmatrix} e_1^{(3)} \\ e_2^{(3)} \\ e_3^{(3)} \end{bmatrix}$$

or

$$\begin{aligned} e_2^{(3)} + e_3^{(3)} &= 2e_1^{(3)} \\ e_1^{(3)} + e_3^{(3)} &= 2e_2^{(3)} \\ e_1^{(3)} + e_2^{(3)} &= 2e_3^{(3)} \end{aligned}$$

which along with the unit vector condition reduces to

$$e_1^{(3)} = e_2^{(3)} = e_3^{(3)} = \frac{1}{\sqrt{3}}$$

which states that the cosine of the angle (54.8°) between the $\tilde{r}_3$ axis and the r_1 axis, between the $\tilde{r}_3$ axis and the r_2 axis, and between the $\tilde{r}_3$ axis and the r_3 axis, is $1/\sqrt{3}$ (Figure 2-22). Additionally, since $\lambda^{(3)}$ is positive while $\lambda^{(1)}$ and $\lambda^{(2)}$ are negative and equal, the fluid is undergoing expansion along the $\tilde{r}_3$ axis and isotropic contraction in the plane normal to this axis. To verify this conclusion, we can make use of the property of the stream

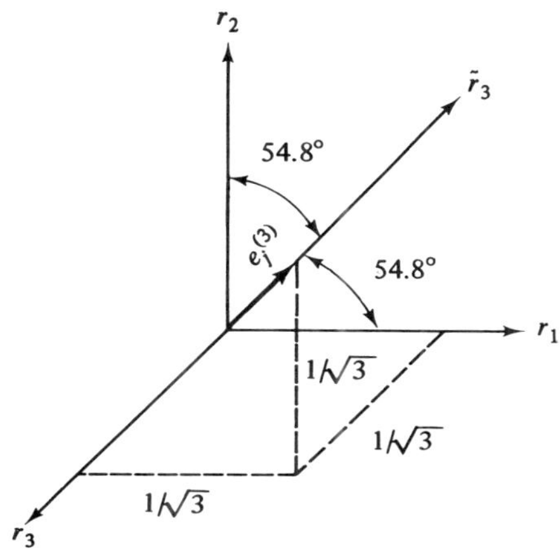

Fig. 2-22. Principal axes for the flow condition of example (b).

function ϕ. As in the previous example, from (2-47),

$$\begin{aligned}\phi &= -\tfrac{1}{2}\tilde{\mathbf{r}}^T\tilde{\mathbf{D}}\tilde{\mathbf{r}} \\ &= -\tfrac{1}{2}[\tilde{r}_1\tilde{r}_2\tilde{r}_3]\begin{bmatrix}-1 & 0 & 0\\ 0 & -1 & 0\\ 0 & 0 & 2\end{bmatrix}\begin{bmatrix}\tilde{r}_1\\ \tilde{r}_2\\ \tilde{r}_3\end{bmatrix} \\ &= \tfrac{1}{2}\tilde{r}_1^2 + \tfrac{1}{2}\tilde{r}_2^2 - \tilde{r}_3^2\end{aligned} \qquad (2\text{-}63)$$

giving

$$-\frac{\partial\phi}{\partial\tilde{r}_1} = -\tilde{r}_1 \qquad -\frac{\partial\phi}{\partial\tilde{r}_2} = -\tilde{r}_2 \qquad -\frac{\partial\phi}{\partial\tilde{r}_3} = 2\tilde{r}_3$$

which are the velocity components in the $\tilde{r}_1$, $\tilde{r}_2$, and $\tilde{r}_3$ directions, respectively; and as asserted, represents isotropic contraction in the $\tilde{r}_1\tilde{r}_2$ plane and expansion along the $\tilde{r}_3$ axis.

The family of quadrics represented by ϕ differ as ϕ is positive or negative. For $\phi > 0$, (2-63) is the equation of a family of hyperboloids of two sheets opening in the positive and negative $\tilde{r}_3$ directions. Sections parallel to the $\tilde{r}_1\tilde{r}_3$ and $\tilde{r}_2\tilde{r}_3$ planes are hyperbolas; the traces in the $\tilde{r}_1\tilde{r}_2$ plane are circles. For $\phi < 0$, (2-63) is the equation of a hyperboloid of one sheet: sections parallel to the $\tilde{r}_1\tilde{r}_3$ and $\tilde{r}_2\tilde{r}_3$ planes are hyperbolas, the traces in the $\tilde{r}_1\tilde{r}_2$ plane again being circles. With $\tilde{r}_3$ taken in the direction of the axis of symmetry of the strain, let $\tilde{r}$ be the radial coordinate in the plane normal to this axis ($\tilde{r}_1$ and $\tilde{r}_2$ are normal to $\tilde{r}_3$ but otherwise indeterminate). The quadric surfaces and the streamlines (which are everywhere normal to the quadric surfaces since $\delta V_i^{(s)}$ is in the direction of the negative of the gradient of ϕ) in the $\tilde{r}_3\tilde{r}$

plane would look as shown in Figure 2-23, where P is the material point of the medium relative to which the streamline pattern applies. In other words, the material near point P appears to be flowing relative to P, as shown. Also illustrated are lines of zero strain rate (or of zero fractional rate of separation) which join points having parallel velocity vectors.

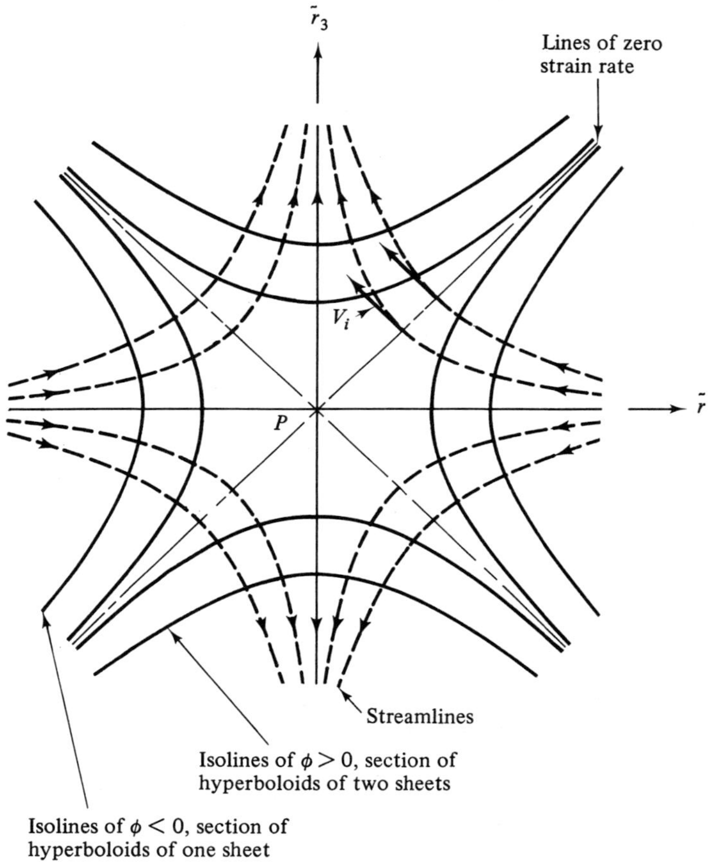

Fig. 2-23. Isolines of the velocity potential and streamlines for the flow field of example (b) as viewed normal to the $\tilde{r}$, $\tilde{r}_3$ plane.

On the other hand, viewed in the plane normal to the $\tilde{r}_3$ axis, the relative streamlines appear, as in Figure 2-24, as a convergent flow. Isolines of ϕ in this special case are circles. For the more general case where $\lambda^{(1)}$ and $\lambda^{(2)}$ are both positive but not equal, they would be ellipses with principal axes in the direction of $\tilde{r}_1$ and $\tilde{r}_2$.

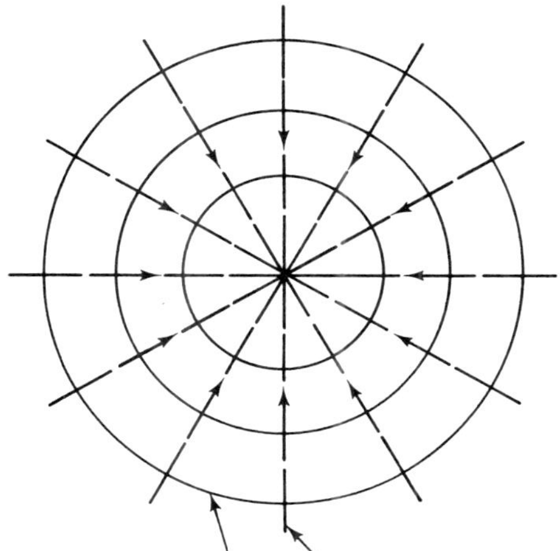

Fig. 2-24. Isolines of the velocity potential and streamlines for the flow field of example (b) as viewed in plane normal to $\bar{r}_3$ axis.

2-12 Kinematic Classification of Fluid Motion

A method of classifying different types of fluid motion is according to kinematic (kinematics: treatment of motion without reference to particular forces or bodies) properties; for example, there are large classes of motion dealing with divergenceless (incompressible) fluids, with irrotational (zero vorticity) flow, and with deformationless (like that of solid bodies) motion. A schematic diagram illustrating various special types of motion is presented in Figure 2-25. The size of the various blocks denotes the overall relative importance; the amount of material available in the literature dealing with a given classification, on the other hand, is in rough measure inversely proportional to the area of the appropriate block. The condition imposed upon each type of motion is stated under each title. Where overlapping occurs, more than one condition is imposed; for example, Laplacian flow represents that for which both $\nabla \times \mathbf{V}$ and $\nabla \cdot \mathbf{V}$ are zero and is more extensively studied than any other in the classical literature of hydrodynamics. It is apparent, however, that it is merely a special case in the general scheme of possible kinematic behavior.

2-13 The Stream Function

If a particular fluid motion is two-dimensional and nondivergent, its velocity field can be represented by a single scalar function, the *stream function.* Letting $\psi(x, y, t)$ represent this function, then

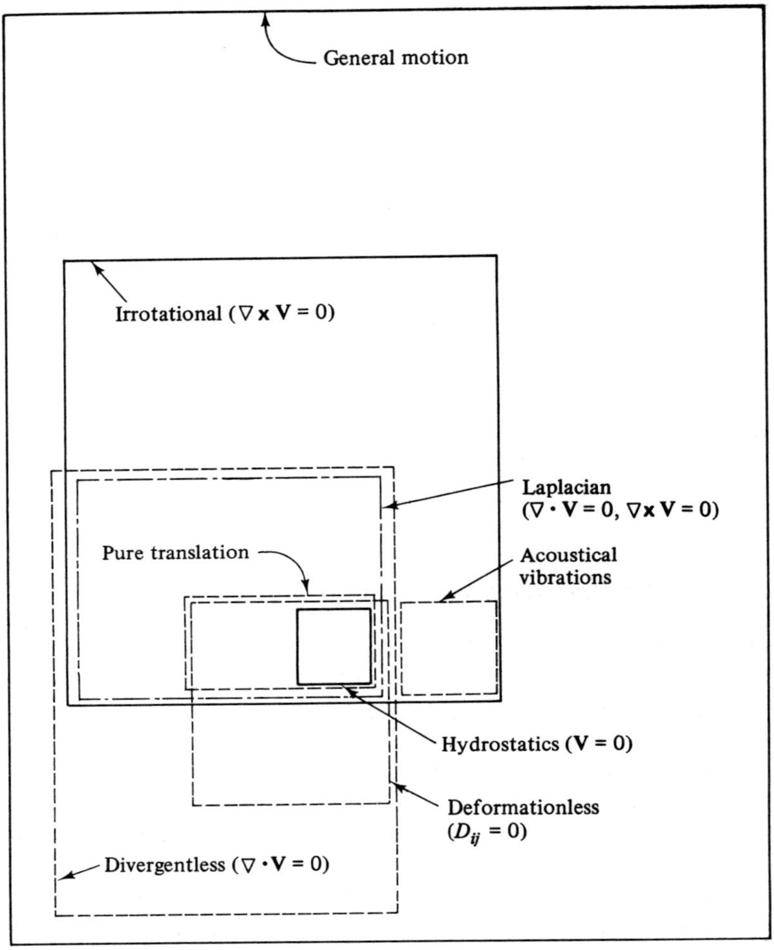

Fig. 2-25. Schematic diagram for the kinematic classification of fluid motions.

$$u = -\frac{\partial \psi}{\partial y} \qquad v = \frac{\partial \psi}{\partial x} \tag{2-64}$$

From this we see that, assuming that ψ has continuous first and second derivatives, a necessary condition for the existence of the stream function is that the fluid motion be nondivergent, since for two-dimensional flow

$$\nabla \cdot \mathbf{V} = \frac{\partial u}{\partial x} + \frac{\partial v}{\partial y} = \frac{\partial}{\partial x}\left(-\frac{\partial \psi}{\partial y}\right) + \frac{\partial}{\partial y}\left(\frac{\partial \psi}{\partial x}\right) = 0$$

where

$$\mathbf{V} = u\mathbf{i} + v\mathbf{j} \tag{2-65}$$

Utilizing (2-64), it is seen that **V** can be written

$$\begin{aligned}\mathbf{V} &= -\frac{\partial \psi}{\partial y}\mathbf{i} + \frac{\partial \psi}{\partial x}\mathbf{j} \\ &= -\boldsymbol{\nabla}\psi \times \mathbf{k}\end{aligned} \tag{2-66}$$

Then

$$\begin{aligned}-\boldsymbol{\nabla}\psi \cdot \mathbf{V} &= -\boldsymbol{\nabla}\psi \cdot (-\boldsymbol{\nabla}\psi \times \mathbf{k}) \\ &= (\boldsymbol{\nabla}\psi \times \boldsymbol{\nabla}\psi) \cdot \mathbf{k} = 0\end{aligned}$$

and from the definition of the dot product, at any one time **V** must be normal to the gradient of ψ. But, since the gradient of ψ is normal to isolines of ψ, we conclude that **V** is *parallel to the isolines of* ψ and that isolines of ψ are therefore streamlines. Additionally, (2-66) states that the magnitude of the velocity is inversely proportional to the spacing of isolines of ψ provided that these isolines are constructed for equal increments of ψ (Figure 2-26). It is not generally true, however, that streamlines are isolines of ψ since streamlines can be constructed for divergent flow, for which case the stream function does not exist.

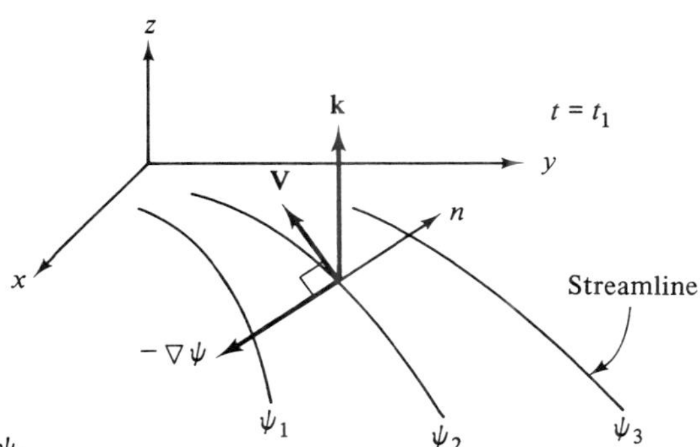

$\psi_3 > \psi_2 > \psi_1$

$\Delta\psi \equiv \psi_2 - \psi_1 = \psi_3 - \psi_2$

$|\mathbf{V}| = \dfrac{\partial \psi}{\partial n} \doteq \dfrac{\Delta\psi}{\Delta n}$ where n = distance measured normal to streamlines

Fig. 2-26. The relation between isolines of the stream function, streamlines, and fluid velocity.

A physical interpretation of the stream function results from the following considerations. The volume transport of fluid per unit z distance (having units of ([length]^3/[time])/[length]) across any cylindrical surface connecting points P and O of Figure 2-27 is given by the line integral

$$Q = \int_{(0,0)}^{(x_1,y_1)} V_n \, ds \qquad t = t_1 \tag{2-67}$$

where V_n is the velocity component normal to the surface. The determination of Q is not dependent on the existence of the stream function; however, in general the quantity Q depends upon the path of integration. Now let Q_1 and Q_2 be the volume transports per unit z distance across the surfaces whose projection on the x, y plane are curve 1 and curve 2, respectively (Figure 2-27). Then

$$Q_1 = \int_{(0,0)}^{(x_1,y_1)} (V_n)_1 \, ds \qquad Q_2 = \int_{(0,0)}^{(x_1,y_1)} (V_n)_2 \, ds \qquad t = t_1$$

and

$$\begin{aligned} Q_2 - Q_1 &= \int_{(0,0)}^{(x_1,y_1)} (V_n)_2 \, ds - \int_{(0,0)}^{(x_1,y_1)} (V_n)_1 \, ds \\ &= \oint V_n \, ds \qquad t = t_1 \end{aligned} \tag{2-68}$$

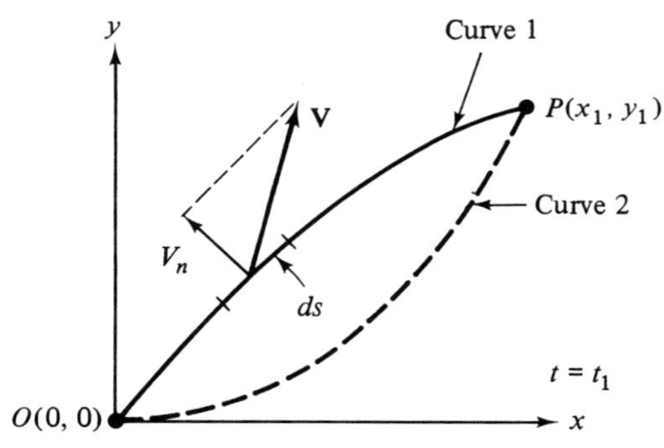

Fig. 2-27. Projection of a right cylindrical surface in the x, y plane.

Using the divergence theorem,

$$\int_v \nabla \cdot \mathbf{V} \, dv = \int_A \mathbf{n} \cdot \mathbf{V} \, dA \tag{2-69}$$

where v and A designate integration over the volume and the area, respectively, the integral on the right of (2-68) can be transformed such that we obtain

$$Q_2 - Q_1 = \int_A \left(\frac{\partial u}{\partial x} + \frac{\partial v}{\partial y}\right) dA \qquad t = t_1 \tag{2-70}$$

In this case the integral, since we are dealing with a unit z distance, is the counterpart of the volume integral of (2-69); that is, for a unit z distance the ds and V_n of (2-68) are equivalent to the dA and $(\mathbf{n} \cdot \mathbf{V})$ of (2-69) and dv of (2-69) is equivalent to the dA of (2-70). In (2-70) the area over which the integration is to be carried out is that enclosed by the curves 1 and 2.

Now it is clear that the condition of no divergence of flow corresponds to the condition of Q_1 being equal to Q_2. Furthermore, since the paths of integration for Q_1 and Q_2 are arbitrary, it may be concluded that for $\nabla \cdot \mathbf{V} = 0$, Q is independent of the path of integration. This implies that for any given time dQ is an exact differential.

Since ψ is a function of only x and y at any given time,

$$d\psi = \frac{\partial \psi}{\partial x} dx + \frac{\partial \psi}{\partial y} dy \tag{2-71}$$

is an exact differential. For $\nabla \cdot \mathbf{V} = 0$, then from (2-64),

$$d\psi = v\,dx - u\,dy \tag{2-72}$$

or

$$\psi(x_1, y_1) = \int_{(0,0)}^{(x_1,y_1)} v\,dx - u\,dy$$

the path from point (0, 0) to point (x_1, y_1) being completely arbitrary. From Figure 2-28 we see that

$$v\,dx - u\,dy = V_n\,ds$$

and, consequently,

$$\psi(x_1, y_1) = \int_{(0,0)}^{(x_1,y_1)} V_n\,ds = Q \tag{2-73}$$

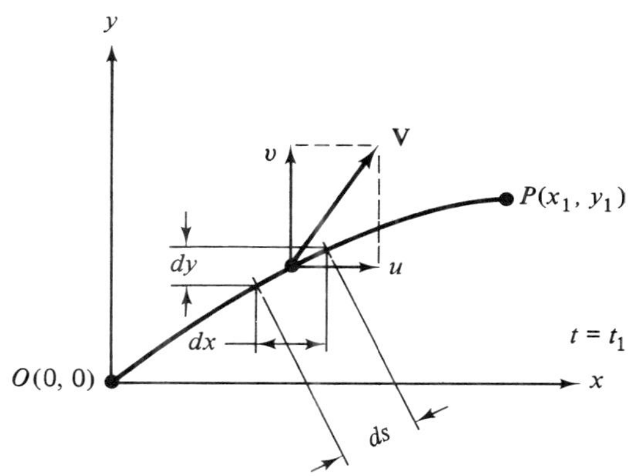

Fig. 2-28. Volume transport across arbitrary surface element.

This asserts that the value of the stream function at the point (x_1, y_1) is associated with the volume transport between that point and a reference point; that the difference between values of ψ of neighboring isolines at any given time represents the volume transport between those isolines per unit distance (for our case, per unit z distance). The volume transport between any two isolines of ψ must therefore be constant, which is consistent with the observation that isolines of ψ are streamlines and as such can have no flow across them (Figure 2-29).

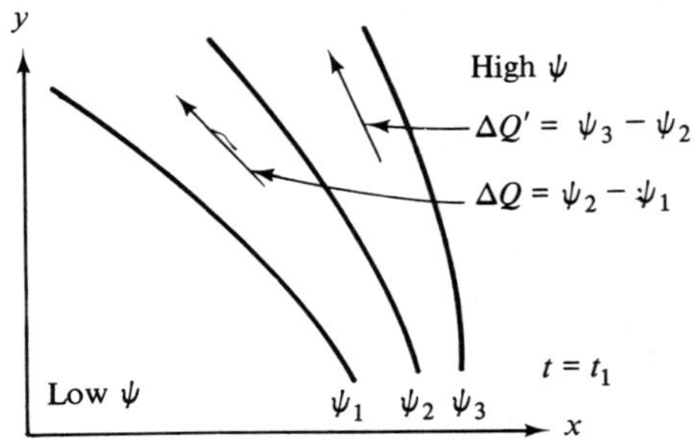

Fig. 2-29. Relation between volume transport and isolines of the stream function.

To summarize:

1. A stream function exists for each and every two-dimensional incompressible flow, whether viscid or inviscid.
2. Lines for which the stream function is a constant are streamlines.
3. At any given time in Cartesian coordinates,

$$u = -\frac{\partial \psi}{\partial y} \qquad v = \frac{\partial \psi}{\partial x}$$

   and, in general,

$$V_s = \frac{\partial \psi}{\partial n}$$

   where s is measured 90° counterclockwise from n, as illustrated in Figure 2-30.

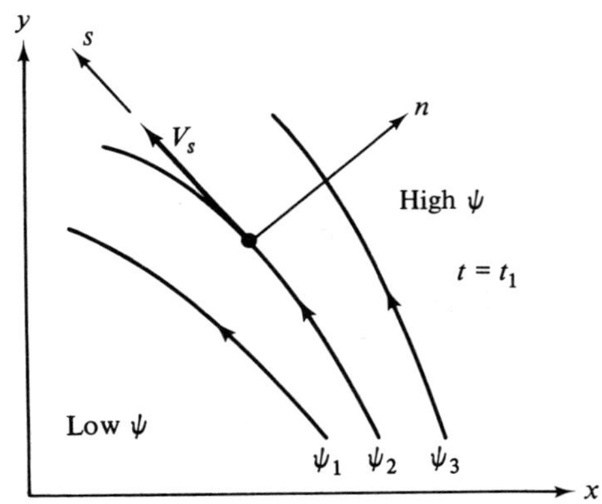

Fig. 2-30. Relation between isolines of the stream function and the fluid velocity in the natural coordinate system.

2-14 The Velocity Potential

If a particular fluid motion is irrotational, it is possible to represent the field of flow in terms of a single scalar function $\phi(x, y, z, t)$ as

$$\mathbf{V} = -\nabla\phi \tag{2-74}$$

or

$$u = -\frac{\partial \phi}{\partial x} \qquad v = -\frac{\partial \phi}{\partial y} \qquad w = -\frac{\partial \phi}{\partial z} \tag{2-75}$$

at any given time t_1, where ϕ is called the *velocity potential.* Unlike the stream function, which is applicable only to two-dimensional, nondivergent flow, the velocity potential is defined for the three-dimensional case. However, the necessary condition for the existence of ϕ that the flow be irrotational requires that

$$\frac{\partial w}{\partial y} = \frac{\partial v}{\partial z} \qquad \frac{\partial u}{\partial z} = \frac{\partial w}{\partial x} \qquad \frac{\partial v}{\partial x} = \frac{\partial u}{\partial y}$$

As is evident from (2-74), the fluid velocity is directed normal to the equiscalar surfaces of ϕ, its sense being toward low values of ϕ and its magnitude given by

$$|\mathbf{V}| = \left|\frac{\partial \phi}{\partial n}\right| \tag{2-76}$$

where n is the distance measured normal to the equiscalar surface at the point in question. If the equiscalar surfaces are constructed for equal intervals of ϕ, then $|\mathbf{V}|$ is inversely proportional to the spacing of the surfaces. Figure 2-31 illustrates these concepts for the special case of two-dimensional flow.

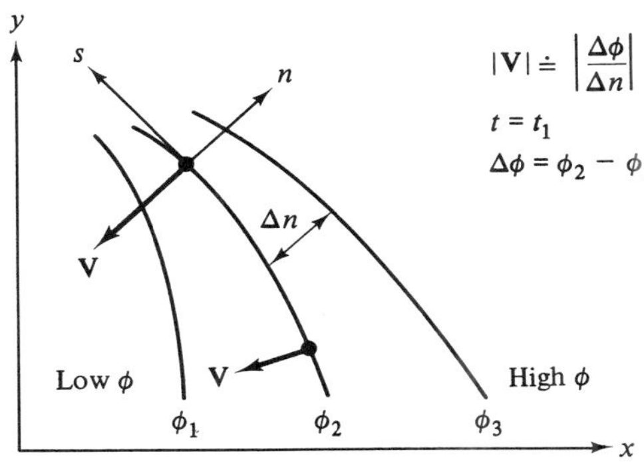

Fig. 2-31. Relation between isolines of the velocity potential and the fluid velocity.

To gain further insight into the character of the velocity potential, as well as to substantiate the assertion of irrotationality, we proceed as follows. At any one time the differential $d\phi$ can be expressed as

$$d\phi = \frac{\partial \phi}{\partial x} dx + \frac{\partial \phi}{\partial y} \partial y + \frac{\partial \phi}{\partial z} dz \tag{2-77}$$

which, from (2-75), can be written

$$d\phi = -(u\, dx + v\, dy + w\, dz) \tag{2-78}$$

Thus, if the velocity field is known, ϕ at a point $P(x_1, y_1, z_1)$ can be evaluated from the line integral

$$\phi(x_1, y_1, z_1) = -\int_{(0,0,0)}^{(x_1,y_1,z_1)} (u\,dx + v\,dy + w\,dz) \tag{2-79}$$

where the origin is taken as the reference point. In the light of the requirement that $\nabla \times \mathbf{V} = 0$, the path of integration is perfectly arbitrary. To prove this we write (2-79) as

$$\phi(x_1, y_1, z_1) = -\int_{(0,0,0)}^{(x_1,y_1,z_1)} \mathbf{V} \cdot d\mathbf{s} = \int_{(0,0,0)}^{(x_1,y_1,z_1)} V_s\,ds \tag{2-80}$$

V_s being the tangential component of $\mathbf{V}$ along the path of integration and $d\mathbf{s}$ the differential line vector

$$d\mathbf{s} = \mathbf{i}\,dx + \mathbf{j}\,dy + \mathbf{k}\,dz$$

If, as in Figure 2-32, we consider two different paths, curve 1 and curve 2, then the difference in the values of ϕ obtained from integration is

$$\begin{aligned}\phi_2 - \phi_1 &= -\left[\int_{(0,0,0)}^{(x_1,y_1,z_1)} (V_s)_2\,ds + \int_{(0,0,0)}^{(x_1,y_1,z_1)} (V_s)_1\,ds\right] \\ &= -\oint V_s\,ds\end{aligned} \tag{2-81}$$

Applying Stokes' theorem,

$$\oint \mathbf{V} \cdot d\mathbf{r} = \int_A (\mathbf{n} \cdot \nabla \times \mathbf{V})\,dA \tag{2-82}$$

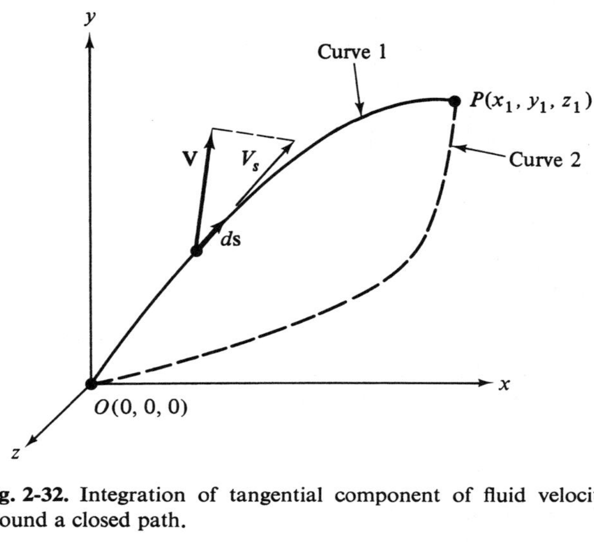

Fig. 2-32. Integration of tangential component of fluid velocity around a closed path.

to the last integral of (2-81),

$$\phi_2 - \phi_1 = -\int_A (\mathbf{n} \cdot \nabla \times \mathbf{V})\, dA \tag{2-83}$$

where A is the area of an arbitrary surface whose boundaries are the curves 1 and 2. In view of the desire that the ϕ obtained by integration along an arbitrary path be unique and single-valued, we require that $\phi_1 = \phi_2$, which from (2-83) necessitates that $\nabla \times \mathbf{V} = 0$, that is, that the fluid flow be irrotational as asserted previously.

To summarize:

1. The velocity potential exists if and only if the flow is irrotation; no other restrictions are required.
2. The fluid velocity is given at any particular time by

$$\mathbf{V} = -\nabla\phi$$

2-15 Laplacian Flow

A two-dimensional field of flow which is both nondivergent and irrotational can be characterized by either (or both) of the scalar functions ψ or ϕ. This type of flow, bearing the name *Laplace*, has been extensively investigated and forms much of the bulk of the classical literature of hydrodynamics.

For two-dimensional flow, the conditions implied by the stipulation that $\nabla \cdot \mathbf{V} = 0$ and $\nabla \times \mathbf{V} = 0$ are

$$\frac{\partial u}{\partial x} = -\frac{\partial v}{\partial y} \qquad \text{and} \qquad \frac{\partial u}{\partial y} = \frac{\partial v}{\partial x} \tag{2-84}$$

We have seen that the relations

$$u = -\frac{\partial \psi}{\partial y} \qquad v = \frac{\partial \psi}{\partial x} \tag{2-85}$$

satisfy the first of these, while

$$u = -\frac{\partial \phi}{\partial x} \qquad v = -\frac{\partial \phi}{\partial y} \tag{2-86}$$

satisfy the second. In Laplacian flow both of (2-84) must be satisfied simultaneously. Consequently, inserting (2-85) into the second part of (2-84) and (2-86) into the first part of (2-84), we find that the functions ψ and ϕ must satisfy the two-dimensional form of Laplace's equation,

$$\begin{aligned} \nabla^2\psi &= 0 \\ \nabla^2\phi &= 0 \end{aligned} \tag{2-87}$$

a second-order, elliptic, partial differential equation. Also, since isolines of ψ are parallel to $\mathbf{V}$ and isolines of ϕ are normal to $\mathbf{V}$, it is clear that isolines of ψ and ϕ must be orthogonal, as shown in Figure 2-33.

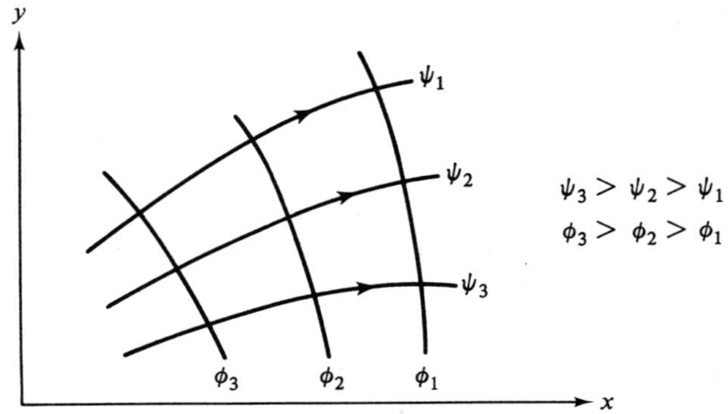

Fig. 2-33. Isolines of the stream function and velocity potential for Laplacian flow.

2-16 The Kinematic Boundary Condition

At an impermeable boundary, a boundary across which there is no flux of matter on a macroscopic scale, the flow of fluid relative to the boundary must be tangential. Examples of impermeable boundaries are the surfaces of solid bodies adjacent to the fluid (the bottom and sides of the ocean when percolation is ignored) and free surfaces (the interface between air and water if we ignore evaporation and condensation or between immiscible fluids). If the boundary is fixed in space, the component of fluid velocity normal to it must be zero,

$$\mathbf{V} \cdot \mathbf{n} = V_n = 0 \tag{2-88}$$

where $\mathbf{n}$ is the unit normal vector to the boundary. However, if the boundary is moving, the normal component of the fluid velocity at the boundary must be equal to the velocity of the boundary normal to itself. This applies whether the interface is solid–fluid or a free surface. If we let $\mathbf{U}$ be the velocity of the boundary and $\mathbf{V}$ be the velocity of the fluid at the boundary, the kinematic boundary condition is

$$(\mathbf{U} - \mathbf{V}) \cdot \mathbf{n} = 0 \tag{2-89}$$

Equation (2-89) represents the kinematic boundary condition for the specific case of an impermeable boundary. In those cases where a change of state of the fluid occurs at the boundary, or where the boundary displays a selective permeability, neither (2-88) nor (2-89) is rigorously applicable. A specific example is the effect of an evaporative flux of water at the sea surface. This process is certainly important in thermodynamic problems; however, in nearly all kinematic or dynamical problems the effect of evaporation can be ignored with regard to the kinematic boundary condition. We merely note here that for evaporation at the sea surface, (2-89) takes the form

$$\rho(\mathbf{U} - \mathbf{V}) \cdot \mathbf{n} = E \tag{2-90}$$

where ρ is the mass density of water and E the magnitude of the mass flux of water vapor emerging from the sea surface (having units of [mass]/([length]^2 [time]).

For completeness we might note that if the fluid is viscous, an additional boundary condition applies, the *no-slip condition* of

$$(\mathbf{U} - \mathbf{V}) \cdot \mathbf{s} = 0 \tag{2-91}$$

where $\mathbf{s}$ is the unit tangent vector to the boundary. Application of the kinematic and no-slip conditions result in the fluid at the boundary having zero velocity with respect to the boundary itself.

Let us now direct our attention specifically to a free surface across which a certain physical or thermal property is continuous. For example, at the sea surface the temperature across the interface between the air and the sea water is assumed to be continuous. Another example is that of pressure for those cases where the effect of surface tension can be ignored. Let $f_1(x, y, z, t)$ represent the field of this continuous property in fluid 1 and $f_2(x, y, z, t)$ represent the field of this same property in fluid 2. Since the property is continuous across the free surface, then

$$f_{1_b}(x, y, z, t) - f_{2_b}(x, y, z, t) = 0 \tag{2-92}$$

where the subscript b indicates that the function is to be evaluated at the boundary between the fluids. If the functions f_1 and f_2 are known, this relation actually defines the equation of the boundary; that is,

$$F(x, y, z, t) = f_{1_b} - f_{2_b} = 0 \tag{2-93}$$

is the implicit equation of the free surface. This equation states that at time t the point whose coordinates (x, y, z) satisfies it lies on the boundary. Since the boundary is moving, at some later time, say $(t + dt)$, the point will have coordinates $(x + dx, y + dy, z + dz)$ such that they again satisfy the relation

$$F(x + dx, y + dy, z + dz, t + dt) = 0 \tag{2-94}$$

which can be written in a Taylor series to the first order as

$$\begin{aligned} F(x + dx, y + dy, z + dz, t + dt) &= F(x, y, z, t) + \frac{\partial F(x, y, z, t)}{\partial x} dx + \frac{\partial F(x, y, z, t)}{\partial y} dy \\ &\quad + \frac{\partial F(x, y, z, t)}{\partial z} dz + \frac{\partial F(x, y, z, t)}{\partial t} dt \\ &= 0 \end{aligned} \tag{2-95}$$

Subtracting (2-94) from (2-95) and dividing the result by dt yields

$$\frac{\partial F}{\partial t} + \frac{dx}{dt}\frac{\partial F}{\partial x} + \frac{dy}{dt}\frac{\partial F}{\partial y} + \frac{dz}{dt}\frac{\partial F}{\partial z} = 0$$

or expressing this result in terms of the material derivative,

$$\frac{DF}{Dt} = \frac{\partial F}{\partial t} + \mathbf{V} \cdot \nabla F = 0 \tag{2-96}$$

This result is not surprising in that it merely states that if we follow a fluid particle on the boundary, it moves with the boundary.

An important application of this result is in the framing of the kinematic free surface boundary condition for gravity waves. We have stated that (2-93) is an implicit equation of the free surface. In wave theory this is generally written in the explicit form

$$z = \eta(x, y, t) \tag{2-97}$$

where η denotes the function representing the elevation of the free surface above some reference plane at position (x, y) and at a time t as shown in Figure 2-34. Now from combining (2-93) and (2-97),

$$g(x, y, z, t) \equiv \eta(x, y, t) - z = 0$$

and applying (2-96),

$$\frac{D}{Dt}[\eta(x, y, t) - z] = 0$$

$$\frac{\partial \eta}{\partial t} + u\Big|_{z=\eta}\frac{\partial \eta}{\partial x} + v\Big|_{z=\eta}\frac{\partial \eta}{\partial y} - w\Big|_{z=\eta} = 0$$

or, rearranging,

$$w\Big|_{z=\eta} = \frac{\partial \eta}{\partial t} + u\Big|_{z=\eta}\frac{\partial \eta}{\partial x} + v\Big|_{z=\eta}\frac{\partial \eta}{\partial y} \tag{2-98}$$

In this, the complete form of the kinematic free surface boundary condition, it is seen that the vertical velocity at the free surface $w|_{z=\eta}$ is not just the local rate of change of elevation $\partial\eta/\partial t$ (as would be measured by a wave staff, for example), but also depends on the advection of the sea surface slopes.

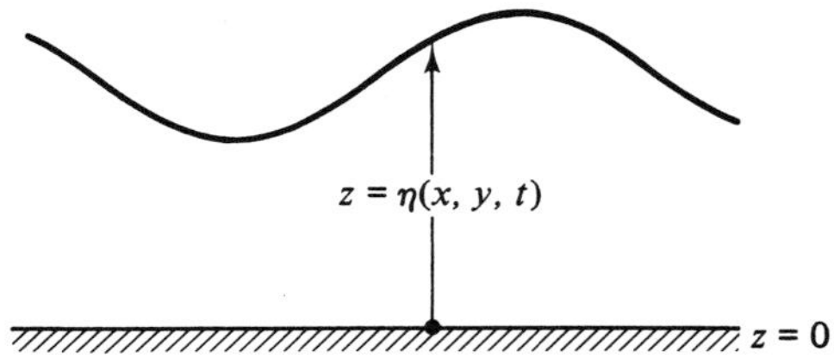

Fig. 2-34. Definition diagram of free surface for gravity waves; explicit form.

As a final point it is observed that in some cases the nonlinear terms of (2-98) are small compared with the linear terms and the approximation

$$w\big|_{z=\eta} \doteq \frac{\partial \eta}{\partial t}$$

is made.

2-17 Distribution of Variables

Let Q be an intensive measure (a quantity expressed per unit volume) of some property q. For example, Q could represent:

1. The density ρ, where the property q is the mass M; that is, the intensive measure of the mass is the density,

$$q \equiv M \qquad Q \equiv \frac{M}{v} \equiv \rho$$

2. The momentum density ρu, ρv, or ρw, where the property q is the momentum Mu, Mv, or Mw; that is, the intensive measure of the momentum is the momentum density,

$$q \equiv Mu \qquad Q \equiv \frac{Mu}{v} = \rho u$$

3. The concentration of salt ρs or the concentration of water $\rho(1 - s)$ in a binary mixture, where s is the salinity expressed in mass of salt per unit mass of seawater (approximating seawater as a binary mixture) while the property q is the total mass of salt or water; that is, the intensive measure of the total salt or total water is the salt per unit volume or the water per unit volume, respectively,

$$q \equiv Ms \qquad Q \equiv \frac{Ms}{v} = \rho s$$

or

$$q \equiv M(1 - s) \qquad Q \equiv \frac{M(1 - s)}{v} = \rho(1 - s)$$

Consider the elemental rectangular parallelopiped of Figure 2-35 with sides of length Δx, Δy, and Δz. We assert that the increase of the intensive measure Q of the property q within the volume over a time interval Δt can only be attributed to:

1. The advection of the property into the volume by the fluid velocity.
2. The diffusion of the property into the volume due to the existence of a gradient of the property.
3. The production of the property within the volume (i.e., the volume acts as a source or sink for the property).

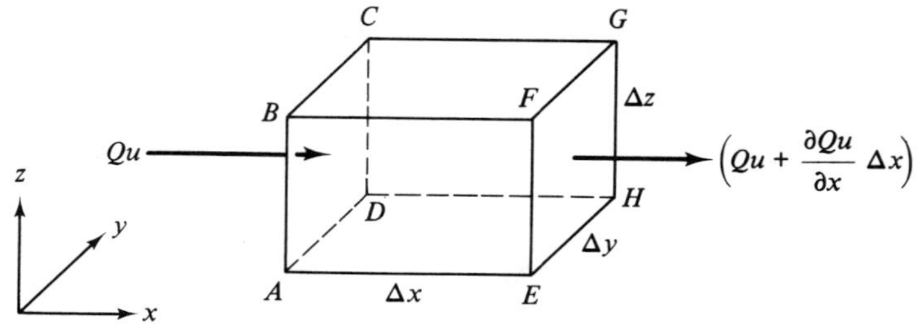

Fig. 2-35. Advective flux of a property in the x direction of an elemental rectangular parallelopiped.

We will examine the effect of each of these individually and then put all three together to form one equation.

Change of the Intensive Property Q Due to Advection

Assume that the components of the fluid velocity in the x, y, and z directions are u, v, and w, respectively. With respect then to the elemental rectangular parallelopiped, we can write

$$\begin{bmatrix} \textit{inflow} \text{ of property} \\ q \text{ across face } ABCD \\ \text{in time } \Delta t \end{bmatrix} = Qu\,\Delta y\,\Delta z\,\Delta t \tag{2-99}$$

and utilizing the Taylor series, to first order,

$$\begin{bmatrix} \textit{inflow} \text{ of property} \\ q \text{ across face } EFGH \\ \text{in time } \Delta t \end{bmatrix} = -\left(Qu + \frac{\partial Qu}{\partial x}\Delta x\right)\Delta y\,\Delta z\,\Delta t \tag{2-100}$$

Adding these last two expressions, we obtain

$$\begin{aligned} \begin{bmatrix} \textit{net inflow} \text{ of prop-} \\ \text{erty } q \text{ in the } x \\ \text{direction in time} \\ \text{interval } \Delta t \end{bmatrix} &= -\frac{\partial Qu}{\partial x}\Delta x\,\Delta y\,\Delta z\,\Delta t \\ &= -\frac{\partial Qu}{\partial x}\Delta \upsilon\,\Delta t \end{aligned} \tag{2-101}$$

where υ is the volume of the parallelopiped.

In a similar fashion, it is seen that

$$\begin{bmatrix} \textit{net inflow} \text{ of prop-} \\ \text{erty } q \text{ in the } y \\ \text{direction in time} \\ \text{interval } \Delta t \end{bmatrix} = -\frac{\partial Qv}{\partial y}\Delta \upsilon\,\Delta t \tag{2-102}$$

and

$$\begin{bmatrix} \textit{net inflow} \text{ of prop-} \\ \text{erty } q \text{ in the } z \\ \text{direction in time} \\ \text{interval } \Delta t \end{bmatrix} = -\frac{\partial Qw}{\partial z}\,\Delta v\,\Delta t \tag{2-103}$$

Summing (2-101), (2-102), and (2-103) yields

$$\begin{bmatrix} \textit{net inflow} \text{ of prop-} \\ \text{erty } q \text{ across all} \\ \text{faces in time } \Delta t \\ \text{due to } \textit{advection} \end{bmatrix} = -\left(\frac{\partial Qu}{\partial x} + \frac{\partial Qv}{\partial y} + \frac{\partial Qw}{\partial z}\right)\Delta v\,\Delta t \tag{2-104}$$

Now if the amount of the property present in the volume at time t is $Q\,\Delta v$, then at time $(t + \Delta t)$, according to a Taylor expansion, the amount of the property present will be

$$Q\,\Delta v + \frac{\partial Q}{\partial t}\,\Delta v\,\Delta t$$

and subtracting we have that the

$$\begin{bmatrix} \textit{net increase} \text{ of property} \\ q \text{ within volume in time} \\ \text{interval } \Delta t \end{bmatrix} = \frac{\partial Q}{\partial t}\,\Delta v\,\Delta t \tag{2-105}$$

The assumption that the only way that the amount of the property q can change within the volume is by advection allows us to equate (2-104) and (2-105). Doing so and canceling common terms results in

$$\frac{\partial Q}{\partial t} + \frac{\partial Qu}{\partial x} + \frac{\partial Qv}{\partial y} + \frac{\partial Qw}{\partial z} = 0$$

alternative forms of which are

$$\frac{\partial Q}{\partial t} + \mathbf{V} \cdot \nabla Q + Q\nabla \cdot \mathbf{V} = 0$$

or

$$\frac{DQ}{Dt} + Q\nabla \cdot \mathbf{V} = 0 \tag{2-106}$$

or

$$\frac{\partial Q}{\partial t} + \nabla \cdot Q\mathbf{V} = 0 \tag{2-107}$$

Change of the Intensive Property Q Due to Diffusion

The quantity Q can vary within a control volume, regardless of the existence of a fluid velocity, as a result of diffusion as long as there is a gradient of the property. It is known that, if one region has a higher concentration of a particular substance, random molecular motion (motion on a microscopic scale) will result in a transport, in a statistical sense, of the substance from regions of higher concentration to regions of lower concentration. This is transport on a smaller-than-continuum scale.

Returning to our elemental rectangular parallelopiped, let us take the fluid velocity $\mathbf{V}$ as zero and define a diffusive flux vector $\mathbf{F}_Q(x, y, z, t)$ as the nonadvective transport of the property q through a unit area normal to the direction of the flux. For the x direction, then,

$$\begin{bmatrix} \textit{inflow} \text{ of property} \\ q \text{ across face } ABCD \\ \text{in time } \Delta t \end{bmatrix} = F_{Q_x} \Delta y \, \Delta z \, \Delta t$$

and via a Taylor expansion

$$\begin{bmatrix} \textit{inflow} \text{ of property} \\ q \text{ across face } EFGH \\ \text{in time } \Delta t \end{bmatrix} = -\left(F_{Q_x} + \frac{\partial F_{Q_x}}{\partial x} \Delta x\right) \Delta y \, \Delta z \, \Delta t$$

where F_{Q_x} is the x component of $\mathbf{F}_Q$. Adding these last two statements, we obtain

$$\begin{bmatrix} \textit{net inflow} \text{ of} \\ \text{property in the } x \\ \text{direction in time} \\ \text{interval } \Delta t \end{bmatrix} = -\frac{\partial F_{Q_x}}{\partial x} \Delta v \, \Delta t \tag{2-108}$$

Similarly, for the y and z directions,

$$\begin{bmatrix} \textit{net inflow} \text{ of} \\ \text{property in the } y \\ \text{direction in time} \\ \text{interval } \Delta t \end{bmatrix} = -\frac{\partial F_{Q_y}}{\partial y} \Delta v \, \Delta t \tag{2-109}$$

and

$$\begin{bmatrix} \textit{net inflow} \text{ of} \\ \text{property in the } z \\ \text{direction in time} \\ \text{interval } \Delta t \end{bmatrix} = -\frac{\partial F_{Q_z}}{\partial z} \Delta v \, \Delta t \tag{2-110}$$

Then, taking (2-108), (2-109), and (2-110) all together yields

$$\begin{bmatrix} \textit{net inflow} \text{ of} \\ \text{property } q \text{ across} \\ \text{all faces in time} \\ \Delta t \text{ due to } \textit{diffusion} \end{bmatrix} = -(\nabla \cdot \mathbf{F}_Q) \, \Delta v \, \Delta t \tag{2-111}$$

Recalling (2-105), and assuming that the only way that the intensive measure of the property can change within the control volume is via diffusion, we obtain that

$$\frac{\partial Q}{\partial t} + \nabla \cdot \mathbf{F}_Q = 0 \qquad (2\text{-}112)$$

Change of the Intensive Property Q Due to Production within Control Volume

If we let $R_Q(x, y, z, t)$ equal the net rate of production of the intensive property and again recall (2-105), but now assume that the only way in which the property can change is via its production within the volume, we obtain

$$\frac{\partial Q}{\partial t} = R_Q \qquad (2\text{-}113)$$

* * *

Now, putting the foregoing together in that we allow for a change in Q to occur due to advection, diffusion, and generation, we obtain from (2-107), (2-112), and (2-113) the most general from of the *distribution of variables equation*,

$$\frac{\partial Q}{\partial t} + \nabla \cdot (Q\mathbf{V} + \mathbf{F}_Q) = R_Q \qquad (2\text{-}114)$$

If the property that concerns us is a conservative one (i.e., one that can be altered through physical processes but not through chemical or biological ones), then no production can occur, R_Q equals zero, and (2-114) reduces to

$$\frac{\partial Q}{\partial t} + \nabla \cdot (Q\mathbf{V} + \mathbf{F}_Q) = 0 \qquad (2\text{-}115)$$

which states that the local rate of change of Q is equal to the convergence of Q due to advection and diffusion.

At this point it seems appropriate to take time to comment briefly on the form of the diffusive flux vector $\mathbf{F}_Q$. An empirical law governing $\mathbf{F}_Q$ that is frequently adopted is of the form

$$\mathbf{F}_Q = -K_Q \nabla Q \qquad (2\text{-}116)$$

where K_Q is the diffusivity coefficient for the particular property under investigation. Several specific but familiar forms of this equation exist. Consider the following examples:

Transfer of Mass Due to a Concentration Gradient

For the case of mass transfer due to a concentration gradient, (2-116) takes the form

$$\mathbf{F}_s = -K_s \nabla(\rho s) \qquad (2\text{-}117)$$

where for this purpose s is the salinity of seawater expressed in mass per unit mass; ρs is the concentration of salt in mass per unit volume; $\nabla(\rho s)$

is the gradient of the concentration, the "driving force" in mass per unit volume per unit distance; K_s is the diffusivity coefficient of salt expressed in area per unit time (physically, K_s is the mass of salt per unit area per unit time diffusing past a point when the concentration gradient is one unit of mass per unit volume per unit distance); and $\mathbf{F}_s$ is the diffusive flux of salt in seawater expressed as mass per unit area per unit time. In general, ρs need not necessarily be the salt concention but could be the mass concentration of any substance. The equation for mass diffusion, of which (2-117) is a specific case, goes under the name of the *Fickian diffusion law*.

When in (2-115) we substitute ρs for Q and $\mathbf{F}_s$ for $\mathbf{F}_Q$, we obtain the *conservative of salt equation*,

$$\frac{\partial \rho s}{\partial t} + \nabla \cdot (\rho s \mathbf{V} + \mathbf{F}_s) = 0 \tag{2-118}$$

which states that the local rate of change of the salt concentration is equal to the convergence of salt due to advection and diffusion. Assuming, as we have been doing that seawater is a binary mixture of salt and water, the concentration of water is $\rho(1 - s)$. Defining $\mathbf{F}_W$ as the diffusive flux of water in seawater and substituting into (2-115) yields the *conservation of water equation*,

$$\frac{\partial \rho(1 - s)}{\partial t} + \nabla \cdot [\rho(1 - s)\mathbf{V} + \mathbf{F}_W] = 0 \tag{2-119}$$

which is physically interpreted in the same manner as (2-118). From these last two conservation equations plus the equation of continuity (which will be developed in Section 2-18) it can readily be shown that the two diffusive fluxes are related by

$$\mathbf{F}_s = -\mathbf{F}_W \tag{2-120}$$

Conductive Transfer of Thermal Energy Due to a Temperature Gradient

For the conductive transfer of thermal energy due to a temperature gradient, (2-116) is written

$$\mathbf{h} = -c\, \nabla T \tag{2-121}$$

where T is the absolute temperature; ∇T is the spatial temperature gradient expressed in absolute degrees per unit length, the "driving force"; c is the coefficient of thermal conductivity expressed as energy per unit time per absolute degree per unit distance; and $\mathbf{h}$ is the energy flux due to thermal conduction expressed as energy per unit time per unit area. Equation (2-121) is the familiar *Fourier law of heat conduction*.

Momentum Transfer Due to Velocity Gradients

An example of the relationship of the transfer of momentum to the velocity gradient is that of *Newton's law of viscosity*, which for fluid flow in the x direction is written

$$\tau_{yx} = \mu \frac{\partial u}{\partial y} \qquad (\mu > 0) \tag{2-122}$$

where $\partial u/\partial y$ is the velocity gradient expressed in units of distance per unit time per unit distance, the "driving force"; μ is the dynamic viscosity expressed as a force per unit area multiplied by time; and τ_{yx} is the shear stress (shearing force per unit area) exerted in the x direction on a surface of constant y by the fluid in the region of greater y. Alternatively, the shear stress can be thought of as the flux of x momentum in the negative y direction and can be expressed equivalently as a force per unit area or as momentum per unit area per unit time. Any fluid that obeys (2-122) is said to be *Newtonian*; we will consider a more general representation of the shear stress in a later section.

2-18 The Continuity Equation

The law of the conservation of mass forms the basis of what is called the *principle of continuity* and is expressed mathematically by the *continuity equation*. The principle states that the rate of increase of the fluid mass contained within a given space must be equal to the difference between the rates of influx and efflux of mass.

To show generality and yet keep the mathematical manipulations from becoming overly tedious, let us again consider seawater to be a binary fluid consisting of salt and water, designated as before by subscripts s and W. Consider a finite volume of the fluid υ of arbitrary but fixed size and shape. Adopting the principle of the conservation of mass, the only way in which the mass of salt and water, M_s and M_W, within the volume can change is by input across its boundaries. M_s and M_W can in general vary, but their variation must be in accord with the equations

$$\frac{dM_s}{dt} = -\int_A \mathbf{n} \cdot \rho_s \mathbf{V}_s \, dA \tag{2-123}$$

and

$$\frac{dM_W}{dt} = -\int_A \mathbf{n} \cdot \rho_W \mathbf{V}_W \, dA \tag{2-124}$$

where $\mathbf{n}$ is the outward directed unit normal to the surface A of the volume υ. In these two equations $\rho_s \mathbf{V}_s$ and $\rho_W \mathbf{V}_W$ are the mass flux vectors of salt and water, ρ_s and ρ_W are the actual densities of the two species in the particular seawater, and $\mathbf{V}_s$ and $\mathbf{V}_W$ are the *total velocities* of salt and water. By total velocity, we mean the diffusion velocity of each species, defined by

$$\mathbf{C}_s \equiv \frac{\mathbf{F}_s}{\rho s} \qquad \mathbf{C}_W \equiv \frac{\mathbf{F}_W}{\rho(1-s)} \tag{2-125}$$

plus what we generally call the *velocity of the fluid* (which is actually the mass average velocity of the individual species). Summing (2-123) and (2-124) yields

$$\frac{dM}{dt} = \frac{dM_s}{dt} + \frac{dM_W}{dt} = -\int_A \mathbf{n} \cdot (\rho_s \mathbf{V}_s + \rho_W \mathbf{V}_W)\, dA \tag{2-126}$$

where M is the total mass of fluid within the volume. However, the mass average velocity or fluid velocity, $\mathbf{V}$, is by definition

$$\mathbf{V} \equiv \frac{1}{\rho}(\rho_s \mathbf{V}_s + \rho_W \mathbf{V}_W) \tag{2-127}$$

Substituting this in (2-127), we have

$$\frac{dM}{dt} = -\int_A \mathbf{n} \cdot \rho \mathbf{V}\, dA \tag{2-128}$$

Alternatively, the time rate of change of mass within the volume can be written

$$\frac{dM}{dt} = \frac{d}{dt} \int_v \rho\, dv \tag{2-129}$$

or, since the volume is fixed, as

$$\frac{dM}{dt} = \int_v \frac{\partial \rho}{\partial t}\, dv \tag{2-130}$$

where Leibnitz's rule has been invoked (see the following section). Equating (2-128) and (2-130) and using the divergence theorem to transform the surface integral results in

$$\int_v \left(\frac{\partial \rho}{\partial t} + \nabla \cdot \rho \mathbf{V} \right) dv = 0 \tag{2-131}$$

Since the volume is arbitrary, the only way that this result can in general be satisfied is that the integrand vanish,

$$\frac{\partial \rho}{\partial t} + \nabla \cdot \rho \mathbf{V} = 0 \tag{2-132}$$

which is the differential form of the continuity equation. We can write this another way by expanding the dot product,

$$\frac{\partial \rho}{\partial t} + \mathbf{V} \cdot \nabla \rho + \rho \nabla \cdot \mathbf{V} = 0$$

and applying the definition of the material derivative,

$$\frac{D\rho}{Dt} + \rho \nabla \cdot \mathbf{V} = 0 \tag{2-133}$$

which states that the fluid can only be divergentless if the material time rate of change of the density is zero.

Equation (2-133) permits a greater insight into the relation between incompressibility and nondivergence. A fluid is said to be incompressible if the density following the fluid motion does not change; that is,

$$\frac{D\rho}{Dt} = 0 \tag{2-134}$$

The continuity equation then reduces to

$$\nabla \cdot \mathbf{V} = 0 \tag{2-135}$$

Incompressibility, as expressed by (2-134), does not imply that the density field of the fluid is both uniform in space and constant in time; that is merely one possibility. It is possible instead that the local rate of change of the density is the negative of its field rate of change,

$$\frac{\partial \rho}{\partial t} = -\mathbf{V} \cdot \nabla \rho$$

2-19 Three-Dimensional Leibnitz Rule and the Reynolds Transport Theorem

Without proof, the Leibnitz rule for the differentiation of an integral is

$$\frac{d}{dt}\int_{\alpha(t)}^{\beta(t)} f(x, t)\, dx = \int_{\alpha}^{\beta} \frac{\partial f}{\partial t}\, dx + f(\beta(t), t)\frac{d\beta}{dt} - f(\alpha(t), t)\frac{d\alpha}{dt} \tag{2-136}$$

One can obtain a useful integral form of the continuity equation by making use of the three-dimensional analogue to (2-136). Again without proof, if $f(x, y, z, t)$, then

$$\frac{d}{dt}\int_{v(t)} f\, dv = \int_{A(t)} \mathbf{n} \cdot \mathbf{U} f\, dA + \int_{v(t)} \frac{\partial f}{\partial t}\, dv \tag{2-137}$$

where $v(t)$ denotes any arbitrary volume whose boundaries move with the arbitrary velocity $\mathbf{U}(x, y, z, t)$.

If we consider an arbitrary volume of fluid whose boundaries move with the fluid velocity, $\mathbf{U} = \mathbf{V}$, and replace f by ρf, then (2-137) becomes

$$\frac{d}{dt}\int_{v(t)} \rho f\, dv = \int_{A(t)} \mathbf{n} \cdot \rho f \mathbf{V}\, dA + \int_{v(t)} \frac{\partial \rho f}{\partial t}\, dv$$

Using the divergence theorem, this converts to

$$\frac{d}{dt}\int_{v(t)} \rho f\, dv = \int_{v(t)} \left(\frac{\partial \rho f}{\partial t} + \nabla \cdot \rho f \mathbf{V}\right) dv \tag{2-138}$$

The integrand on the right-hand side upon expansion can be written

$$f\left(\frac{\partial \rho}{\partial t} + \nabla \cdot \rho \mathbf{V}\right) + \rho\left(\frac{\partial f}{\partial t} + \mathbf{V} \cdot \nabla f\right)$$

The first term in parentheses vanishes in view of the continuity equation, while the second can be expressed as a material derivative. Thus, (2-138) reduces to

$$\frac{d}{dt}\int_{v(t)} \rho f\, dv = \int_{v(t)} \rho \frac{Df}{Dt}\, dv \qquad (2\text{-}139)$$

remembering here that the boundaries of the volume *move with the fluid velocity*. Furthermore, under this condition there can be no net mass transport across the boundaries, thus conserving mass within the volume. Finally, identifying $\rho\, dv$ as identical to dM, an alternative form of (2-139) is that of

$$\frac{d}{dt}\int_{M} f\, dM = \int_{M} \frac{Df}{Dt}\, dM \qquad (2\text{-}140)$$

for a volume of mass M whose boundaries move with the fluid velocity. Equations (2-139) and (2-140) are two forms of the *Reynolds transport theorem*.

2-20 Moving Cartesian Coordinate Systems

In this section we will examine in detail the relations between inertial and noninertial Cartesian coordinate systems. Under the latter type we first consider only translation, second only rotation, and third the combination of translation and rotation.

Translating Cartesian Coordinate Systems

With reference to Figure 2-36, consider a point in space P which is located by the two vectors $\mathbf{r}$ and $\mathbf{r}'$ with respect to the origins of two Cartesian coordinate systems O and O'; and let O' be located by the vector $\mathbf{h}$ with

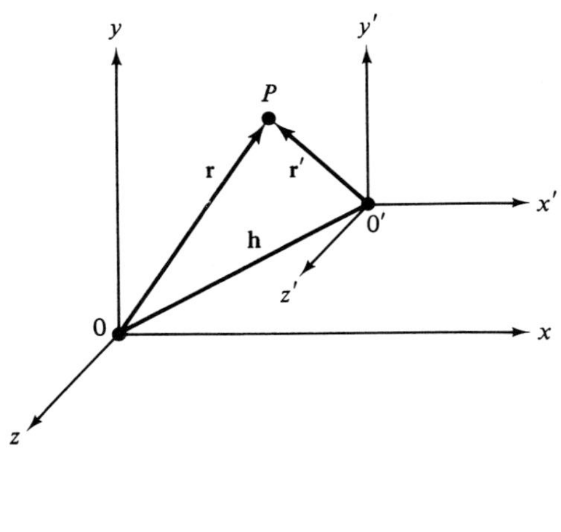

Fig. 2-36. Definition diagram for translating Cartesian coordinate system.

respect to O. Then

$$\mathbf{r} = \mathbf{r}' + \mathbf{h} \tag{2-141}$$

which can be written in terms of the Cartesian components of $\mathbf{r}, \mathbf{r}'$, and $\mathbf{h}$ as

$$x = x' + h_x \qquad y = y' + h_y \qquad z = z' + h_z \tag{2-142}$$

Now if we regard the origin O as fixed (inertial), the origin O' as translating (no rotation) with respect to origin O, and the point P as moving with respect to both coordinate systems, the velocity of P given by differentiation of (2-141) is

$$\mathbf{V} \equiv \frac{d\mathbf{r}}{dt} = \frac{d\mathbf{r}'}{dt} + \frac{d\mathbf{h}}{dt} \tag{2-143}$$

or

$$\mathbf{V} = \mathbf{V}' + \mathbf{V}_h \tag{2-144}$$

where $\mathbf{V}$ is the velocity of P relative to O, $\mathbf{V}'$ is the velocity of P relative to O', and $\mathbf{V}_h$ is the velocity of O' relative to O. Differentiating a second time, we obtain the relation between the relative accelerations,

$$\mathbf{a} \equiv \frac{d\mathbf{V}}{dt} = \frac{d^2\mathbf{r}}{dt^2} = \frac{d^2\mathbf{r}'}{dt^2} + \frac{d^2\mathbf{h}}{dt^2} \tag{2-145}$$

or

$$\mathbf{a} = \mathbf{a}' + \mathbf{a}_h$$

in which $\mathbf{a}$ is the acceleration of P relative to O, $\mathbf{a}'$ is the acceleration of P relative to O', and $\mathbf{a}_h$ is the acceleration of O' relative to O.

Newton's equation of motion holds in the inertial coordinate system, O, and if we designate the force per unit mass acting as $\mathbf{F}_m$, then

$$\mathbf{F}_m = \frac{d^2\mathbf{r}}{dt^2} = \frac{d^2\mathbf{r}'}{dt^2} + \mathbf{a}_h \tag{2-146}$$

If O' is moving at a constant velocity relative to O, then $\mathbf{a}_h$ equals zero, and

$$\mathbf{F}_m = \frac{d^2\mathbf{r}'}{dt^2} \tag{2-147}$$

which states that Newton's equation of motion holds in any coordinate system moving with a uniform velocity relative to an inertial or fixed coordinate system; that both coordinate systems are then inertial.

Rotating Cartesian Coordinate Systems

Consider the two coordinate systems (x, y, z) and (x', y', z') of Figure 2-37 whose axes are rotated relative to one another and whose origins coincide. The units vectors $\mathbf{i}, \mathbf{j}$, and $\mathbf{k}$ are associated with the unprimed system

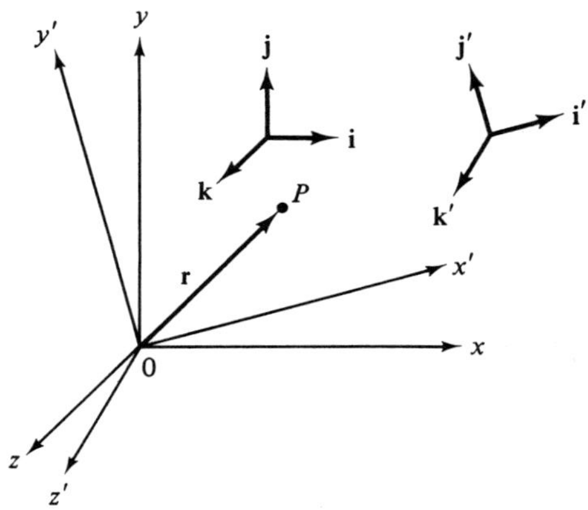

Fig. 2-37. Definition diagram for rotated Cartesian coordinate system.

and the unit vectors $\mathbf{i}', \mathbf{j}'$, and $\mathbf{k}'$ are associated with the primed system. A position vector $\mathbf{r}$ can be expressed as either

$$\mathbf{r} = x\mathbf{i} + y\mathbf{j} + z\mathbf{k} \tag{2-148}$$

or

$$\mathbf{r} = x'\mathbf{i}' + y'\mathbf{j}' + z'\mathbf{k}' \tag{2-149}$$

The equations relating the two coordinate systems are obtained by taking the dot product of either the primed or unprimed unit vectors with (2-148) and (2-149) and equating the corresponding results. For example, working with the dot product of the unprimed vectors; $(\mathbf{i} \cdot \mathbf{r})$, $(\mathbf{j} \cdot \mathbf{r})$, and $(\mathbf{k} \cdot \mathbf{r})$, we get

$$\begin{aligned} x &= x'(\mathbf{i}' \cdot \mathbf{i}) + y'(\mathbf{j}' \cdot \mathbf{i}) + z'(\mathbf{k}' \cdot \mathbf{i}) \\ y &= x'(\mathbf{i}' \cdot \mathbf{j}) + y'(\mathbf{j}' \cdot \mathbf{j}) + z'(\mathbf{k}' \cdot \mathbf{j}) \\ z &= x'(\mathbf{i}' \cdot \mathbf{k}) + y'(\mathbf{j}' \cdot \mathbf{k}) + z'(\mathbf{k}' \cdot \mathbf{k}) \end{aligned} \tag{2-150}$$

The dot products of the unit vectors are the cosine of the angles between the corresponding axes; for example,

$$(\mathbf{i}' \cdot \mathbf{j}) = |\mathbf{i}'||\mathbf{j}| \cos \theta = \alpha_{x'y}$$

where $\alpha_{x'y}$ is the cosine of the angle between the x' axis and the y axis. Equations (2-148), (2-149), and (2-150) do not require that $\mathbf{r}$ be drawn from the origin of the coordinates; since we are dealing with free vectors, analogous formulas apply in terms of the components of any other free vector, $\mathbf{A}$.

Next we consider the case where the two coordinate systems are rotating with respect to each other. Since the time derivative of a given vector $\mathbf{A}$ will be different in the two systems, we denote d/dt as the time derivative with respect to the unprimed coordinate system which we will regard as fixed and thus inertial, and d'/dt as the time derivative with respect to the primed coordinate system which we will consider as rotating and thus noninertial.

This distinction is made for vectors only; there is no corresponding ambiguity associated with the time derivatives of scalars (numerical quantities)—their time derivatives will be denoted by d/dt, which has the same meaning in all coordinate systems. Let **A** be given by

$$\mathbf{A} = A_x\mathbf{i} + A_y\mathbf{j} + A_z\mathbf{k} \tag{2-151}$$

or

$$\mathbf{A} = A'_x\mathbf{i}' + A'_y\mathbf{j}' + A'_z\mathbf{k}' \tag{2-152}$$

Differentiating each of these,

$$\frac{d\mathbf{A}}{dt} = \frac{dA_x}{dt}\mathbf{i} + \frac{dA_y}{dt}\mathbf{j} + \frac{dA_z}{dt}\mathbf{k} \tag{2-153}$$

and

$$\frac{d'\mathbf{A}}{dt} = \frac{dA'_x}{dt}\mathbf{i}' + \frac{dA'_y}{dt}\mathbf{j}' + \frac{dA'_z}{dt}\mathbf{k}' \tag{2-154}$$

From (2-152), we can also obtain the formula for d/dt in terms of the primed components, but now the unit vectors $\mathbf{i}'$, $\mathbf{j}'$, and $\mathbf{k}'$ have time derivatives since they are moving relative to the unprimed system. Thus,

$$\frac{d\mathbf{A}}{dt} = \frac{dA'_x}{dt}\mathbf{i}' + \frac{dA'_y}{dt}\mathbf{j}' + \frac{dA'_z}{dt}\mathbf{k}' + A'_x\frac{d\mathbf{i}'}{dt} + A'_y\frac{d\mathbf{j}'}{dt} + A'_z\frac{d\mathbf{k}'}{dt} \tag{2-155}$$

Now let us designate as OQ the axis, through the origin, about which the primed coordinate system rotates with an angular velocity $\boldsymbol{\omega}$ (Figure 2-38). The primed derivative of a vector **B** at rest in the primed system is

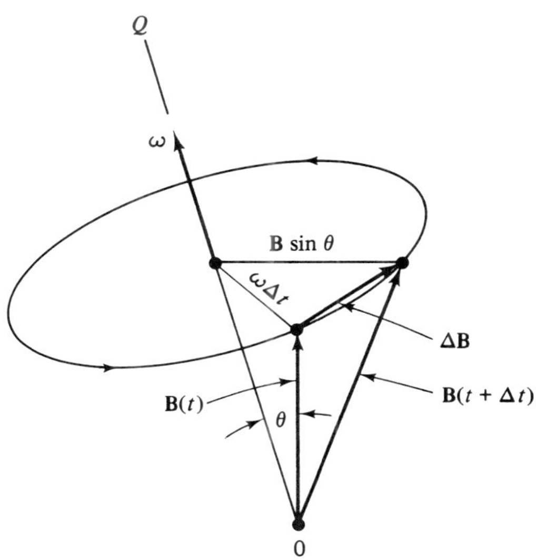

Fig. 2-38. Definition diagram for a rotating vector.

zero. From the figure it is seen that for the unprimed system,

$$\left|\frac{\Delta \mathbf{B}}{\Delta t}\right| = \frac{(B \sin \theta)(\omega\, \Delta t)}{\Delta t}$$

In the limit, and considering the definition of the vector product,

$$\frac{d\mathbf{B}}{dt} = \boldsymbol{\omega} \times \mathbf{B} \tag{2-156}$$

Substituting (2-154) for the first three right-hand terms in (2-155) and then applying (2-156) to the time derivatives of the unit vectors $\mathbf{i}', \mathbf{j}'$, and $\mathbf{k}'$, we obtain

$$\frac{d\mathbf{A}}{dt} = \frac{d'\mathbf{A}}{dt} + A'_x(\boldsymbol{\omega} \times \mathbf{i}') + A'_y(\boldsymbol{\omega} \times \mathbf{j}') + A'_z(\boldsymbol{\omega} \times \mathbf{k}')$$

or

$$\frac{d\mathbf{A}}{dt} = \frac{d'\mathbf{A}}{dt} + \boldsymbol{\omega} \times \mathbf{A} \tag{2-157}$$

This is the fundamental relation between time derivatives for rotating coordinate systems. To find the second time derivative, we take the derivative of (2-157) and apply (2-157) to $\mathbf{A}$ and $d'\mathbf{A}/dt$. Thus,

$$\begin{aligned}\frac{d^2\mathbf{A}}{dt^2} &= \frac{d}{dt}\left(\frac{d'\mathbf{A}}{dt}\right) + \boldsymbol{\omega} \times \frac{d\mathbf{A}}{dt} + \frac{d\boldsymbol{\omega}}{dt} \times \mathbf{A} \\ &= \frac{d'^2\mathbf{A}}{dt^2} + \boldsymbol{\omega} \times \frac{d'\mathbf{A}}{dt} + \boldsymbol{\omega} \times \left(\frac{d'\mathbf{A}}{dt} + \boldsymbol{\omega} \times \mathbf{A}\right) \\ &\quad + \frac{d\boldsymbol{\omega}}{dt} \times \mathbf{A}\end{aligned}$$

or, finally,

$$\frac{d^2\mathbf{A}}{dt^2} = \frac{d'^2\mathbf{A}}{dt^2} + 2\boldsymbol{\omega} \times \frac{d'\mathbf{A}}{dt} + \boldsymbol{\omega} \times (\boldsymbol{\omega} \times \mathbf{A}) + \frac{d\boldsymbol{\omega}}{dt} \times \mathbf{A} \tag{2-158}$$

Note from (2-157) that the primed and unprimed derivatives of any vector parallel to the axis of rotation are the same; specifically,

$$\frac{d\boldsymbol{\omega}}{dt} = \frac{d'\boldsymbol{\omega}}{dt} \tag{2-159}$$

where as previously the vector $\boldsymbol{\omega}$ is the angular velocity of the primed system relative to the unprimed system.

It is appropriate now to apply the results of (2-157) and (2-158) to the position vector $\mathbf{r}$ of (2-148) and (2-149),

$$\frac{d\mathbf{r}}{dt} = \frac{d'\mathbf{r}}{dt} + \boldsymbol{\omega} \times \mathbf{r} \tag{2-160}$$

$$\frac{d^2\mathbf{r}}{dt^2} = \frac{d'^2\mathbf{r}}{dt^2} + \boldsymbol{\omega} \times (\boldsymbol{\omega} \times \mathbf{r}) + 2\boldsymbol{\omega} \times \frac{d\mathbf{r}}{dt} + \frac{d\boldsymbol{\omega}}{dt} \times \mathbf{r} \tag{2-161}$$

where $d'^2\mathbf{r}/dt^2$ is the acceleration of the point P relative to the primed system; $\boldsymbol{\omega} \times (\boldsymbol{\omega} \times \mathbf{r})$ is the centripetal (toward the center) acceleration of the point P in rotation about an axis (the magnitude of the acceleration is $\omega^2 r \sin\theta$; Figure 2-39); $2\boldsymbol{\omega} \times d'\mathbf{r}/dt$ is the Coriolis acceleration, which is present only when the point P and thus position vector $\mathbf{r}$ is moving in the primed system; and $d\boldsymbol{\omega}/dt \times \mathbf{r}$ is the acceleration resulting from a change of direction or magnitude of $\boldsymbol{\omega}$ and is zero for a constant angular velocity.

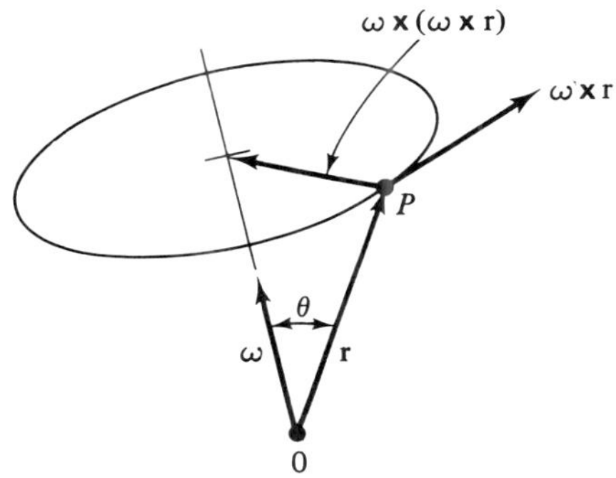

Fig. 2-39. Orientation of the centripetal acceleration vector.

If we suppose that the unprimed coordinate system is inertial, and that therefore Newton's second law holds, we have in the primed system that

$$\frac{d'^2\mathbf{r}}{dt^2} = \mathbf{F}_m - \boldsymbol{\omega} \times (\boldsymbol{\omega} \times \mathbf{r}) - 2\boldsymbol{\omega} \times \frac{d\mathbf{r}}{dt} - \frac{d\boldsymbol{\omega}}{dt} \times \mathbf{r} \qquad (2\text{-}162)$$

where $\mathbf{F}_m$ is, as before, the force acting per unit mass. This result is an equation of motion similar in form to that of Newton. The second term on the right, $-\boldsymbol{\omega} \times (\boldsymbol{\omega} \times \mathbf{r})$, is now referred to as the centrifugal (away from the center) acceleration. In summary, we see from (2-162) that we can treat a rotating coordinate system as if it were fixed through the introduction of the centrifugal acceleration, the Coriolis acceleration, and the acceleration arising from changes in the direction or magnitude of the rotation.

Translating and Rotating Cartesian Coordinate Systems

Coordinate systems in simultaneous translation and rotation relative to each other can be treated by using (2-141) to represent the relation between the vectors $\mathbf{r}$ and $\mathbf{r}'$ relative to the origins O and O', not necessarily coincident. In substituting $\mathbf{r}'$ for $\mathbf{r}$ in (2-160) and (2-161) and the result for $d\mathbf{r}'/dt$ and $d^2\mathbf{r}'/dt^2$ in (2-143) and (2-145), we obtain for the position, velocity, and acceleration of a point P with respect to coordinate systems in both relative translation and rotation the following:

$$\mathbf{r} = \mathbf{r}' + \mathbf{h} \tag{2-163}$$

$$\frac{d\mathbf{r}}{dt} = \frac{d'\mathbf{r}'}{dt} + \boldsymbol{\omega} \times \mathbf{r}' + \frac{d\mathbf{h}}{dt} \tag{2-164}$$

and

$$\begin{aligned} \frac{d^2\mathbf{r}}{dt^2} = \frac{d'^2\mathbf{r}'}{dt^2} &+ \boldsymbol{\omega} \times (\boldsymbol{\omega} \times \mathbf{r}') + 2\boldsymbol{\omega} \times \frac{d'\mathbf{r}'}{dt} \\ &+ \frac{d\boldsymbol{\omega}}{dt} \times \mathbf{r}' + \frac{d^2\mathbf{h}}{dt^2} \end{aligned} \tag{2-165}$$

2-21 Newton's Law of Motion on a Rotating Earth

Of the various forms of the equation of motion derived in the previous section, that of (2-165) applies most closely to our interest concerning relative motion over the earth referred to a coordinate system *fixed relative to the earth*. Substituting $\mathbf{F}_m$ for $d^2\mathbf{r}/dt^2$ in this last result and rearranging, we obtain

$$\begin{aligned} \frac{d'^2\mathbf{r}'}{dt^2} = \mathbf{F}_m &- \boldsymbol{\omega} \times (\boldsymbol{\omega} \times \mathbf{r}') - 2\boldsymbol{\omega} \times \frac{d'\mathbf{r}'}{dt} \\ &- \frac{d\boldsymbol{\omega}}{dt} \times \mathbf{r}' - \frac{d^2\mathbf{h}}{dt^2} \end{aligned} \tag{2-166}$$

With reference to the coordinate system of Figure 2-40, for geophysical purposes, what we must consider is the rotation of a relative coordinate system at a constant angular velocity $\boldsymbol{\Omega}$ about an axis located at a distance

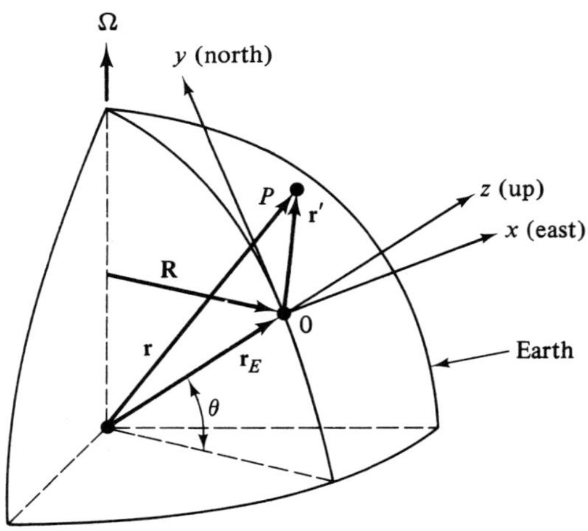

Fig. 2-40. Noninertial coordinate system fixed relative to the earth.

R from the center of the relative system. Now we can make the following physical identifications of the terms in (2-166): $d'\mathbf{r}'/dt$ is equal to $\mathbf{V}$, the velocity of point P as measured relative to the x, y, z coordinate system (a noninertial system) with origin O and fixed with respect to the earth; $d'^2\mathbf{r}'/dt^2$ is then equal to $d'\mathbf{V}/dt$, the acceleration of point P as measured in the (x, y, z) coordinate system; $\boldsymbol{\omega}$ is the angular velocity of rotation of the earth on its axis, this will specifically be assigned the symbol $\boldsymbol{\Omega}$; $d\boldsymbol{\omega}/dt$ is $d\boldsymbol{\Omega}/dt$, the time rate of change of the earth's diurnal rotation, taken for our purposes at this point as being equal to zero; and $d^2\mathbf{h}/dt^2$ is the acceleration of the origin O resulting from its motion in a circle of radius $|\mathbf{R}|$ and is therefore equivalent to $\boldsymbol{\Omega} \times (\boldsymbol{\Omega} \times \mathbf{R})$. The quantity $\boldsymbol{\Omega} \times \mathbf{R}$, the peripheral velocity of the point O, is directed toward the east. It is readily verified that $\boldsymbol{\Omega} \times \mathbf{R}$ is identical to $\boldsymbol{\Omega} \times \mathbf{r}_E$. The magnitude of $\boldsymbol{\Omega} \times \mathbf{r}_E$ is $\Omega r_E \sin(\pi/2 - \theta)$, which is simply $\Omega r_E \cos\theta$, but $r_E \cos\theta$ is equal to R. Thus, the magnitude of $\boldsymbol{\Omega} \times \mathbf{r}_E$ is equal to that of $\boldsymbol{\Omega} \times \mathbf{R}$ and the direction is the same. In view of the foregoing, we can rewrite (2-166) as

$$\frac{d'\mathbf{V}}{dt} = \mathbf{F}_m - \boldsymbol{\Omega} \times (\boldsymbol{\Omega} \times \mathbf{r}) - 2\boldsymbol{\Omega} \times \mathbf{V} \tag{2-167}$$

where since

$$\mathbf{r}_E + \mathbf{r}' = \mathbf{r}$$

we have replaced the two terms $-\boldsymbol{\Omega} \times (\boldsymbol{\Omega} \times \mathbf{r}')$ and $-\boldsymbol{\Omega} \times (\boldsymbol{\Omega} \times \mathbf{r}_E)$ by its equivalent $-\boldsymbol{\Omega} \times (\boldsymbol{\Omega} \times \mathbf{r})$. In the future we will, as a matter of convenience, drop the prime from $d'\mathbf{V}/dt$ and write, instead, $d\mathbf{V}/dt$ with the understanding that by this we mean the acceleration of point P as measured in the (x, y, z) coordinate system.

For our particular interests in fluid motion, since $d\mathbf{V}/dt$ is the acceleration of a particular particle of fluid, it can be replaced by the material derivative $D\mathbf{V}/Dt$, thus making it possible to write (2-167) as

$$\frac{D\mathbf{V}}{Dt} = \mathbf{F}_m - \boldsymbol{\Omega} \times (\boldsymbol{\Omega} \times \mathbf{r}) - 2\boldsymbol{\Omega} \times \mathbf{V} \tag{2-168}$$

or, equivalently,

$$\frac{\partial \mathbf{V}}{\partial t} + (\mathbf{V} \cdot \nabla)\mathbf{V} = \mathbf{F}_m - \boldsymbol{\Omega} \times (\boldsymbol{\Omega} \times \mathbf{r}) - 2\boldsymbol{\Omega} \times \mathbf{V} \tag{2-169}$$

or in indicial notation as

$$\frac{\partial V_i}{\partial t} + V_j \frac{\partial V_i}{\partial x_j} = F_{m_i} - \epsilon_{ijk}\epsilon_{kmn}\Omega_j\Omega_m r_n - 2\epsilon_{ijk}\Omega_j V_k \tag{2-170}$$

These results will be temporarily put aside while we investigate the form of the $\mathbf{F}_m$ term, that is, its representation as the sum of volume and surface forces per unit mass.

2-22 Volume Forces

The forces acting on an element of a continuous medium may be of two kinds: (1) *volume* (or *external* or *body*) *forces* are those which can be regarded as reaching into the medium and acting throughout the volume, and (2) *surface* (or *internal* or *contact*) *forces* are those which are to be regarded as acting on an element of volume through its bounding surface. In this section we will discuss the character of the former and in the next section that of the latter.

In general, volume forces are long range in nature and are proportional to the amount of mass present. Frequently, they are expressed as forces per unit mass or accelerations. The only volume force that concerns us here is that of gravitational or Newtonian attraction. Other volume forces exist; those of an electromagnetic nature, which may act when the medium carries an electric charge or an electric current, and centrifugal and Coriolis forces associated with motion referred to accelerating coordinate systems as in (2-168).

If we let $\mathbf{g}_E$, $\mathbf{g}_M$, and $\mathbf{g}_S$ be the absolute gravitational attractions (as given by Newton's law of universal gravitation) on a unit mass due to the earth, moon, and sun, respectively, then the differential gravitational force $d\mathbf{F}$ acting on the differential mass dM having a volume dv is

$$d\mathbf{F} = \rho(\mathbf{g}_E + \mathbf{g}_M + \mathbf{g}_S)\, dv$$

or

$$d\mathbf{F} = \rho\mathbf{g}_a\, dv \tag{2-171}$$

where $\mathbf{g}_a$ is the vector sum of $\mathbf{g}_E$, $\mathbf{g}_M$, and $\mathbf{g}_S$. Writing the force on a per unit volume basis and aligning the z axis of our coordinate system along the direction of $-\mathbf{g}_a$ (this condition will be slightly modified a little later), then the gravitational force per unit volume, $d\mathbf{F}/dv$, is

$$\frac{d\mathbf{F}}{dv} = \rho\mathbf{g}_a = -\rho g_a\mathbf{k} \equiv \mathbf{F}_g \tag{2-172}$$

or, in indicial notation,

$$(F_g)_i = -\rho g_a \delta_{i3} \tag{2-173}$$

A force field is said to be *conservative* if the net work done on an object when moved through a closed path is zero; that is, the force field $\mathbf{F}(x, y, z)$ is conservative if

$$\oint \mathbf{F} \cdot d\mathbf{r} = 0 \tag{2-174}$$

for any closed path contained in the field of $\mathbf{F}$. A property of a conservative force is that it can always be expressed in terms of the gradient of a scalar function of position. It can be shown that $\mathbf{F}_g$, the gravitational force per unit volume, or $\mathbf{g}_a$, the gravitational force per unit mass, are conservative and as such can be expressed as

$$\mathbf{g}_a = -\nabla \chi_a \tag{2-175}$$

or since our coordinate z axis is aligned with $\mathbf{g}_a$, the magnitude of $\mathbf{g}_a$ is given by

$$g_a = \frac{d\chi_a}{dz} \tag{2-176}$$

and from (2-172),

$$\mathbf{F}_g = -\rho \nabla \chi_a \tag{2-177}$$

χ_a is called the *absolute gravitational potential* and has units appropriate to those of energy per unit mass. Equation (2-176) can be integrated to obtain

$$(\chi_a)_2 - (\chi_a)_1 = \int_{z_1}^{z_2} g_a \, dz$$

Taking $(\chi_a)_1$ arbitrarily as zero at z_1, also equal to zero (say mean sea level), gives us

$$\chi_a = \int_0^z g_a \, dz$$

which permits χ_a to be physically interpreted simply as the potential energy per unit mass by virtue of an object's relative position in the earth's gravitational (Newtonian) force field.

2-23 Surface Forces

Next we examine the character of the surface forces acting on the boundaries of a volume of fluid υ. These forces are distinct from the volume forces in that the latter act at long range while the former act at short range; they are forces acting between the fluid particles immediately adjacent to one side of the surface and those particles immediately adjacent to the other side. Consider a small area increment of a surface, ΔA (Figure 2-41). The force exerted by the fluid particles exterior to the volume upon this element will be denoted by $\mathbf{P}\,\Delta A$. The vector $\mathbf{P}$ is the stress (expressed as a force per unit

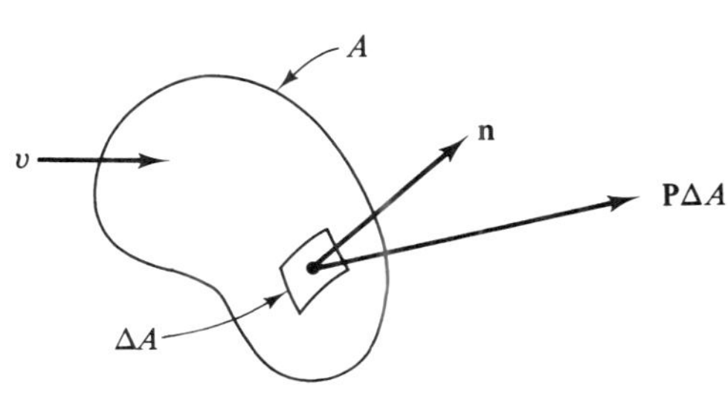

Fig. 2-41. Definition sketch for surface forces.

area) at the point in question. This stress is in general inclined with respect to a normal to the surface; it will have both normal and tangential components. An example of a possible normal component is that which we refer to as the fluid pressure. In fact, this is the only surface force acting under hydrostatic conditions ($\mathbf{V} = 0$). However, if the fluid is in motion, then shearing stresses can exist and $\mathbf{P}$ is no longer normal to ΔA. In that which follows, we will adopt the sign convention with reference to the normal component of the stress $\mathbf{P}$ such that tension is taken as positive and compression as negative.

Hydrostatic pressure is independent of the orientation of the surface upon which it acts. The stress $\mathbf{P}$, however, does depend in general on both the location and the orientation of the surface ΔA as well as on time. Let $\mathbf{n}$ be the outward unit normal to the small element of surface. Then

$$\mathbf{P} = \mathbf{P}(x, y, z, \mathbf{n}, t) = \mathbf{P}(\mathbf{r}, \mathbf{n}, t)$$

To determine the form of the dependence of $\mathbf{P}$ on $\mathbf{n}$, consider the volume, shown in Figure 2-42, shaped in the form of a tetrahedron, three of whose faces are normal to the three coordinate axes and whose fourth face is determined by the unit normal $\mathbf{n}$ and whose area is ΔA. There exist three compo-

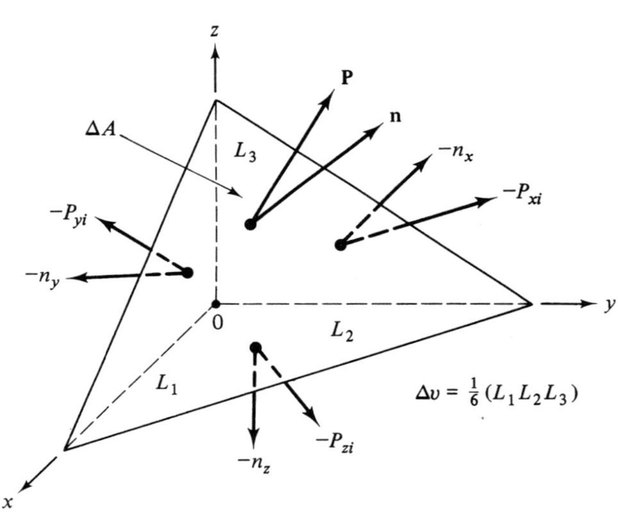

P_{xi}, P_{yi}, P_{zi} ≡ Stress (a vector acting on the positive side of the coordinate surface normal to the x-, y-, z-axis, respectively)

$-n_x, -n_y, -n_z$ ≡ Outward normal to the coordinate surface normal to the x-, y-, z-axis, respectively (they are the negative components of the outward unit normal $\mathbf{n}$)

Fig. 2-42. Stress acting on a tetrahedral element.

nent stresses acting on each face. Assuming equilibrium, the total of the nine *coordinate stresses* acting on the *coordinate surfaces* along with the force $\mathbf{P}\,\Delta A$ and the body force $\rho g_a\,\Delta v$ must sum to zero. The component stresses acting on the positive sides of the three coordinate surfaces define a stress tensor P_{ij}, where i and j take on the values x, y, and z independently. We will not take the time to prove it, but it can readily be shown that P_{ij} obeys the transformation law with respect to coordinate axes for a Cartesian tensor of order two. The components of the stresses P_{xi}, P_{yi}, and P_{zi} are taken as positive in the direction of the positive normals n_x, n_y, and n_z to the coordinate surfaces.

We can now write that the

$$\begin{bmatrix}\text{total force on the}\\ \text{coordinate face}\\ \text{normal to the } x \text{ axis}\end{bmatrix} = -\tfrac{1}{2}L_2L_3P_{xi}$$

However, from geometrical considerations,

$$\tfrac{1}{2}L_2L_3 = n_x\,\Delta A$$

where n_x is given by

$$\mathbf{n} = n_x\mathbf{i} + n_y\mathbf{j} + n_z\mathbf{k}$$

and ΔA being the area of the inclined face. Similar expressions hold for the other two coordinate surfaces, thus permitting us to write the total forces acting on the surfaces as $-n_xP_{xi}\,\Delta A$, $-n_yP_{yi}\,\Delta A$, and $-n_zP_{zi}\,\Delta A$, all of which are vectors and whose sum is simply $-n_iP_{ij}\,\Delta A$, following the summation convention of indicial notation. Now summing all the forces acting on the tetrahedron,

$$P_j\,\Delta A - n_iP_{ij}\,\Delta A + \rho(g_a)_j\,\Delta v = 0 \qquad (2\text{-}178)$$

If we designate by L some typical linear dimension of the element, say the average length of its sides, then the volume is of the order of the magnitude of L^3 or of $O(L^3)$. On the other hand, ΔA is of $O(L^2)$. If we allow the element to become arbitrarily small, we have from (2-178),

$$P_j - n_iP_{ij} - \rho(g_a)_jO(L)$$

which as $L \longrightarrow 0$ reduces to

$$P_j = n_iP_{ij} \qquad (2\text{-}179)$$

at the point (x, y, z). This equation relates the stress vector P_j at a point corresponding to the direction of n_j to the stress components P_{ij} at the point; it tells us that the system of stresses in a fluid is not so complicated as to require the compilation of a table of the function $\mathbf{P}(\mathbf{r}, \mathbf{n})$ at any given instant of time, but that the stresses depend rather simply on $\mathbf{n}$ through the nine components of $P_{ij}(\mathrm{r}_i)$. Moreover, since the quantities are components of a tensor, any equation we derive with them will be true under any rotation of the coordinate axes.

In the special case of hydrostatic conditions, **P** is always normal to the surface ΔA, and is independent of the orientation of the surface. Consequently, the hydrostatic stress $(\mathbf{P})_0$ can be expressed as

$$(\mathbf{P})_0 = -p\mathbf{n} \tag{2-180}$$

or

$$(P_i)_0 = pn_i \tag{2-181}$$

where the negative sign is introduced since the hydrostatic pressure p is to be taken as a positive quantity and our previously adopted convention for **P** was that tension is to be positive. From (2-179) and (2-181), we can derive a *hydrostatic stress tensor*, $(P_{ij})_0$,

$$(P_j)_0 = n_i(P_{ij})_0 = -pn_j$$

Expanding this yields

$$[n_1 n_2 n_3]\begin{bmatrix}(P_{11})_0 & (P_{12})_0 & (P_{13})_0\\ (P_{21})_0 & (P_{22})_0 & (P_{23})_0\\ (P_{31})_0 & (P_{32})_0 & (P_{33})_0\end{bmatrix} = [-pn_1 - pn_2 - pn_3]$$

and thus obtaining

$$(P_{ij})_0 = \begin{bmatrix}-p & 0 & 0\\ 0 & -p & 0\\ 0 & 0 & -p\end{bmatrix} \tag{2-182}$$

It is convenient to introduce the Kronecker delta and write (2-182) as

$$(P_{ij})_0 = -p\delta_{ij} \tag{2-183}$$

Now, defining τ_{ij} as the anomaly of the stress tensor from the hydrostatic condition, we assert that

$$P_{ij} \equiv (P_{ij})_0 + \tau_{ij} \tag{2-184}$$

or

$$P_{ij} \equiv -p\delta_{ij} + \tau_{ij} \tag{2-185}$$

The anomaly τ_{ij} is known as the *viscous stress tensor.*

Finally, returning to the arbitrary volume v of Figure 2-41, we see that acting on the area A of the volume the net surface force is given by

$$[\text{net surface force}] = \int_A \mathbf{P}\, dA$$

and substituting from (2-179) yields

$$\int_A P_j\, dA = \int_A n_i P_{ij}\, dA$$

Using the divergence theorem, we obtain

$$\int_A P_j \, dA = \int_v \frac{\partial P_{ij}}{\partial x_i} \, dv$$

and therefore

$$\begin{bmatrix} \text{net surface force} \\ \text{per unit volume} \end{bmatrix} = \frac{\partial P_{ij}}{\partial x_i} \tag{2-186}$$

or, in terms of the hydrostatic pressure and the viscous stress tensor,

$$\begin{bmatrix} \text{net surface force} \\ \text{per unit volume} \end{bmatrix} = -\frac{\partial p}{\partial x_j} + \frac{\partial \tau_{ij}}{\partial x_i} \tag{2-187}$$

2-24 Symmetry of the Viscous Stress Tensor

It is our purpose at this point to show that the nine components of the viscous stress tensor are not all independent. Figures 2-43 and 2-44 show elemental cubical parallelopipeds at whose centers, point O, the viscous stresses τ_{ij} act. For reasons of clarity, the first of these figures shows only the viscous stresses acting on the positive faces of the parallelopiped while the second shows all the viscous stresses exerted on the element, but only those which act in the x direction are labeled. Examining now the cross section of the

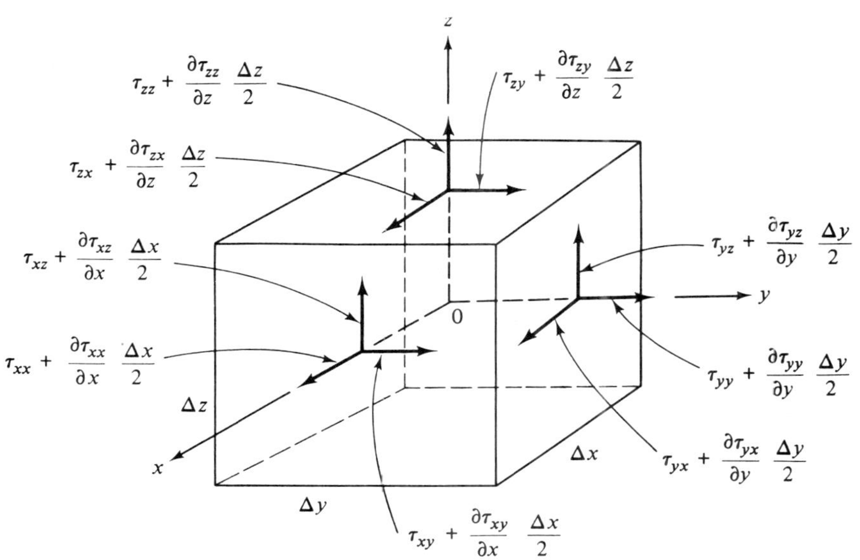

Fig. 2-43. Components of the viscous stress tensor acting on the positive surfaces of an elemental parallelopiped.

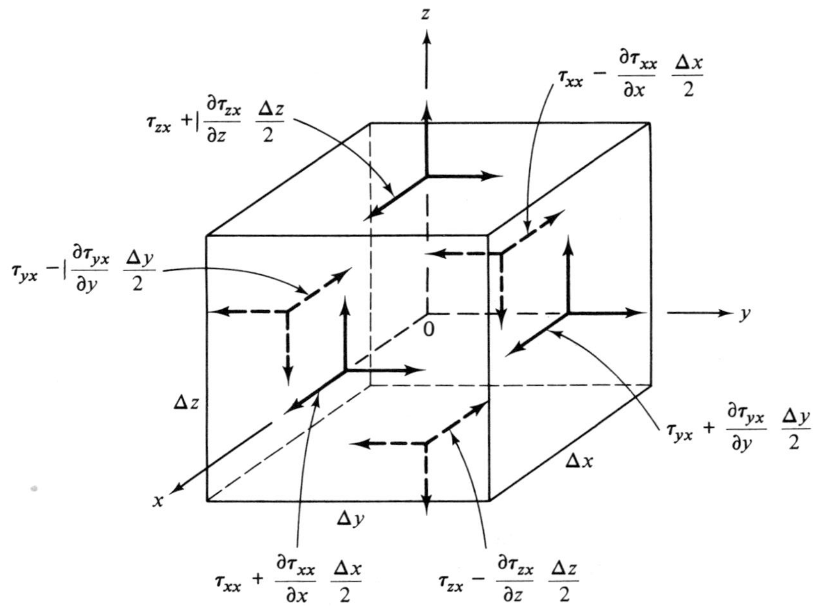

Fig. 2-44. Components of the viscous stress tensor acting on all the surfaces of an elemental parallelopiped.

parallelopiped in the plane $z = 0$, we see that since the normal stresses and gravity forces act through the center of the element, only the shear stresses act so as to produce a torque about the z axis. The force couples shown in Figure 2-45 will tend to rotate the element about the z axis. If we let r be the

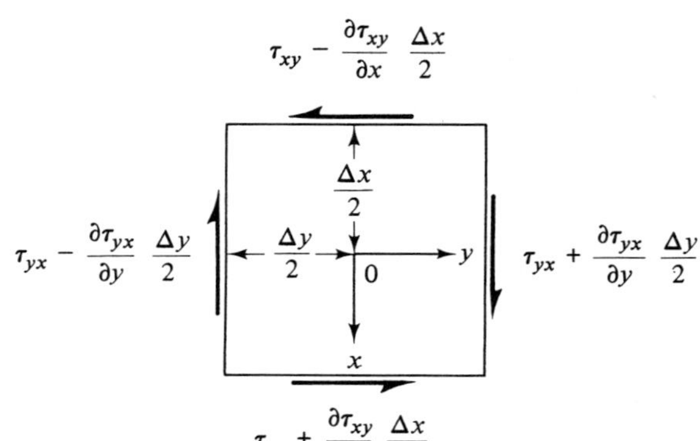

Fig. 2-45. Shear stresses acting on the surfaces of an elemental parallelopiped which produce a torque about the z axis.

radius of gyration of the parallelopiped about the z axis, then

$$I = mr^2$$

and from Newton's second law for rotation,

$$\Gamma = I\alpha = mr^2\alpha \tag{2-188}$$

in which I is the moment of inertia, Γ the resultant torque, m the mass of the body, and α the angular displacement in the counterclockwise sense, so (2-188) for our particular element is written

$$mr^2 \frac{d^2\theta}{dt^2} = \left(\tau_{xy} \frac{\partial \tau_{xy}}{\partial x}\frac{\Delta x}{2} + \tau_{xy} - \frac{\partial \tau_{xy}}{\partial x}\frac{\Delta x}{2}\right)(\Delta y\, \Delta z)\frac{\Delta x}{2}$$
$$-\left(\tau_{yx} + \frac{\partial \tau_{yx}}{\partial y}\frac{\Delta y}{2} + \tau_{yx} - \frac{\partial \tau_{yx}}{\partial y}\frac{\Delta y}{2}\right)(\Delta x\, \Delta z)\frac{\Delta y}{2}$$

which since

$$m = \rho\, \Delta x\, \Delta y\, \Delta z$$

reduces to

$$\rho r^2 \frac{d^2\theta}{dt^2} = \tau_{xy} - \tau_{yx}$$

Now, r is of the order of magnitude Δx and Δy, and if the angular acceleration is to be finite, then in the limit as $\Delta x \longrightarrow 0$ and $\Delta y \longrightarrow 0$, we must have that

$$\tau_{xy} = \tau_{yx}$$

In a similar manner, by examining the torques in the y, z plane (i.e., about the x axis) and in the x, z plane (about the y axis) we would find that in general

$$\tau_{ij} = \tau_{ji} \tag{2-189}$$

that is, that the viscous stress tensor is symmetrical and therefore consists of only six different components, three of which are of a normal type and three of which are of a shearing type.

2-25 The Momentum Equation

The Eulerian equation of motion giving the relationship between the field variables ρ and $\mathbf{V}$ is obtained directly from Newton's second law. This equation is merely a statement of the principle of the conservation of linear momentum and is frequently referred to as the *momentum equation*. We have at hand from the previous discussions all the parts required to formulate this conservation statement; it remains now only to assemble them.

We have as the second law, equation (2-169),

$$\frac{\partial \mathbf{V}}{\partial t} + (\mathbf{V} \cdot \nabla)\mathbf{V} + \mathbf{\Omega} \times (\mathbf{\Omega} \times \mathbf{r}) + 2\mathbf{\Omega} \times \mathbf{V} = \mathbf{F}_m \tag{2-169}$$

Multiplying through by the density ρ yields

$$\rho\frac{\partial \mathbf{V}}{\partial t} + \rho(\mathbf{V}\cdot\nabla)\mathbf{V} + \rho\mathbf{\Omega}\times(\mathbf{\Omega}\times\mathbf{r}) + 2\rho\mathbf{\Omega}\times\mathbf{V} = \mathbf{F}_v \qquad (2\text{-}190)$$

where $\mathbf{F}_v$ is now the net force per unit volume. The contributors to the net force per unit volume are the body force per unit volume, which from (2-177) can be written in terms of the absolute gravitational potential as $-\rho\nabla\chi_a$, and, from (2-186), the surface forces per unit volume, $\partial P_{ij}/\partial x_i$, or, alternatively, $\nabla\cdot\mathcal{P}$. Now putting all together into (2-190),

$$\rho\frac{\partial \mathbf{V}}{\partial t} + \rho(\mathbf{V}\cdot\nabla)\mathbf{V} + \rho\mathbf{\Omega}\times(\mathbf{\Omega}\times\mathbf{r}) + 2\rho\mathbf{\Omega}\times\mathbf{V} = \nabla\cdot\mathcal{P} - \rho\nabla\chi_a \qquad (2\text{-}191)$$

Some simplification results from combining the centripetal acceleration and the Newtonian attraction in the form

$$\mathbf{g} \equiv -\nabla\chi_a - \mathbf{\Omega}\times(\mathbf{\Omega}\times\mathbf{r}) \qquad (2\text{-}192)$$

where $\mathbf{g}$ is the acceleration of gravity (or just plain "gravity") of the earth at the latitude θ. The absolute gravitational vector $\mathbf{g}_a$, for an assumed spherical earth, is directed toward the exact center of the earth, while the centrifugal acceleration vector $-\mathbf{\Omega}\times(\mathbf{\Omega}\times\mathbf{r})$ is directed outward and normal to the earth's axis of rotation (Figure 2-46). The direction of the resultant vector $\mathbf{g}$ is that assumed by a plumb line at the latitude θ; it is the "local down." It can be shown that the centrifugal force is also a conservative force and as such it too can, like the Newtonian attraction, be represented as the gradient of a scalar force potential. Further, the sum of two or more conservative force fields is also a conservative force field, which means that the acceleration of gravity can be written

$$\mathbf{g} = -\nabla\chi \qquad (2\text{-}193)$$

where $\chi(x, y, z, t)$ is called the *geopotential*. Additionally, from now on we

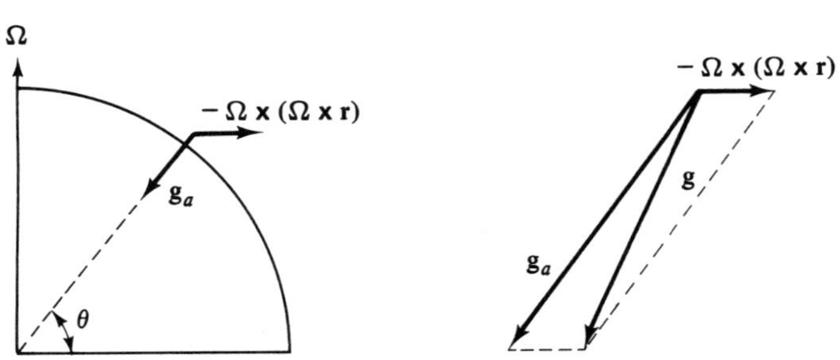

Fig. 2-46. The relationship between Newtonian acceleration and the acceleration of gravity.

will take the direction of the z axis of our relative coordinate system as being in the direction of $-\mathbf{g}$, that is, along the direction of the local vertical. This does not constitute a serious compromise since $\mathbf{g}$ and $\mathbf{g}_a$ are almost collinear due to $|\mathbf{\Omega} \times (\mathbf{\Omega} \times \mathbf{r})|$ being much smaller in magnitude than $|\mathbf{g}_a|$. Using (2-192) and (2-193) in (2-191), we obtain

$$\rho \frac{\partial \mathbf{V}}{\partial t} + \rho(\mathbf{V} \cdot \nabla)\mathbf{V} + 2\rho\mathbf{\Omega} \times \mathbf{V} = \nabla \cdot \mathcal{P} - \rho\nabla\chi \tag{2-194}$$

The result may also be written in terms of the hydrostatic pressure p and the viscous stress tensor τ_{ij} ($\equiv \mathfrak{T}$) from (2-185),

$$\rho \frac{\partial \mathbf{V}}{\partial t} + \rho(\mathbf{V} \cdot \nabla)\mathbf{V} + 2\rho\mathbf{\Omega} \times \mathbf{V} = -\nabla p + \nabla \cdot \mathfrak{T} - \rho\nabla\chi \tag{2-195}$$

If we multiply the continuity equation of (2-132) by $\mathbf{V}$ and add the result to (2-194), we obtain Newton's law in its most basic form, which states that the time rate of change of momentum is equal to the net force acting. For our particular case all the quantities are expressed as being per unit volume. Proceeding,

$$\frac{\partial \rho\mathbf{V}}{\partial t} + \nabla \cdot \rho\mathbf{V}\mathbf{V} = -2\rho\mathbf{\Omega} \times \mathbf{V} - \nabla p + \nabla \cdot \mathfrak{T} + \rho\mathbf{g} \tag{2-196}$$

The two terms $\nabla \cdot \rho\mathbf{V}\mathbf{V}$ and $\nabla \cdot \mathfrak{T}$ are, of course, not simple divergences due to the tensorial nature of $\rho\mathbf{V}\mathbf{V}$ and $\boldsymbol{\tau}$; however, the physical interpretation of these terms is analogous to that of $\nabla \cdot \rho\mathbf{V}$ in the equation of continuity. Recall that $\nabla \cdot \rho\mathbf{V}$ represents the rate of loss of mass (a scalar quantity) per unit volume of fluid due to advection by the fluid flow. Similarly, the quantity $\nabla \cdot \rho\mathbf{V}\mathbf{V}$ is the rate of loss of momentum (a vector quantity) per unit volume of fluid due to advection. In detail, then, we can physically interprete the individual terms of (2-196) as follows:

1. $\partial\rho\mathbf{V}/\partial t$ is the local time rate of increase of momentum, per unit volume of fluid.
2. $-\nabla \cdot \rho\mathbf{V}\mathbf{V}$ is the time rate of momentum gain due to advection, per unit volume.
3. $-2\rho\mathbf{\Omega} \times \mathbf{V}$ is the time rate of momentum gain due to the action of the Coriolis force, per unit volume.
4. $-\nabla p$ is the time rate of momentum gain due to the action of the pressure force, per unit volume.
5. $\nabla \cdot \mathfrak{T}$ is the time rate of momentum gain due to the action of viscous forces, per unit volume.
6. $\rho\mathbf{g}$ is the time rate of momentum gain due to the action of the gravity force, per unit volume.

It is possible, of course, to write any of the foregoing results as three component scalar equations or as one vector equation in indicial form. For example, since

$$\begin{aligned}\mathbf{\Omega} \times \mathbf{V} &= \begin{vmatrix} \mathbf{i} & \mathbf{j} & \mathbf{k} \\ 0 & \Omega_y & \Omega_z \\ u & v & w \end{vmatrix} \\ &= (w\Omega_y - u\Omega_z)\mathbf{i} + u\Omega_z\mathbf{j} \\ &\quad -u\Omega_y\mathbf{k} \\ &= (w\Omega\cos\theta - v\Omega\sin\theta)\mathbf{i} \\ &\quad + (u\Omega\sin\theta)\mathbf{j} - (u\Omega\cos\theta)\mathbf{k}\end{aligned}$$

we can write the three component scalar equations of (2-195) as follows:

1. x-component equation,

$$\begin{aligned}\frac{\partial u}{\partial t} + u\frac{\partial u}{\partial x} + v\frac{\partial u}{\partial y} + w\frac{\partial u}{\partial z} + 2w\,\Omega\cos\theta - 2v\Omega\sin\theta \\ = -\frac{1}{\rho}\frac{\partial p}{\partial x} + \frac{1}{\rho}\left(\frac{\partial \tau_{xx}}{\partial x} + \frac{\partial \tau_{yx}}{\partial y} + \frac{\partial \tau_{zx}}{\partial x}\right)\end{aligned} \tag{2-197}$$

2. y-component equation,

$$\begin{aligned}\frac{\partial v}{\partial t} + u\frac{\partial v}{\partial x} + v\frac{\partial v}{\partial y} + w\frac{\partial v}{\partial z} + 2u\Omega\sin\theta \\ = -\frac{1}{\rho}\frac{\partial p}{\partial y} + \frac{1}{\rho}\left(\frac{\partial \tau_{xy}}{\partial x} + \frac{\partial \tau_{yy}}{\partial y} + \frac{\partial \tau_{zy}}{\partial z}\right)\end{aligned} \tag{2-198}$$

3. z-component equation,

$$\begin{aligned}\frac{\partial w}{\partial t} + u\frac{\partial w}{\partial x} + v\frac{\partial w}{\partial y} + w\frac{\partial w}{\partial z} - 2u\Omega\cos\theta \\ = -\frac{1}{\rho}\frac{\partial p}{\partial z} + \frac{1}{\rho}\left(\frac{\partial \tau_{xz}}{\partial x} + \frac{\partial \tau_{yz}}{\partial y} + \frac{\partial \tau_{zz}}{\partial z}\right) - g\end{aligned} \tag{2-199}$$

Similarly, since

$$\mathbf{VV} = V_i V_j$$

and

$$\nabla \cdot \rho\mathbf{VV} = \frac{\partial}{\partial x_j}(\rho V_i V_j)$$

(2-196) in indicial form is

$$\rho\frac{\partial V_i}{\partial t} + \frac{\partial \rho V_i V_j}{\partial x_j} = -2\rho\epsilon_{ijk}\Omega_j V_k - \frac{\partial p}{\partial x_i} + \frac{\partial \tau_{ij}}{\partial x_j} - \rho g\delta_{i3} \tag{2-200}$$

2-26 Representation of the Viscous Stress Tensor and the Navier–Stokes Equation

Empirically, it is known that the viscous stress is proportional to the local velocity gradient. Thus, for the case of simple shear where, for example, $v = w = 0$ and at most $u = u(y, t)$ we have Newton's law of viscosity, equation (2-122),

$$\tau_{yx} = \mu \frac{\partial u}{\partial y} \tag{2-122}$$

where the proportionality constant μ is called the *dynamic viscosity*. Fluids that obey this law are said to be Newtonian. This result, however, is not general enough in that it applies only to the flow of a thin lamina of fluid contained between two flat, parallel plates. For complex flows, with the fluid moving in all directions, something more general is required. A general form for the tensor τ_{ij} can be established in the following way. Only when there is a relative motion between various parts of the fluid, when different fluid particles move with different velocities, do processes involving internal friction occur. It is reasonable, then, that τ_{ij} depends on the space derivatives of the velocity and that there be no terms in τ_{ij} which are independent of these derivatives, since it is obvious that τ_{ij} must vanish for V_i uniform. Second, τ_{ij} must vanish when the entire fluid is uniformly rotating; in such motion no internal friction acts. For uniform rotation with an angular velocity $\boldsymbol{\omega}$, the velocity $\mathbf{V}$ is equal to the vector product $\boldsymbol{\omega} \times \mathbf{r}$. Thus,

$$\begin{aligned} \mathbf{V} = \boldsymbol{\omega} \times \mathbf{r} &= \begin{vmatrix} \mathbf{i} & \mathbf{j} & \mathbf{k} \\ \omega_x & \omega_y & \omega_z \\ x & y & z \end{vmatrix} \\ &= (\omega_y z - \omega_z y)\mathbf{i} + (\omega_z x - \omega_x z)\mathbf{j} + (\omega_x y - \omega_y x)\mathbf{k} \end{aligned} \tag{2-201}$$

or

$$u = \omega_y z - \omega_z y$$
$$v = \omega_z x - \omega_x z$$
$$w = \omega_x y - \omega_y x$$

Now, then, the sums $(\partial V_i/\partial x_j + \partial V_j/\partial x_i)$ are linear combinations of the derivatives $\partial V_i/\partial x_j$ and vanish when (2-201) applies. Hence, τ_{ij} must contain just these symmetrical combinations of the space derivatives of the velocity. Following Stokes and others, it is apparent that the most general tensor of order two satisfying all the conditions above is that of a linear combination of the deformation tensor and the divergence: namely,

$$\tau_{ij} = \rho\nu\left(\frac{\partial V_i}{\partial x_j} + \frac{\partial V_j}{\partial x_i}\right) + \rho\nu'\,\delta_{ij}\frac{\partial V_k}{\partial x_k} \tag{2-202}$$

where the coefficients ν and ν', having dimensions of [length]^2/[time], are called the *kinematic shear viscosity* and the *kinematic bulk viscosity*. These kinematic viscosities are related to the dynamic viscosities through

$$\nu = \frac{\mu}{\rho} \qquad \nu' = \frac{\mu'}{\rho} \tag{2-203}$$

Sometimes ν, μ and ν', μ' are referred to as the first and second coefficients of viscosity. The combination of (2-202) of course satisfies the requirements that τ_{ij} be symmetrical and reduce to (2-122) for the case of simple shearing flow.

Turning to ν and ν', it turns out that the value of one is not completely independent of the other. One possible relation between them was given by Navier and Stokes, but Eckart (1947) has shown that the Navier–Stokes relation is only a limiting case of a more general expression. Specifically, through application of the second law of thermodynamics, Eckart has shown that

$$\nu \geq 0 \qquad \nu' \geq -\tfrac{2}{3}\nu \tag{2-204a}$$

The limiting condition on ν',

$$\nu' = -\tfrac{2}{3}\nu \tag{2-204b}$$

represents the celebrated *Navier–Stokes relation*, and substitution of (2-204b) into (2-202) and then in turn into the momentum equation gives the result known as the *Navier–Stokes equation*. Proceeding in this fashion, we find from (2-202) that

$$\tau_{ij} = \rho\nu\left(\frac{\partial V_i}{\partial x_j} + \frac{\partial V_j}{\partial x_i}\right) - \frac{2}{3}\rho\nu\,\delta_{ij}\frac{\partial V_k}{\partial x_k} \tag{2-205}$$

Differentiating yields

$$\frac{\partial \tau_{ij}}{\partial x_i} = \rho\nu\frac{\partial^2 V_j}{\partial x_i^2} + \frac{1}{3}\rho\nu\frac{\partial}{\partial x_j}\left(\frac{\partial V_i}{\partial x_i}\right) \tag{2-206}$$

which is written in vector notation as

$$\frac{1}{\rho}\nabla \cdot \mathfrak{T} = \nu\nabla^2 \mathbf{V} + \frac{1}{3}\nu\nabla(\nabla \cdot \mathbf{V}) \tag{2-207a}$$

Using the vector identity

$$\nabla \times (\nabla \times \mathbf{V}) = \nabla(\nabla \cdot \mathbf{V}) - \nabla^2\mathbf{V}$$

we find that (2-207a) can be rendered as

$$\frac{1}{\rho}\nabla \cdot \mathfrak{T} = \frac{4}{3}\nu\nabla^2\mathbf{V} + \frac{1}{3}\nu\nabla \times (\nabla \times \mathbf{V}) \tag{2-207b}$$

These two forms, (2-207a) and (2-207b), are particularly useful since we see from the former that for nondivergent flow the viscous force per unit volume reduces to $\mu\nabla^2\mathbf{V}$; while for irrotational flow the latter reduces to $\frac{4}{3}(\mu\nabla^2\mathbf{V})$. Returning to (2-206), substitution into the indicial form of (2-195), the momentum equation, yields

$$\rho\frac{\partial V_i}{\partial t} + \rho V_j\frac{\partial V_i}{\partial x_j} + 2\rho\epsilon_{ijk}\Omega_j V_k = -\frac{\partial p}{\partial x_i} + \mu\frac{\partial^2 V_i}{\partial x_j^2} + \frac{1}{3}\mu\frac{\partial^2 V_i}{\partial x_j^2} - \rho g\delta_{i3} \tag{2-208}$$

or, using (2-207a) in the vector form of the momentum equation,

$$\rho\frac{\partial \mathbf{V}}{\partial t} + \rho(\mathbf{V}\cdot\nabla)\mathbf{V} + 2\rho\,\mathbf{\Omega}\times\mathbf{V} = -\nabla p + \mu\nabla^2\mathbf{V} + \tfrac{1}{3}\mu\nabla(\nabla\cdot\mathbf{V}) + \rho\mathbf{g} \tag{2-209}$$

Equations (2-208) and (2-209) are two forms of the *general Navier–Stokes equation.*

At this point, having arrived at many mathematical formulations, we might take some time to see how close we are to having a deterministic set of equations, a set consisting of as many equations as there are unknowns. In general, our independent variables are the three space coordinates and time. We are required to find the dependent variables u, v, w, ρ, and p and as such we will need a total of five equations for the five unknowns. In physical oceanography it is frequently assumed that the fluid is incompressible. Under this assumption, the Navier–Stokes equation of (2-209) yields the following three component equations:

1. x-component equation,

$$\frac{\partial u}{\partial t} + u\frac{\partial u}{\partial x} + v\frac{\partial u}{\partial y} + w\frac{\partial u}{\partial z} + 2w\Omega\cos\theta - 2v\Omega\sin\theta = -\frac{1}{\rho}\frac{\partial p}{\partial x} + \nu\left(\frac{\partial^2 u}{\partial x^2} + \frac{\partial^2 u}{\partial y^2} + \frac{\partial^2 u}{\partial z^2}\right) \tag{2-210}$$

2. y-component equation,

$$\frac{\partial v}{\partial t} + u\frac{\partial v}{\partial x} + v\frac{\partial v}{\partial y} + w\frac{\partial v}{\partial z} + 2u\Omega\sin\theta = -\frac{1}{\rho}\frac{\partial p}{\partial y} + \nu\left(\frac{\partial^2 v}{\partial x^2} + \frac{\partial^2 v}{\partial y^2} + \frac{\partial^2 v}{\partial z^2}\right) \tag{2-211}$$

3. z-component equation,

$$\frac{\partial w}{\partial t} + u\frac{\partial w}{\partial x} + v\frac{\partial w}{\partial y} + w\frac{\partial w}{\partial z} - 2u\Omega\sin\theta = -\frac{1}{\rho}\frac{\partial p}{\partial z} + \nu\left(\frac{\partial^2 w}{\partial x^2} + \frac{\partial^2 w}{\partial y^2} + \frac{\partial^2 w}{\partial z^2}\right) - g \tag{2-212}$$

Also available are continuity equation expressing the conservation of mass, which we have seen for incompressible flow is

$$\frac{\partial \rho}{\partial t} + u\frac{\partial \rho}{\partial x} + v\frac{\partial \rho}{\partial y} + w\frac{\partial \rho}{\partial z} = 0 \tag{2-134}$$

and the statement of the conservation of volume,

$$\frac{\partial u}{\partial x} + \frac{\partial v}{\partial y} + \frac{\partial w}{\partial z} = 0 \tag{2-135}$$

thus giving us five equations in five unknowns.

If incompressibility is not assumed, then additional independent relations must be called upon, since (2-134) and (2-135) will not be true. In that event we may have the Navier–Stokes form of the momentum equations,

$$\rho\frac{D\mathbf{V}}{Dt} + 2\rho\mathbf{\Omega} \times \mathbf{V} = -\mathbf{\nabla} p + \mu\nabla^2\mathbf{V} + \tfrac{1}{3}\mu\mathbf{\nabla}(\mathbf{\nabla} \cdot \mathbf{V}) + \rho\mathbf{g} \tag{2-209}$$

and the general form of the continuity equation,

$$\frac{D\rho}{Dt} + \rho\mathbf{\nabla} \cdot \mathbf{V} = 0 \tag{2-133}$$

giving a total of four equations with five unknowns. The inclusion of some form of an equation of state,

$$\rho = \rho(s, T, p) \tag{2-213}$$

is possible, but the price we must pay is the introduction of the two additional independent variables, s (salinity) and T (temperature). The score at this point is equations 5, independent variables 7. One additional independent equation, which fortunately introduces no additional variables, is that of the conservation of salt,

$$\frac{\partial \rho s}{\partial t} + \mathbf{\nabla} \cdot \rho s\mathbf{V} + \mathbf{\nabla} \cdot \mathbf{F}_s = 0 \tag{2-118}$$

where the diffusive flux vector $\mathbf{F}_s$ can be replaced using the Fickian diffusion equation of (2-117). Score: equations 6, independent variables 7. As a seventh expression, we cannot call upon the conservation of water equation since it is not independent of (2-118) and (2-133); however, it is possible to apply the distribution of variables equation to the conservation of thermal energy.

The heat "content" of a mass of fluid m can be written as mc_sT, where c_s is the specific heat and T the absolute temperature, and the heat content per unit volume is ρc_sT. If this heat content per unit volume is treated as a conservative property, that is, we assume (1) no thermal sources or sinks and (2) that the only way thermal energy can be transferred is via conduction and advection, then in (2-115) Q can be replaced by ρc_sT to give

$$\frac{\partial}{\partial t}(\rho c_sT) + \mathbf{\nabla} \cdot (\rho c_sT\mathbf{V}) - \mathbf{\nabla} \cdot (c\mathbf{\nabla} T) = 0 \tag{2-214}$$

where Fourier's law of heat conduction, (2-121), has replaced $\mathbf{F}_Q$. Treating c_s as a constant, expanding, rearranging, and employing the continuity equation results in

$$\rho\left(\frac{\partial T}{\partial t} + \mathbf{V} \cdot \nabla T\right) - c'\nabla^2 T = 0 \tag{2-215}$$

where $c' \equiv c/c_s$. Score: equations 7, independent variables 7.

These few relations—(1) (2-134), (2-135), and (2-210) through (2-212) for incompressible flow, and (2) (2-118), (2-133), (2-209), (2-213), and (2-215) for compressible flow—constitute what we understand to be the basic equations of dynamical oceanography. In theory, given proper initial and boundary conditions they can be solved for the field variables as functions of the Eulerian variables x, y, z, and t.

2-27 The Hydrostatic Approximation

Under certain conditions the z-component equation of the equation of motion (2-212) can be greatly simplified. First, it may be possible to drop the frictional terms, especially for large or synoptic times scales, since it is thought that for such circumstances frictional dissipation is of a secondary importance. Thus, we write for the z component,

$$\frac{Dw}{Dt} - 2u\Omega \cos\theta = -\frac{1}{\rho}\frac{\partial p}{\partial z} - g \tag{2-216}$$

A convenient technique for estimating the magnitudes of the various terms in a governing equation for a particular type of motion is that of scale analysis or scaling. In scaling, the typical expected values of (1) the magnitudes of the field variables, u, v, w, ρ, p, T, s, for example; (2) the amplitudes of the fluctuations in the field variables; and (3) the characteristic length, depth, and time scales over which these fluctuations occur, are specified. These typical values are then used to compare the magnitudes of the various terms in the governing equations.

For oceanographic, midlatitude, synoptic scales the characteristic value of the field variables can be taken as follows:

Horizontal velocity scale, $U \sim 10^2$ cm/sec
Vertical velocity scale, $W \sim 10^{-1}$ cm/sec
Length scale, $L \sim 10^8$ cm
Depth scale, $H \sim 10^6$ cm
Horizontal pressure fluctuation scale, $\Delta P \sim 10^6$ dynes/cm^2
Vertical pressure fluctuation scale, $\Delta P_z/H \sim 10^2$ dynes/cm^3
Time scale, $L/U \sim 10^6$ sec

Considering disturbances centered at a latitude θ of 45°, we find that

$$f \equiv 2\Omega \sin \theta = 2\Omega \cos \theta \sim 10^{-4} \text{ sec}^{-1}$$

Substituting these scales into (2-216) yields

$$\begin{array}{ccccccc} \dfrac{Dw}{Dt} & - & 2u\Omega \cos \theta & = & -\dfrac{1}{\rho}\dfrac{\partial p}{\partial z} & - & g \\ \dfrac{UW}{L} & & fU & & \dfrac{\nabla P_z}{\rho H} & & g \\ 10^{-7} & & 10^{-2} & & 10^{3} & & 10^{3} \end{array}$$

which shows that to a high degree of accuracy the pressure field is in hydrostatic equilibrium. In summary, then, for midlatitude, synoptic scale motions we can use as the vertical equation of motion the *hydrostatic equation* as given by

$$\frac{\partial p}{\partial z} = -\rho g \tag{2-217}$$

2-28 The Vorticity Equation

Let us return now to the Stokes form of the viscous stress tensor as given by (2-202),

$$\tau_{ij} = \rho\nu\left(\frac{\partial V_i}{\partial x_j} + \frac{\partial V_j}{\partial x_i}\right) + \rho\nu'\delta_{ij}\frac{\partial V_k}{\partial x_k} \tag{2-202}$$

the space derivative of which yields

$$\frac{\partial \tau_{ij}}{\partial x_j} = \rho\nu\frac{\partial}{\partial x_i}\left(\frac{\partial V_j}{\partial x_j}\right) + \rho\nu\frac{\partial}{\partial x_j}\left(\frac{\partial V_i}{\partial x_j}\right) + \rho\nu'\delta_{ij}\frac{\partial}{\partial x_j}\left(\frac{\partial V_k}{\partial x_k}\right)$$

or, equivalently,

$$\nabla \cdot \mathfrak{T} = \rho(\nu + \nu')\nabla(\nabla \cdot \mathbf{V}) + \rho\nu\nabla^2\mathbf{V}$$

Substituting this result into (2-195), the equation of motion, gives

$$\frac{D\mathbf{V}}{Dt} + 2\mathbf{\Omega} \times \mathbf{V} + \alpha\nabla p + \nabla\chi = \nu\nabla^2\mathbf{V} + (\nu + \nu')\nabla(\nabla \cdot \mathbf{V}) \tag{2-218}$$

where α is the specific volume, $1/\rho$, and recalling that ν and ν' are the first and second coefficients of the kinematic viscosity whose values are restricted by the conditions that $\nu \geq 0$ and $\nu' \geq -\frac{2}{3}\nu$ for any real fluid satisfying the second law of thermodynamics. Equation (2-218) is basically a statement relating the time rate of change of the velocity to all the things that can effect such a change, such as the Coriolis force, the pressure force, the gravitational force, and the viscous forces. We have seen that this equation is merely a form of Newton's second law or of the principle of the conservation of linear momentum. We might logically expect that a similar equation could be formulated that would express the principle of the conservation of angular momentum in terms of the applied torques. Such a result can be obtained by taking the curl of the equation of motion; but first it is expedient to do a small amount of rewriting.

The material derivative, through use of the vector identity

$$(\nabla \times \mathbf{V}) \times \mathbf{V} = (\mathbf{V} \cdot \nabla)\mathbf{V} - \nabla(\tfrac{1}{2}|\mathbf{V}|^2)$$

can be rendered in the equivalent form

$$\frac{D\mathbf{V}}{Dt} = \frac{\partial \mathbf{V}}{\partial t} + + (\mathbf{V} \cdot \nabla)\mathbf{V} = \frac{\partial \mathbf{V}}{\partial t} + \nabla(\tfrac{1}{2}|\mathbf{V}|^2) + \boldsymbol{\zeta} \times \mathbf{V} \qquad (2\text{-}219)$$

where $\boldsymbol{\zeta}$ is the vorticity (i.e., the curl of the velocity). Substituting into (2-218),

$$\begin{aligned}\frac{\partial \mathbf{V}}{\partial t} + \nabla(\chi + \tfrac{1}{2}|\mathbf{V}|^2) + (\boldsymbol{\zeta} + 2\boldsymbol{\Omega}) \times \mathbf{V} + \alpha\nabla p& \\ = \nu\nabla^2\mathbf{V} + (\nu + \nu')\nabla(\nabla \cdot \mathbf{V})&\end{aligned} \qquad (2\text{-}220)$$

Before taking the curl of this result, we note several things that will simplify the process. First, the curl operation is commutative with scalar operators such as the partial derivative with respect to time, $\partial/\partial t$, and the Laplacian, ∇^2. Second, it will be recalled that

$$\nabla \times \nabla B = 0$$

where B is any scalar function; and third, that

$$\nabla \times (\alpha\nabla p) = \nabla\alpha \times \nabla p + \alpha\nabla \times \nabla p = \nabla\alpha \times \nabla p$$

since $\nabla p \times \nabla p$ is equal to zero. Applying these facts, the curl of (2-220) is

$$\frac{\partial \boldsymbol{\zeta}}{\partial t} + \nabla \times (\boldsymbol{\zeta}_a \times \mathbf{V}) + \nabla\alpha \times \nabla p = \nu\nabla^2\boldsymbol{\zeta} \qquad (2\text{-}221)$$

where $\boldsymbol{\zeta}_a$, the *absolute vorticity*, is taken as the vector sum of the *relative vorticity* $\boldsymbol{\zeta}$ and the *planetary vorticity* $2\boldsymbol{\Omega}$; that is,

$$\boldsymbol{\zeta}_a \equiv \boldsymbol{\zeta} + 2\boldsymbol{\Omega} \qquad (2\text{-}222)$$

Equation (2-221) is one form, albeit not the most useful form, of the *general vorticity equation*. To put it in a more useful form, we begin by expanding the second term on the left,

$$\nabla \times (\boldsymbol{\zeta}_a \times \mathbf{V}) = (\mathbf{V} \cdot \nabla)\boldsymbol{\zeta}_a - (\boldsymbol{\zeta}_a \cdot \nabla)\mathbf{V} - \mathbf{V}(\nabla \cdot \boldsymbol{\zeta}_a) + \boldsymbol{\zeta}_a\nabla \cdot \mathbf{V}$$

We see that the third term on the right is zero since

$$\nabla \cdot \boldsymbol{\zeta}_a = \nabla \cdot (\boldsymbol{\zeta} + 2\boldsymbol{\Omega}) = 0$$

owing to the twofold fact that $\boldsymbol{\Omega}$ is taken as being invariant in space and time and the divergence of the curl of a vector, in this instance $\nabla \cdot (\nabla \times \mathbf{V})$, is zero. Additionally, we can, because of the invariant character of $\boldsymbol{\Omega}$, in (2-221) substitute $\partial\boldsymbol{\zeta}_a/\partial t$ for $\partial\boldsymbol{\zeta}/\partial t$ and $\nabla^2\boldsymbol{\zeta}_a$ for $\nabla^2\boldsymbol{\zeta}$. Accordingly, the vorticity equation may be rendered in the form

$$\frac{\partial \boldsymbol{\zeta}_a}{\partial t} + (\mathbf{V} \cdot \nabla)\boldsymbol{\zeta}_a = (\boldsymbol{\zeta}_a \cdot \nabla)\mathbf{V} - \boldsymbol{\zeta}_a(\nabla \cdot \mathbf{V}) - \nabla\alpha \times \nabla p + \nu\nabla^2\boldsymbol{\zeta}_a$$

or

$$\frac{D\boldsymbol{\zeta}_a}{Dt} = (\boldsymbol{\zeta}_a \cdot \nabla)\mathbf{V} - \boldsymbol{\zeta}_a(\nabla \cdot \mathbf{V}) - \nabla\alpha \times \nabla p + \nu\nabla^2\boldsymbol{\zeta}_a \qquad (2\text{-}223)$$

a more useful form of the *general vorticity equation*. Physically, (2-223) asserts that the rate of change of the absolute vorticity following a fluid particle is equal to the sum of the four terms on the right, each of which may be regarded as a torque (actually as a torque per unit of inertia, this being equivalent in rotation to a force per unit mass in translation). The first of these exists if variations in the fluid velocity (speed or direction) occurs along vortex lines—a vortex line being a line that is everywhere parallel to the local vorticity vector (the vortex line is to vorticity what a streamline is to the velocity). This term is called the *tipping and stretching term*. The second term of (2-223) occurs only if the fluid is compressible and is called, fittingly enough, the *divergence term*. The third term, called the *solenoidal term*, exists only for baroclinic conditions (that is, the fluid density must be a function of properties other than just pressure; seawater, for example, has a density which in general depends upon salinity and temperature as well as pressure). The last term, the *viscous torque term*, results from the torque produced by the viscous stresses within the fluid. We will examine each of these in a little more detail so as to arrive at a truly physical understanding of their nature.

The Tipping and Stretching Term, $(\boldsymbol{\zeta}_a \cdot \nabla)\mathbf{V}$

This term gives vorticity changes a distinctive character in that no counterpart to it exists in the equation of motion. It, like all of those in (2-223), is expressed as a time rate of change of vorticity and its meaning becomes evident when we write

$$(\boldsymbol{\zeta}_a \cdot \nabla)\mathbf{V} = |\boldsymbol{\zeta}_a| \lim_{PQ \to 0} \frac{\delta \mathbf{V}}{PQ}, \qquad (2\text{-}224)$$

where P and Q are two neighboring points on a local vortex line and $\delta\mathbf{V}$ is the velocity of the fluid at Q relative to that at P (Figure 2-47). Either term in this equation, when divided by $|\boldsymbol{\zeta}_a|$, becomes a fractional time rate of change of vorticity. The corresponding contribution then to the fractional time rate of change of vorticity on the left of (2-223), that is, either $(1/|\boldsymbol{\zeta}_a|)\partial\boldsymbol{\zeta}_a/\partial t$ or $(1/|\boldsymbol{\zeta}_a|)D\boldsymbol{\zeta}_a/Dt$, is identical with the fractional rate of change of the material line element vector extending from point P to Q, which is the meaning of $\lim_{PQ\to 0} \delta\mathbf{V}/PQ$, P and Q being regarded now as material points. We can

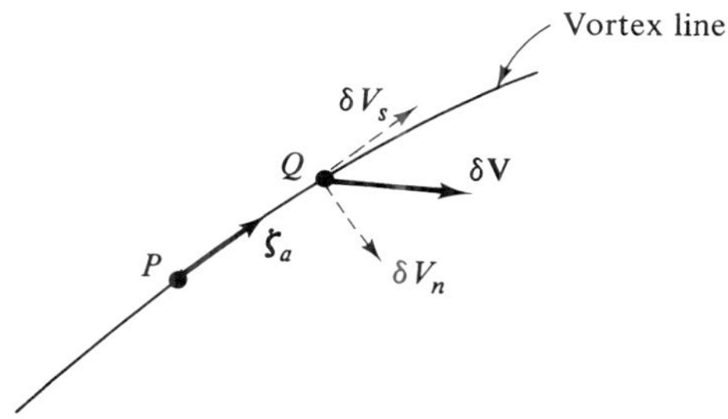

Fig. 2-47. Schematic for the vorticity tipping and stretching term.

think of ζ_a behaving like a material line element coinciding for an instant with a section of the vortex line. A portion of the change in ζ_a, its direction, coming from the rigid rotation of the line element due to the component of $\delta\mathbf{V}$ normal to ζ_a, the tipping part of the term, and a portion being a change in ζ_a, its magnitude, coming from the extension or contraction of the line element due to the component of $\delta\mathbf{V}$ parallel to ζ_a, the stretching part of the term.

The Divergence Term, $-\zeta_a(\nabla \cdot \mathbf{V})$

The term $-\zeta_a(\nabla \cdot \mathbf{V})$ accounts for changes in the moment of inertia due to changes in the density (distribution of mass) of the fluid. If $\nabla \cdot \mathbf{V} > 0$, then we have positive divergence and the fractional rate of change of the density following the fluid is negative, since from the continuity equation

$$\frac{1}{\rho}\frac{D\rho}{Dt} = -\nabla \cdot \mathbf{V}$$

A decrease in the mass per unit volume for the parcel results in an increase in its radius of gyration, which, in turn, since $I = mr^2$, increases the moment of inertia. If the angular momentum is to be conserved, the angular velocity (or vorticity) must decrease. We see that positive divergence results in a decrease of the absolute vorticity. For a negative divergence (i.e., a convergence), we obtain through this term an increase in the absolute vorticity. This same effect is employed by a skater doing a spin. The skater starts turning with arms extended and then pulls them in close, whereupon the spinning increases. By drawing in the arms, the skater is actually decreasing the moment of inertia; then to keep the angular momentum the same, the angular velocity increases.

The Solenoidal Term, $-\nabla\alpha \times \nabla p$

A possible arrangement of level surfaces, isobars, and isosteres (surfaces along which α, the specific volume, is a constant; equivalent to the surface along which the density is a constant, an isopycnal) for a baroclinic

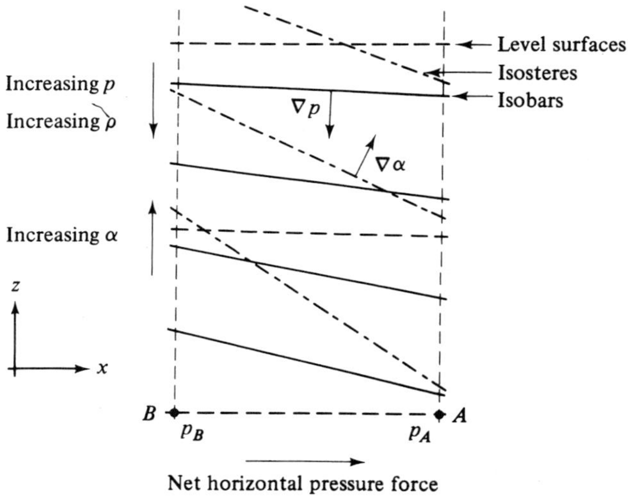

Fig. 2-48. Distribution of level surfaces, isobars, and isosteres in a baroclinic fluid.

fluid is shown in Figure 2-48. For the baroclinic situation, the slope of the isobars and the isosteres are not the same as they would be in the purely barotropic case, where the specific volume is a function of only the pressure. Thus, we have the result that the solenoidal term (a solenoid traditionally is a cell bounded in two dimensions by two isobars and two isosteres), $\nabla\alpha \times \nabla p$, equals zero for a barotropic fluid. For the two typical points A and B on a level surface, we note that the pressure at B, p_B, is greater than the pressure at A, p_A. If a vertical plot of the horizontal pressure force distribution were constructed, it might appear as in Figure 2-49. As shown, the downward-increasing net pressure force will tend to produce a vertical velocity profile

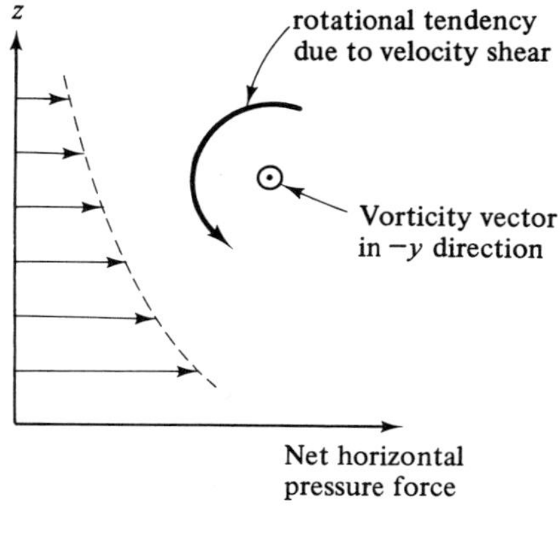

Fig. 2-49. Vertical distribution of the net horizontal pressure force for the baroclinic fluid of Fig. 2-48.

with shear. This velocity gradient will tend to produce vorticity (or rotation). The vorticity vector in this particular case will be directed in the $-y$ direction as a result of the counterclockwise rotational tendency.

The Viscous Torque Term, $\nu\nabla^2\boldsymbol{\zeta}_a$

The term $\nu\nabla^2\boldsymbol{\zeta}_a$ represents the rate of change of vorticity due to the molecular diffusion of vorticity. This term is exactly analogous to that of $\nu\nabla^2\mathbf{V}$ in the equation of motion, in which it represents the contribution to the acceleration due to the diffusion of velocity or linear momentum. Of similar form is the term in any equation representing he balance of a conservative property. For example, if in the conservation of salt equation of (2-118) we substitute the Fickian diffusion law of (2-117), we obtain the term $-K_s\nabla^2(\rho s)$, which represents the rate of change of salt due to the molecular diffusion of salt and which is analogous to the viscous torque term. Finally, we can state that since all components of the velocity $\mathbf{V}$ are transportable quantities, so are all the spatial derivatives of the velocity $\mathbf{V}$, and since the vorticity is a function of these spatial derivatives, so also the vorticity is transportable.

2-29 Conservation of Potential Vorticity

From the results of Section 2-28 it is possible to derive an important equation first developed and fruitfully employed by Rossby in 1940. As a starting point, we have the vorticity equation,

$$\frac{D\boldsymbol{\zeta}_a}{Dt} = (\boldsymbol{\zeta}_a \cdot \nabla)\mathbf{V} - \boldsymbol{\zeta}_a(\nabla \cdot \mathbf{V}) - \nabla\alpha \times \nabla p + \nu\nabla^2\boldsymbol{\zeta}_a \qquad (2\text{-}223)$$

Consider this result in the light of the following initial assumptions:

1. That the fluid is homogeneous throughout so that α is a constant, resulting in the solenoidal and divergence terms of (2-223) being zero.
2. That the fluid is frictionless so that ν is zero, resulting in the viscous torque term being zero.

Under these assumptions, (2-223) becomes

$$\frac{\partial\boldsymbol{\zeta}_a}{\partial t} + (\mathbf{V} \cdot \nabla)\boldsymbol{\zeta}_a - (\boldsymbol{\zeta}_a \cdot \nabla)\mathbf{V} = 0$$

and from the definition of the absolute vorticity can be written

$$\frac{\partial\boldsymbol{\zeta}}{\partial t} + (\mathbf{V} \cdot \nabla)(\boldsymbol{\zeta} + 2\boldsymbol{\Omega}) - [(\boldsymbol{\zeta} + 2\boldsymbol{\Omega}) \cdot \nabla]\mathbf{V} = 0 \qquad (2\text{-}225)$$

where, as previously, $\boldsymbol{\zeta}$ is the relative vorticity and $2\boldsymbol{\Omega}$ the planetary vorticity. Recalling that we have defined the Coriolis parameter f as being equal to

$2\Omega \sin \theta$, we can write the vertical component equation of this result as

$$\frac{\partial \zeta_z}{\partial t} + u\frac{\partial}{\partial x}(\zeta_z + f) + v\frac{\partial}{\partial y}(\zeta_z + f) + w\frac{\partial}{\partial z}(\zeta_z + f) - \zeta_x \frac{\partial w}{\partial x} - (\zeta_y + 2\Omega \cos \theta)\frac{\partial w}{\partial y} - (\zeta_z + f)\frac{\partial w}{\partial z} = 0 \tag{2-226}$$

Now, under the additional assumptions that:

3. Since the fluid is homogeneous and frictionless, the horizontal velocities are independent of the depth, yielding that $\partial u/\partial z$ and $\partial v/\partial z$ are both zero.

4. The vertical velocity is small and the horizontal scale is large, so that the spatial derivatives of the vertical velocity, $\partial w/\partial x$ and $\partial w/\partial y$, are small compared to the other remaining terms and therefore can be approximated as zero, we see that ζ_y and ζ_x are zero.

Then (2-226) reduces to

$$\frac{D(\zeta_z + f)}{Dt} - (\zeta_z + f)\frac{\partial w}{\partial z} = 0$$

Integrating this result in the vertical yields

$$\int_{z=-h}^{z=0} \frac{D}{Dt}(\zeta_z + f)\,dz - \int_{z=-h}^{z=0} (\zeta_z + f)\frac{\partial w}{\partial z}\,dz = 0$$

where $z = 0$ is the top and $z = -h$ is the bottom of a vertical column. Since the integrand of the first term is not a function of z, this reduces to

$$h\frac{D}{Dt}(\zeta_z + f) - (\zeta_z + f)(w|_{z=0} - w|_{z=-h}) = 0$$

and following the fluid column,

$$h\frac{D}{Dt}(\zeta_z + f) - (\zeta_z + f)\frac{Dh}{Dt} = 0$$

Dividing by h^2 and combining terms we finally obtain that

$$\frac{D}{Dt}\left(\frac{\zeta_z + f}{h}\right) = 0 \tag{2-227}$$

where the quantity in parentheses is called the *potential vorticity* and the equation expresses the principle of the *conservation of potential vorticity*; specifically, that following a fluid column potential vorticity is conserved.

As an illustration of the usefulness of (2-227), we can point to the following examples.

Stationary Column of Fluid with Vertical Stretching and Shrinking

From (2-227), for a stationary column with f constant, we see that a change in the thickness of the column h will produce a corresponding change in the relative vorticity component ζ_z. Thus, positive stretching results, in the Northern Hemisphere, in an increase in both h and ζ_z, thereby producing cyclonic or positive relative vorticity (counterclockwise rotation). This cyclonic vorticity is produced by the deflection of the inflowing fluid by the Coriolis force (see Figure 2-50). Similarly, shrinking of the column results in the outflowing fluid being deflected so as to produce clockwise rotation and therefore anticyclonic or negative relative vorticity. This effect is confirmed by (2-227) in that a decrease in h will produce a proportional decrease in ζ_z so that overall the potential vorticity will be conserved.

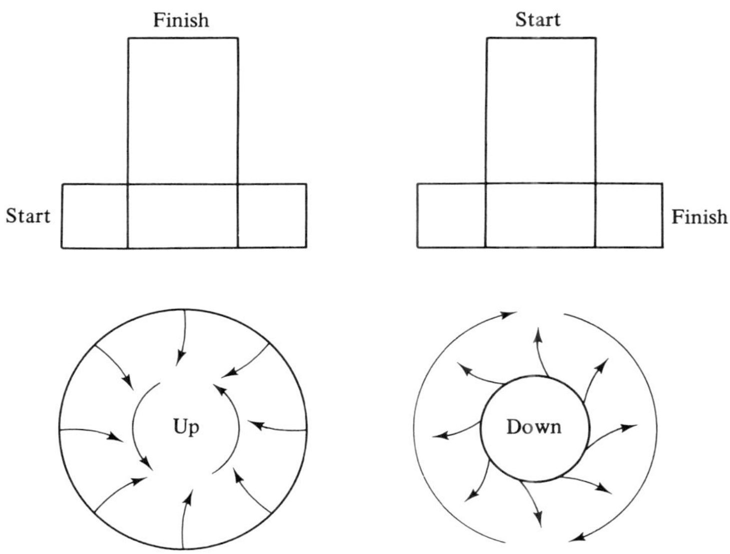

Fig. 2-50. Production of relative vorticity in a stationary column of fluid by stretching and shrinking (for the Northern Hemisphere).

Column of Constant Thickness but with Changing Latitude

Consider a column of fluid of constant thickness in the Northern Hemisphere and for convenience having a relative vorticity ζ_z of zero. If the column moves north, the planetary vorticity f increases and to conserve potential vorticity the column must begin to acquire clockwise rotation and thus anticyclonic or negative relative vorticity. In effect, the fluid column of constant thickness in moving poleward is moving into a region having a greater planetary vorticity and thus appears to develop clockwise rotation. For displacement equatorward, the fluid column moves into a region of decreasing planetary vorticity and in so doing appears to acquire counterclockwise rotation, or positive vorticity. Again this result is confirmed by application of (2-227).

Zonal Flow over an Infinite Ridge

This example of the application of the principle of conservation of potential vorticity is considerably more complicated than the two previous examples, but both the techn que of solution and the solution itself are deemed to be sufficiently valuable to dwell on a bit. Suppose initially that we have pure zonal flow from west to east over a ridge of infinite horizontal extent north and south (Figure 2-51). Under assumed steady-state conditions,

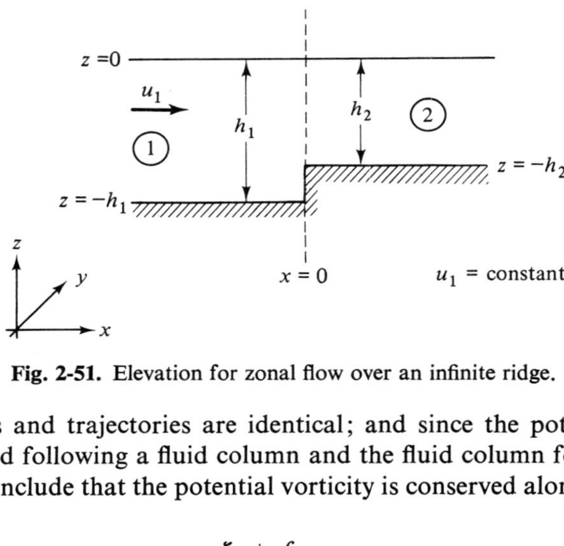

Fig. 2-51. Elevation for zonal flow over an infinite ridge.

streamlines and trajectories are identical; and since the potential vorticity is conserved following a fluid column and the fluid column follows a trajectory, we conclude that the potential vorticity is conserved along a streamline. That is,

$$\frac{\zeta_z + f}{h} = F(\psi) \tag{2-228}$$

where $F(\psi)$ is a to be determined function of the stream function ψ, at a distance from the ridge such that w can be taken as essentially zero. This relation merely states that for ψ a constant (i.e., along a streamline), the potential vorticity equals a constant. The derivation of (2-227) assumed that the fluid is incompressible and assures the existence of ψ. The relative vorticity ζ_z can be written as

$$\zeta_z = \frac{\partial v}{\partial x} - \frac{\partial u}{\partial y} = \frac{\partial^2 \psi}{\partial x^2} + \frac{\partial^2 \psi}{\partial y^2} = \nabla_H^2 \psi$$

in which we take the Laplacian as being two-dimensional; substitution into (2-228) yields

$$\frac{\nabla_H^2 \psi + f}{h} = F(\psi) \tag{2-229}$$

In order to represent the Coriolis parameter f as a function of our independent variables x and y, we make what is known as the *beta-plane approximation*. It is assumed that on the tangent plane f is a linear function of y,

$$f = f_0 + \beta y \tag{2-230}$$

where f_0 is the value of f at the latitude θ_0 of the origin of the relative coordinate system. Differentiating the Coriolis parameter, defined as

$$f \equiv 2\Omega \sin \theta$$

we have that

$$\frac{df}{dy} = \frac{df}{d\theta}\frac{d\theta}{dy} = \frac{d(2\Omega \sin \theta)}{d\theta}\frac{d\theta}{dy} = 2\Omega \cos \theta \frac{d\theta}{dy}$$

But, from Figure 2-52, we can write that

$$r_E d\theta = dy$$

and therefore

$$\beta = \frac{2\Omega \cos \theta_0}{r_E}$$

where r_E is the radius of the earth. As an example, at a latitude of 45°, β is 1.62×10^{-13} cm^{-1}/sec.

Now in the upstream flow region designated by the number 1, where $u = u_1$ and $v = 0$, integration of

$$u = -\frac{\partial \psi}{\partial y} \qquad v = \frac{\partial \psi}{\partial x}$$

yields

$$\psi = -u_1 y \tag{2-231}$$

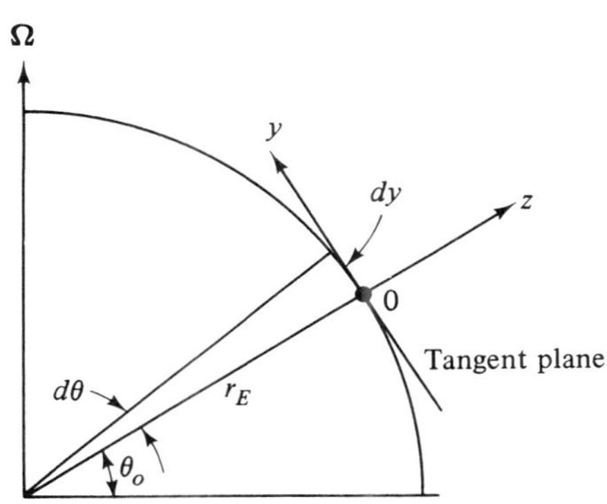

Fig. 2-52. Definition diagram for making the beta-plane approximation.

and upon substitution in (2-229),

$$F(\psi) = \frac{f_0 + \beta y}{h_1} = \frac{f_0 - \beta\psi/u_1}{h_1} \tag{2-232}$$

Thus, $f(\psi)$ has been expressed in terms of the parameters of the problem, and additionally we have obtained an express on that is applicable over region 2, downstream of the ridge. For region 2, then, combining this result with (2-229), one obtains

$$\frac{\nabla_H^2\psi + (f_0 + \beta y)}{h_2} = \frac{f_0 - \beta\psi/u_1}{h_1}$$

or the second-order, nonhomogeneous, linear, partial differential equation

$$\nabla_H^2\psi + k^2\psi = f_0\left(\frac{h_2 - h_1}{h_1}\right) - \beta y \tag{2-233}$$

where

$$k^2 \equiv \frac{h_2}{h_1}\frac{\beta}{u_1}$$

A particular solution, ψ_2, to (2-233) is clearly

$$\psi_2 = \frac{1}{k^2}\left[f_0\left(\frac{h_2 - h_1}{h_1}\right) - \beta y\right] \tag{2-234}$$

The homogeneous equivalent of (2-233),

$$\nabla_H^2\psi + k^2\psi = 0 \tag{2-235}$$

is the two-dimensional form of the well-known Helmholtz equation, solutions of which are generally found through the separation of variables. Alternatively, a shortcut can be utilized due to the observation that ψ should be linearly dependent upon y; that is, streamlines should essentially repeat themselves in the y direction. Proceeding, then, it is assumed that the solution, ψ_1, to (2-235) is of the form

$$\psi_1 = (y + c)X(x) \tag{2-236}$$

where c is a constant to be determined. Substituting back into (2-235),

$$(y + c)\frac{d^2X}{dx^2} + k^2(y + c)X = 0$$

or

$$\frac{d^2X}{dX^2} + k^2X = 0$$

whose solution is of the form

$$X = A\cos kx + B\sin kx$$

where A and B are constants. Using this result in (2-236), we obtain

$$\psi_1 = (y + c)(A\cos kx + B\sin kx)$$

and combining this with (2-234) we have the general solution to (2-233),

$$\psi = \psi_2 + \psi_1 = (y + c)(A \cos kx + B \sin kx) + \frac{1}{k^2}\left[f_0\left(\frac{h_2 - h_1}{h_1}\right) - \beta y\right] \tag{2-237}$$

To evaluate the constants A and B we avail ourselves of the requirement that there be continuity at the ridge:

$$u_1 = -\frac{\partial \psi}{\partial y} \qquad \text{at } x = 0$$

and (2-238)

$$\frac{\partial \psi}{\partial x} = 0 \qquad \text{at } x = 0$$

Application of the first of these to (2-237) gives

$$u_1 = -\left(A \cos kx + B \sin kx - \frac{\beta}{k^2}\right) \qquad \text{at } x = 0$$

and therefore

$$A = \frac{\beta}{k^2} - u_1$$

Application of the second continuity condition yields

$$k(y + c)(-A \sin kx + B \cos kx) = 0 \qquad \text{at } x = 0$$

which requires that

$$B = 0$$

Inserting these results into (2-237),

$$\psi = (y + c)\left[\left(\frac{\beta}{k^2} - u_1\right)\cos kx\right] + \frac{1}{k^2}\left[f_0\left(\frac{h_2 - h_1}{h_1}\right) - \beta y\right] \tag{2-239}$$

To find the value of the constant c, which merely establishes the relative numerical value of the streamlines, we can require that $\psi = 0$ at $x = 0$, $y = 0$. In so doing, we find that

$$c = \frac{f_0}{\beta}$$

and therefore

$$\psi = \left(y + \frac{f_0}{\beta}\right)\left[\left(\frac{\beta}{k^2} - u_1\right)\cos kx\right] + \frac{1}{k^2}\left[f_0\left(\frac{h_2 - h_1}{h_1}\right) - \beta y\right]$$

Rearranging yields

$$\psi = -u_1 y - \left[u_1\left(\frac{h_2 - h_1}{h_1}\right)(f_0 + \beta y)\right]\frac{\cos kx - 1}{\beta} \tag{2-240}$$

The essential features of the streamlines, as shown in Figure 2-53 for the

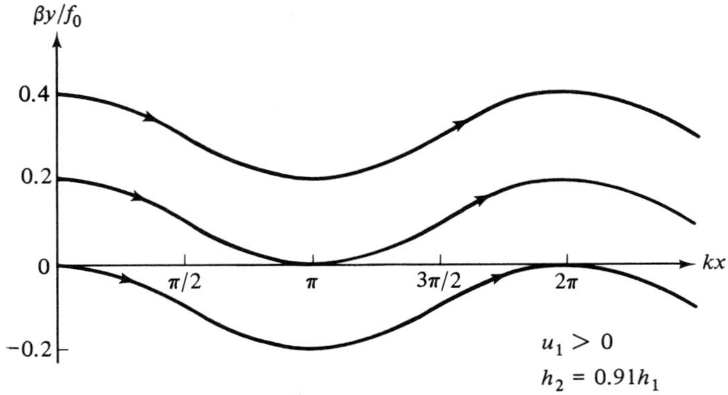

Fig. 2-53. Streamlines for uniform eastward flow over a north-south ridge.

case of $h_2 = 0.91h_1$, can be obtained readily when one approximates (2-240) as

$$\frac{\beta y}{f_0} \doteq -\frac{\psi\beta}{u_1 f_0} - \frac{\bar{f}}{f_0}\left(\frac{h_2 - h_1}{h_1}\right)(\cos kx - 1) \tag{2-241}$$

where

$$\bar{f} \equiv f_0 + \beta\bar{y} \tag{2-242}$$

$\bar{y}$ being the average value of y along any given streamline. The streamlines then differ in shape from each other only to the extent of the change in scale induced by the ratio $\bar{f}/f_0$.

It is possible to physically interpret the behavior of the fluid in the following manner. Assuming that upstream to the ridge the relative vorticity of the fluid is zero, the bottom topography provides relative vorticity at the ridge itself by shrinking the fluid column, thereby causing the outflow to be deflected by the Coriolis force so as to produce anticyclonic rotation. This induced anticyclonic rotation is the cause of the flow turning toward the south (Figure 2-54). The wavy character of the streamlines downstream of the ridge is due to the nonuniformity of the Coriolis parameter. In traveling south, the fluid moves through a region of decreasing f and to conserve potential vorticity, ζ_z must become more positive (less anticyclonic). As the relative vorticity decreases in magnitude, the curvature of the streamline decreases until at the latitude θ_2, corresponding to $f_2 = ch_2$, the relative vorticity and the curvature are both zero. Still continuing, because of its momentum, to move into a region of decreasing f, the fluid starts to acquire cyclonic relative vorticity and consequently starts to curve to the left, eventually moving due east, at which point its time rate of change of relative vorticity, $D\zeta_z/Dt$, is zero and ζ_z is a maximum positive. Continuing to curve to the left, the

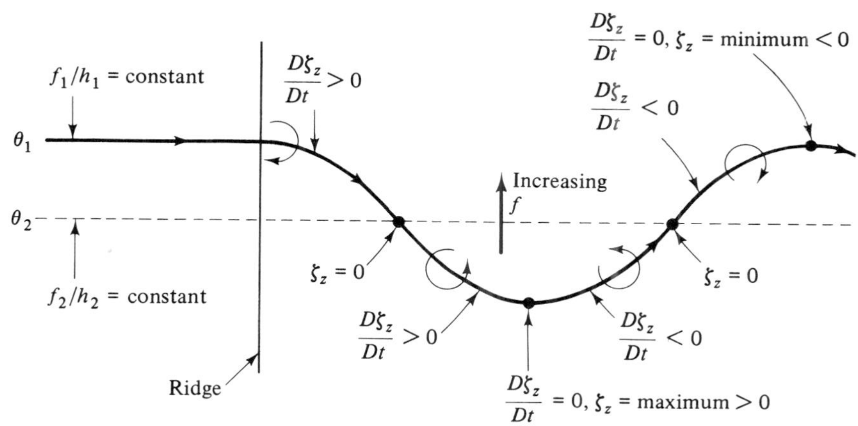

Fig. 2-54. Variation of relative vorticity along a streamline for uniform eastward flow over a north-south ridge, plan view.

fluid now begins to move with a northward component of velocity and thus in the direction of increasing planetary vorticity. To conserve potential vorticity, ζ_z begins to decrease, with the result that the curvature of the streamline decreases. As earlier on the way southward, when the flow reaches the latitude θ_2, the relative vorticity and curvature are both zero. Continuing its northward motion, anticyclonic vorticity increases along with the resulting curvature to the right until the flow is once again due east at latitude θ_1, at which point the relative vorticity has achieved its maximum anticyclonic value and the curvature is maximum in magnitude. The flow continues then to oscillate around the parallel of latitude, θ_2, corresponding to f_2 such that $f_2/h_2 = f_1/h_1 = \text{constant}$.

Finally, returning to (2-240) and the definition of the wave number k, we see that the wavelength L of the oscillation is given by

$$L \equiv \frac{2\pi}{k} = 2\pi\left(\frac{u_1 h_1}{\beta h_2}\right)^{1/2}$$

which for a latitude of 45°, a u_1 of 1 m/sec, and $(h_1 - h_2) \ll h_1$ comes out to approximately 1600 km. Also from (2-240), the southward excursion of the streamline passing through the origin is found by solving for y when $\psi = 0$ and $kx = \pi$, yielding

$$\text{amplitude} = \frac{2f_0}{\beta}\left(\frac{h_1 - h_2}{2h_1 - h_2}\right) = 2r_E \tan\theta_0 \left(\frac{h_1 - h_2}{2h_1 - h_2}\right)$$

For a streamline originating at a latitude of 45°, this excursion is equivalent to a range of latitude of $2(h_1 - h_2)/(2h_1 - h_2)$ radians, or approximately 1050 km or 9.5° of latitude for the case of $h_2 = 0.91h_1$.

2-30 The Bernoulli Equation

Clearly, the complete form of the equation of motion is mathematically very complicated; general solutions to it are difficult to come by. As a matter of fact, it is not possible to generate a general solution to it and then evaluate the resulting constants by application of a set of boundary conditions. Solutions can be found only by considering special cases and utilizing special techniques. In this section we will examine one such special case.

We can start with just about any of the previously obtained forms of the equation of motion. Taking, for example (2-195), and assuming that the flow is frictionless, we get

$$\frac{\partial \mathbf{V}}{\partial t} + (\mathbf{V} \cdot \nabla)\mathbf{V} + 2\boldsymbol{\Omega} \times \mathbf{V} = -\alpha \nabla p - \nabla \chi \tag{2-243}$$

Proceeding now to integrate this statement along a streamline using the differential vector element $d\mathbf{r}$, as shown in Figure 2-55 [recalling that the dot product of the terms of (2-243) with $d\mathbf{r}$ gives their value along $d\mathbf{r}$],

$$\oint \frac{\partial \mathbf{V}}{\partial t} \cdot d\mathbf{r} + \oint (\mathbf{V} \cdot \nabla)\mathbf{V} \cdot d\mathbf{r} + 2\oint (\boldsymbol{\Omega} \times \mathbf{V}) \cdot d\mathbf{r} = -\oint \alpha \nabla p \cdot d\mathbf{r} - \oint \nabla \chi \cdot d\mathbf{r} \tag{2-244}$$

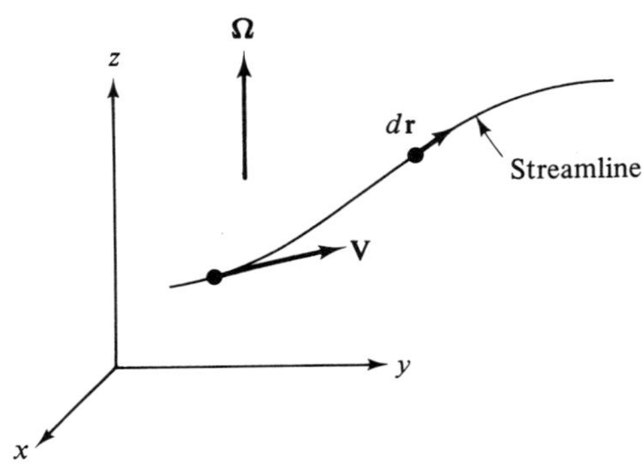

Fig. 2-55. Schematic for integration along a streamline.

To put this in a simpler form, we observe first that

$$2\oint (\boldsymbol{\Omega} \times \mathbf{V}) \cdot d\mathbf{r} = 0$$

since the vector $\boldsymbol{\Omega} \times \mathbf{V}$ is perpendicular to the vector $d\mathbf{r}$, $d\mathbf{r}$ and $\mathbf{V}$ always being coincident. Second, it is possible to write $(\mathbf{V} \cdot \nabla)\mathbf{V}$ as

$$(\mathbf{V} \cdot \nabla)\mathbf{V} = \tfrac{1}{2}\nabla |\mathbf{V}|^2 - \mathbf{V} \times (\nabla \times \mathbf{V})$$

and finally, all terms of the form $\nabla B \cdot d\mathbf{r}$ can be transformed as

$$\nabla B \cdot d\mathbf{r} = \left(\frac{\partial B}{\partial x}\mathbf{i} + \frac{\partial B}{\partial y}\mathbf{j} + \frac{\partial B}{\partial z}\mathbf{k}\right) \cdot (\mathbf{i}\,dx + \mathbf{j}\,dy + \mathbf{k}\,dz)$$
$$= \frac{\partial B}{\partial x}\,dx + \frac{\partial B}{\partial y}\,dy + \frac{\partial B}{\partial z}\,dz$$
$$= dB$$

for any particular time if B is time-dependent.

Utilizing all of this in (2-244), one obtains

$$\oint \frac{\partial \mathbf{V}}{\partial t} \cdot d\mathbf{r} + \frac{1}{2}\oint \nabla |\mathbf{V}|^2 \cdot d\mathbf{r} - \oint [\mathbf{V} \times (\nabla \times \mathbf{V})] \cdot d\mathbf{r}$$
$$= -\oint \alpha \nabla p \cdot d\mathbf{r} - \oint \nabla \chi \cdot d\mathbf{r}$$

or

$$\oint \frac{\partial \mathbf{V}}{\partial t} \cdot d\mathbf{r} + \frac{1}{2}\oint dV^2 - \oint [\mathbf{V} \times (\nabla \times \mathbf{V}) \cdot d\mathbf{r}$$
$$= -\oint \alpha\,dp - \oint d\chi \qquad (2\text{-}245)$$

Now $\mathbf{V} \times (\nabla \times \mathbf{V})$ is actually $\mathbf{V} \times \boldsymbol{\zeta}$, and from the definition of the cross product, is a vector perpendicular to the plane of $\mathbf{V}$ and $\boldsymbol{\zeta}$ and as such has no component along the streamline; therefore,

$$\mathbf{V} \times (\nabla \times \mathbf{V}) \cdot d\mathbf{r} = (\mathbf{V} \times \boldsymbol{\zeta}) \cdot d\mathbf{r} = 0$$

Additionally, from (2-193),

$$\mathbf{g} = -\nabla \chi$$

or

$$-g\mathbf{k} = -\frac{d\chi}{dz}\mathbf{k}$$

All of this results in (2-245), reducing to, assuming that g is constant,

$$\oint \frac{\partial \mathbf{V}}{\partial t} \cdot d\mathbf{r} + \tfrac{1}{2}V^2 + \oint \alpha\,dp + gz = F(t) \qquad (2\text{-}246)$$

along a streamline, the most general form of the *Bernoulli equation*.

Special forms of the Bernoulli equation that might be more familiar are discussed next.

Steady-State Flow

For steady-state conditions, $\partial \mathbf{V}/\partial t = 0$ and (2-246) becomes

$$\tfrac{1}{2}V^2 + \oint \alpha\,dp + gz = c' \qquad (2\text{-}247)$$

where c' is a constant having a particular value along any given streamline.

Steady Flow and Uniform Density

If the flow is steady and the density is constant, one obtains

$$\tfrac{1}{2}V^2 + \alpha p + gz = c' \tag{2-248}$$

Irrotational Flow

If the flow is irrotational, a velocity potential ϕ exists and the first term of Bernoulli's equation can be rewritten as

$$\oint \frac{\partial \mathbf{V}}{\partial t} \cdot d\mathbf{r} = -\oint \frac{\partial \nabla\phi}{\partial t} \cdot d\mathbf{r} = -\oint \nabla\left(\frac{\partial \phi}{\partial t}\right) \cdot d\mathbf{r} = -\oint d\left(\frac{\partial \phi}{\partial t}\right) = -\frac{\partial \phi}{\partial t}$$

Substituting in (2-246) yields

$$-\frac{\partial \phi}{\partial t} + \frac{1}{2}V^2 + \int \alpha\, dp + gz = F(t) \tag{2-249}$$

where $F(t)$ is a function of time for time-dependent flow or a constant, c, for the non-time-dependent case. In either case the function or the constant is the *same for every* point within the fluid. We will presently show this to be true.

Irrotational, Steady Flow of a Homogeneous Fluid

If the fluid is homogeneous and the flow is steady and irrotational, we obtain the most familiar form of Bernoulli's equation, what might be called the engineering form,

$$\tfrac{1}{2}V^2 + \alpha p + gz = c \tag{2-250}$$

where each term is expressed as energy per unit mass. Dividing through by the acceleration of gravity, g, the terms will all have units of length and as such are referred to as "heads," being, from the left, the velocity head, the pressure head, and the elevation head, the sum of which equal a constant for all points in the fluid

In the remaining portion of this section, it will be shown that the $F(t)$ of (2-249) is the same throughout the fluid. Assuming that the fluid is incompressible and inviscid and neglecting the Coriolis acceleration (the latter is not essential but simplifies the manipulations), the equations of motion from (2-210) through (2-212) are

$$\begin{aligned}
\frac{\partial u}{\partial t} + u\frac{\partial u}{\partial x} + v\frac{\partial u}{\partial y} + w\frac{\partial u}{\partial z} &= -\frac{1}{\rho}\frac{\partial p}{\partial x} \\
\frac{\partial v}{\partial t} + u\frac{\partial v}{\partial x} + v\frac{\partial v}{\partial y} + w\frac{\partial v}{\partial z} &= -\frac{1}{\rho}\frac{\partial p}{\partial y} \\
\frac{\partial w}{\partial t} + u\frac{\partial w}{\partial x} + v\frac{\partial w}{\partial y} + w\frac{\partial w}{\partial z} &= -\frac{1}{\rho}\frac{\partial p}{\partial z} - g
\end{aligned} \tag{2-251}$$

Since the motion is irrotational,

$$\frac{\partial u}{\partial z} - \frac{\partial w}{\partial x} = 0 \qquad \frac{\partial w}{\partial y} - \frac{\partial v}{\partial z} = 0 \qquad \frac{\partial v}{\partial x} - \frac{\partial u}{\partial y} = 0 \tag{2-252}$$

and a velocity potential ϕ exists such that

$$u = -\frac{\partial \phi}{\partial x} \qquad v = -\frac{\partial \phi}{\partial y} \qquad w = -\frac{\partial \phi}{\partial z} \tag{2-253}$$

From the fact that

$$|\mathbf{V}|^2 = V^2 = \mathbf{V} \cdot \mathbf{V} = u^2 + v^2 + w^2$$

then

$$\frac{1}{2}\frac{\partial V^2}{\partial x} = \frac{1}{2}\frac{\partial}{\partial x}(u^2 + v^2 + w^2) = u\frac{\partial u}{\partial x} + v\frac{\partial v}{\partial x} + w\frac{\partial w}{\partial x}$$

and similarly,

$$\begin{aligned} \frac{1}{2}\frac{\partial V^2}{\partial y} &= u\frac{\partial u}{\partial y} + v\frac{\partial v}{\partial y} + w\frac{\partial w}{\partial y} \\ \frac{1}{2}\frac{\partial V^2}{\partial z} &= u\frac{\partial u}{\partial z} + v\frac{\partial v}{\partial z} + w\frac{\partial w}{\partial z} \end{aligned} \tag{2-254}$$

Substituting (2-252) and (2-253) into (2-251), we obtain

$$\begin{aligned} -\frac{\partial}{\partial t}\left(\frac{\partial \phi}{\phi x}\right) + u\frac{\partial u}{\partial x} + v\frac{\partial v}{\partial x} + w\frac{\partial w}{\partial x} &= -\frac{1}{\rho}\frac{\partial p}{\partial x} \\ -\frac{\partial}{\partial t}\left(\frac{\partial \phi}{\partial y}\right) + u\frac{\partial u}{\partial y} + v\frac{\partial v}{\partial y} + w\frac{\partial w}{\partial y} &= -\frac{1}{\rho}\frac{\partial p}{\partial y} \\ -\frac{\partial}{\partial t}\left(\frac{\partial \phi}{\partial z}\right) + u\frac{\partial u}{\partial z} + v\frac{\partial v}{\partial z} + w\frac{\partial w}{\partial z} &= -\frac{1}{\rho}\frac{\partial p}{\partial z} - g \end{aligned}$$

and application of the three equations of (2-254) to this result gives us

$$\begin{aligned} -\frac{\partial}{\partial t}\left(\frac{\partial \phi}{\partial x}\right) + \frac{1}{2}\frac{\partial V^2}{\partial x} &= -\frac{1}{\rho}\frac{\partial p}{\partial x} \\ -\frac{\partial}{\partial t}\left(\frac{\partial \phi}{\partial y}\right) + \frac{1}{2}\frac{\partial V^2}{\partial y} &= -\frac{1}{\rho}\frac{\partial p}{\partial y} \\ -\frac{\partial}{\partial t}\left(\frac{\partial \phi}{\partial z}\right) + \frac{1}{2}\frac{\partial V^2}{\partial z} &= -\frac{1}{\rho}\frac{\partial p}{\partial z} - g \end{aligned}$$

If the first of these is multiplied by dx, the second by dy, and the third by dz and added, we have

$$-\frac{\partial}{\partial t}\left(\frac{\partial \phi}{\partial x}dx + \frac{\partial \phi}{\partial y}dy + \frac{\partial \phi}{\partial z}dz\right) + \left(\frac{1}{2}\frac{\partial V^2}{\partial x}dx + \frac{1}{2}\frac{\partial V^2}{\partial y}dy + \frac{1}{2}\frac{\partial V^2}{\partial z}dz\right)$$
$$= -\frac{1}{\rho}\left(\frac{\partial p}{\partial x}dx + \frac{\partial p}{\partial y}dy + \frac{\partial p}{\partial z}dz\right) - g\,dz$$

For any *one given time* ϕ, V, and p are functions of only the position coordinates, and thus the three terms in parentheses in the preceding equation are the perfect differentials of ϕ, $\frac{1}{2}V^2$, and p. Rewriting, then,

$$-\frac{\partial}{\partial t}(d\phi) + d\left(\frac{1}{2}V^2\right) + \frac{1}{\rho}(dp) + g\,dz = 0$$

and upon integrating with respect to distance in any direction,

$$-\int d\left(\frac{\partial \phi}{\partial t}\right) + \int d\left(\frac{1}{2}V^2\right) + \int \frac{1}{\rho}\,dp + \int g\,dz = F(t)$$

where $F(t)$ is some arbitrary function of time and plays a role similar to that of a constant of integration. The final result can be written

$$-\frac{\partial \phi}{\partial t} + \frac{1}{2}V^2 + \int \alpha\,dp + gz = F(t)$$

which for any given time t is identical to (2-249).

2-31 Elementary Applications of the Horizontal Equations of Motion

The equations of motion have, up to this point, been derived and discussed in a general way. As a demonstration of their utility, in this section we will apply our results to solve some special, but useful, oceanographic flow problems of both a frictionless and frictional nature. These special cases successfully explain certain typical features of large- and small-scale motions of different types found in the ocean. To recapitulate, the equation motion for incompressible flow from (2-218) is

$$\frac{D\mathbf{V}}{Dt} + 2\mathbf{\Omega} \times \mathbf{V} + \alpha\nabla p + \nabla\chi = \nu\nabla^2\mathbf{V} \tag{2-255}$$

If it is assumed that the fluid motion is essentially horizontal (two-dimensional), we can adopt the hydrostatic equation in the vertical and the scalar component equations of (2-255) are

$$\frac{Du}{Dt} - fv = -\frac{1}{\rho}\frac{\partial p}{\partial x} + \nu\left(\frac{\partial^2 u}{\partial x^2} + \frac{\partial^2 u}{\partial y^2} + \frac{\partial^2 u}{\partial z^2}\right) \tag{2-256}$$

$$\frac{Dv}{Dt} + fu = -\frac{1}{\rho}\frac{\partial p}{\partial y} + \nu\left(\frac{\partial^2 v}{\partial x^2} + \frac{\partial^2 v}{\partial y^2} + \frac{\partial^2 v}{\partial z^2}\right) \tag{2-257}$$

$$\frac{\partial p}{\partial z} = -\rho g \tag{2-258}$$

These three equations are the starting point for our consideration of inertia currents, geostrophic currents, and Ekman wind-drift currents.

Inertia Currents

If the flow is considered to be frictionless, (2-256) and (2-257) indicate that the accelerations on the left are balanced by the acceleration resulting from the horizontal pressure gradients. These gradients may result, for example, from an inclination of the sea surface in a homogeneous ocean due to the effect of a wind stress. Let us assume that, once having been set up, they vanish. The accelerations on the left of (2-256) and (2-257) then sum to zero, giving

$$\frac{Du}{Dt} - fv = 0 \tag{2-259}$$

and

$$\frac{Dv}{Dt} + fu = 0 \tag{2-260}$$

That is, the acceleration of any water particle is balanced by the Coriolis acceleration; flow under such a condition is said to be an *inertia current.*

To determine the form of the velocities u and v, and thus the character of inertial flow, we can combine (2-259) and (2-260) by differentiating the former with respect to t,

$$\frac{D^2u}{Dt^2} - f\frac{Dv}{Dt} = 0$$

Assuming f constant, and substituting this result in (2-260) we obtain the differential equation in u,

$$\frac{D^2u}{Dt^2} + f^2u = 0$$

whose solution is of the form

$$u = A\cos ft + B\sin ft$$

where A and B are constants. After imposing the initial condition that $u = 0$ at $t = 0$, we see that $A = 0$ and therefore

$$u = B\sin ft \tag{2-261}$$

Substitution of this into (2-259) yields

$$v = B\cos ft \tag{2-262}$$

These two results tell us that the current velocity $\mathbf{V}$ is of constant magnitude since

$$|\mathbf{V}| = (\mathbf{V}\cdot\mathbf{V})^{1/2} = V = (u^2 + v^2)^{1/2} = B \tag{2-263}$$

The trajectory of a fluid particle is determined by integrating (2-261),

$$x = \int (B \sin ft)\, dt + B'$$

$$= -\frac{1}{f} B \cos ft + B' = -\frac{v}{f} + B'$$

and (2-262),

$$y = \int (B \cos ft)\, dt + B''$$

$$= \frac{1}{f} B \sin ft + B'' = \frac{u}{f} + B''$$

and eliminating t by squaring and summing the result,

$$(x - B')^2 + (y - B'')^2 = \left(\frac{V}{f}\right)^2 \tag{2-264}$$

This is the equation of the *circle of inertia* with center at $x = B'$, $y = B''$, and of radius V/f. Because the force that produces the radial acceleration is the Coriolis force, which in the Northern Hemisphere acts to the right of the velocity, the motion around the inertia circle is clockwise or *cum sole* ("with the sun"), as shown in Figure 2-56. Taking T_I as the time required by a particle, traveling with the speed V, to complete one full circle of radius r,

$$T_I = \frac{2\pi r}{V} = \frac{2\pi}{f} = \frac{\pi}{\Omega \sin \theta}$$

where T_I is called the *inertial period.* At the equator, $\theta = 0$ and T_I is infinity; at the poles, $\theta = \pi/2$ and $T_1 = 12$ hours. Also, since $2\pi/(\Omega \sin \theta)$ is the period of revolution of a Foucault pendulum at latitude θ, the inertial period is, therefore, one-half of a pendulum day. Finally, for purposes of comparison, for a current of 2 knots we obtain the following:

Latitude (deg)	*Period (hr)*	*Diameter of Inertial Circle (nautical mi)*
90	12	7.6
60	14	8.2
30	24	15.3
0	∞	∞

Geostrophic Currents (Nonhomogeneous Ocean)

For a frictionless, unaccelerated current flowing horizontally with the only external force being that of gravity, it follows from (2-256) and (2-257)

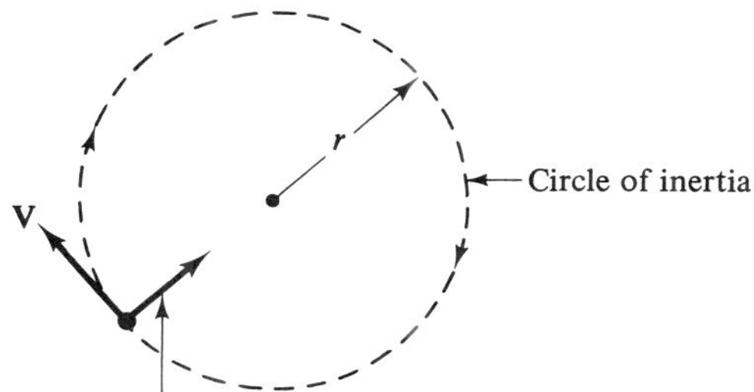

Fig. 2-56. Inertia current in the Northern Hemisphere.

that for the horizontal coordinate directions the components of the Coriolis force balance the components of the pressure gradient force; that is,

$$fv = \frac{1}{\rho}\frac{\partial p}{\partial x} \tag{2-265}$$

and

$$fu = -\frac{1}{\rho}\frac{\partial p}{\partial y} \tag{2-266}$$

If these are squared and added, the result can be written as

$$\frac{\partial p}{\partial n} = f\rho V \tag{2-267}$$

where

$$V \equiv (u^2 + v^2)^{1/2}$$

and

$$\frac{\partial p}{\partial n} \equiv \left[\left(\frac{\partial p}{\partial x}\right)^2 + \left(\frac{\partial p}{\partial y}\right)^2\right]^{1/2}$$

The equilibrium of forces expressed by (2-267) states that the Coriolis force must be exactly opposite to the horizontal pressure gradient force, which in turn requires that the horizontal current vector **V** be parallel to the isobars and, in the Northern Hemisphere, in such a direction that, facing in the direction of the flow, the higher pressure is to the right. Taking θ as negative in the Southern Hemisphere, the higher pressure is then to the left when facing downstream (see Figure 2-57). This type of current is called a *geostrophic current* and the equilibrium of forces expressed by (2-267) is called *geostrophic equilibrium*.

A formal solution to the differential geostrophic equations can be obtained in the following way. Integrating the hydrostatic equation of (2-258) from the surface down to some depth z, we obtain

$$p_z - p_0 = -\int_0^z \rho g\, dz = -g\int_0^z \rho\, dz$$

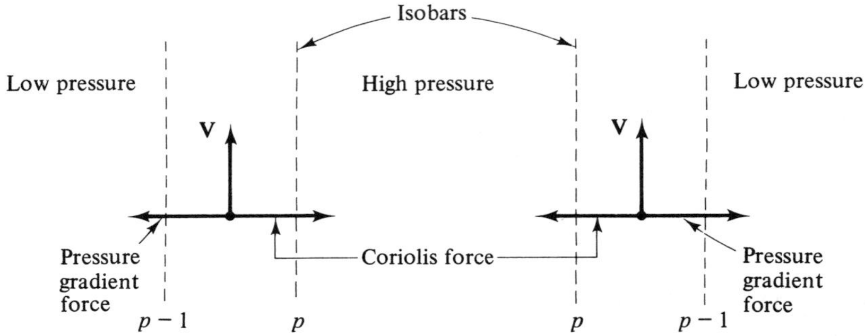

Fig. 2-57. Distribution of forces and current for geostrophic flow.

where p_z is the pressure at depth z, p_0 the pressure at the surface ($z = 0$), and g has been taken as constant. No loss of generality results by taking p_0 as zero; therefore, we may write the last equation as

$$p_z = -g \int_0^z \rho \, dz$$

Now, with reference to Figure 2-58 and through application of a Taylor expansion, it is possible to write p_z at a point x and p_z at a point $x + dx$ as, respectively,

$$(p_z)_x = -g \int_0^z \rho \, dz$$

and

$$(p_z)_{x+dx} = -g \int_0^z \rho \, dz + \left(-g \frac{\partial}{\partial x} \int_0^z \rho \, dz\right) dx + \rho_0 g \, d\eta$$

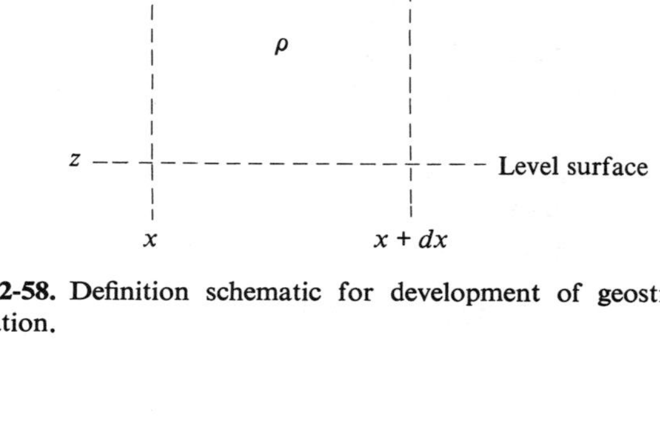

Fig. 2-58. Definition schematic for development of geostrophic equation.

where ρ_0, the density above the level surface $z = 0$, is taken as a constant. Subtracting yields

$$(p_z)_{x+dx} - (p_z)_x = \left(g \frac{\partial}{\partial x} \int_z^0 \rho \, dz\right) dx + \rho_0 g \, d\eta$$

or

$$\frac{\partial p_z}{\partial x} = g \frac{\partial}{\partial x} \int_z^0 \rho \, dz + \rho_0 g \frac{\partial \eta}{\partial x} \tag{2-268}$$

where the first term on the right represents the baroclinic contribution to the horizontal pressure gradient and the second term represents the barotropic contribution. Inserting (2-268) into (2-265), we find that

$$v = \frac{g}{\rho f} \frac{\partial}{\partial x} \int_z^0 \rho \, dz + \frac{\rho_0}{\rho} \frac{g}{f} \frac{\partial \eta}{\partial x}$$

where for all practical purposes we can take $\rho_0/\rho \simeq 1$ for all depths. Letting $z = 0$, this result then reduces to

$$v_0 = \frac{g}{f} \frac{\partial \eta}{\partial x}$$

where v_0 is the current at $z = 0$ and is referred to as the *slope current* or the *barotropic current*. It is merely that portion of the absolute current which results from sea surface being inclined with respect to the level surfaces. We can now write that the *relative* or *baroclinic current*, that portion of the absolute current whose source is the variation of the fluid density due to temperature and salinity differences, is given by

$$v_z - v_0 = \frac{g}{\rho f} \frac{\partial}{\partial x} \int_z^0 \rho \, dz \tag{2-269}$$

In a similar way, for a sea surface having a slope component in the y direction, it is readily shown that

$$u_z - u_0 = -\frac{g}{\rho f} \frac{\partial}{\partial y} \int_z^0 \rho \, dz \tag{2-270}$$

where the left-hand side is the y component of the relative or baroclinic current. Equations (2-269) and (2-270) then permit computation of the relative geostrophic current at any depth within the fluid from a knowledge of the density field.

It is possible to reduce these last two equations to the more practical form, at least as far as actual computations are concerned, discussed in many introductory oceanography textbooks. We can begin to do so by writing (2-269) as

$$v_z - v_0 = \frac{g}{\bar{\rho} f} \int_z^0 \frac{\partial \rho}{\partial x} \, dz$$

where $\bar{\rho}$ is taken as the average density in the vertical column. Introducing the specific volume α,

$$v_z - v_0 = \frac{g}{\bar{\rho} f} \int_z^0 \frac{\partial \alpha^{-1}}{\partial x} dz = -\frac{g}{\bar{\rho} f} \int_z^0 \rho^2 \frac{\partial \alpha}{\partial x} dz$$

and applying the hydrostatic equation, (2-258), to the integral, we obtain

$$\begin{aligned} v_z - v_0 &= \frac{g}{\bar{\rho} f} \int_{p_z}^{p_0} \rho \frac{\partial \alpha}{\partial x} dp \\ &\doteq \frac{1}{f} \int_{p_z}^{p_0} \frac{\partial \alpha}{\partial x} dp \doteq \frac{1}{f} \frac{\partial}{\partial x} \int_{p_z}^{p_0} \alpha \, dp \end{aligned} \tag{2-271}$$

in which integration and differentiation were rearranged via the assumption that the slope of the isobars are so small that they can be neglected. The integral on the right side of (2-271) is the known expression for the dynamic height D, defined as

$$D \equiv \int_{p_z}^{p_0} \alpha \, dp$$

substitution of which results in the familiar form

$$v_z - v_0 = \frac{1}{f} \frac{\partial D}{\partial x} \tag{2-272}$$

and similarly, starting with (2-270),

$$u_z - u_0 = -\frac{1}{f} \frac{\partial}{\partial y} \int_{p_z}^{p_0} \alpha \, dp = -\frac{1}{f} \frac{\partial D}{\partial y} \tag{2-273}$$

When written and applied in finite-difference form, these equations permit calculation of the relative or baroclinic component of the geostrophic current.

Consider briefly the case, shown in Figure 2-59, of two isobars p_1 and p_2 within the fluid having an inclination of γ_1 and γ_2 with respect to level surfaces. Then, from application of (2-272) we can write that

$$v_1 - v_2 = \frac{1}{f} \frac{\Delta D}{\Delta x} = \frac{1}{fL} \left[\left(\int_{p_2}^{p_1} \alpha \, dp \right)_A - \left(\int_{p_2}^{p_1} \alpha \, dp \right)_B \right] \tag{2-274}$$

where v_1 and v_2 are the y components of the fluid velocity at depths z_1 and z_2, L is the horizontal distance between the two hydrographic stations A and B, and the subscripts on the integrals refer to the two stations. This relation was first derived by Sandstörm and Helland-Hansen in 1903 and is known as the *Helland-Hansen formula.* The physical meaning of (2-274) is readily apparent if the integrals are converted back to integration with respect to elevation through application of the hydrostatic equation to give, on the right,

$$f \frac{1}{L} \left[\left(\int_{z_2}^{z_1} - g \, dz \right)_A - \left(\int_{z_2}^{z_1} - g \, dz \right)_B \right]$$

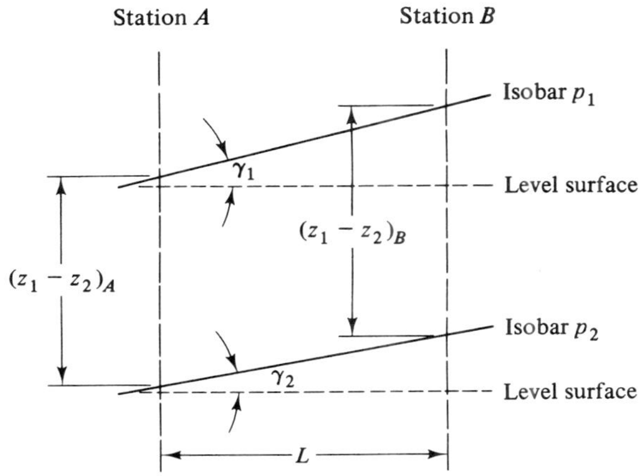

Fig. 2-59. Schematic diagram for application of the Helland-Hansen formula.

which for g a constant reduces to

$$\frac{g}{fL}[(z_1 - z_2)_A - (z_1 - z_2)_B]$$

This is merely equal to g/f times the slope of the isobar p_1 relative to the isobar p_2. Observe also that the quantity in the square brackets of (2-274) is the negative of the area between the α_A and α_B curves for the stations in the interval from p_1 and p_2 (Figure 2-60).

In concluding this discussion, it should again be noted that the geostrophic current can be considered as being composed of two parts: one that is dependent on the slope of the free surface and is called the slope

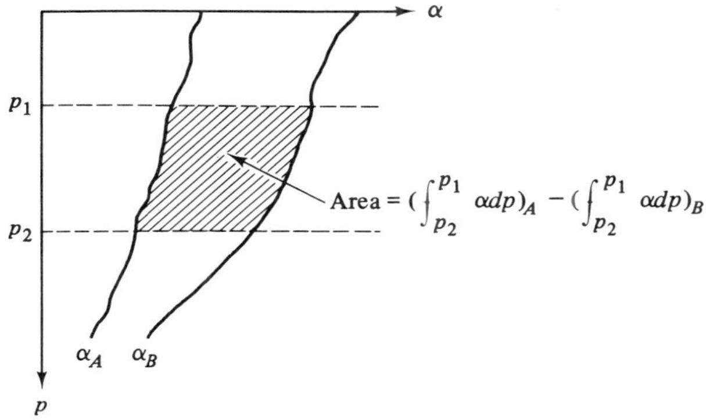

Fig. 2-60. Relationship between α–p plot and the Helland-Hansen formula.

current, and one that is dependent on the internal density distribution and is called the relative current. Of these two, only the relative current can be determined from observations of the density distribution, since in general the added slope of the isobaric surfaces due to the piling up of water cannot be directly observed. To arrive at the absolute current, either the absolute pressure field or the absolute velocity must be known at one level between two adjacent hydrographic stations. The possibility exists of (1) finding the depth if it exists, where the currents become zero, called the *depth of no motion*, through direct current measurements, or (2) making a direct determination of surface currents. Either of these methods encounters considerable difficulty.

Ekman Wind-Drift Currents (the Ekman Layer Equations for a Homogeneous Ocean)

Pure wind-drift currents are the result of the drag exerted by the wind in passing over homogeneous water. Horizontal pressure gradients are not considered; the only driving force is that of friction in the form of the wind stress applied at the sea surface and the frictional coupling between individual layers of water. The necessary conditions for pure wind-drift currents is that the ocean must be assumed infinitely deep and unbounded in the horizontal. Additionally, it is taken that the wind field is uniform in that it is of constant speed and direction everywhere. Under these conditions, no water will pile up near coasts or in the open sea; no slope of the sea surface will therefore develop. It was Ekman in 1902 who first considered the nonaccelerated (stationary) drift currents in this idealized ocean.

The equations of motion for Ekman wind-drift currents from (2-256) and (2-257) take the form

$$-fv = \nu \frac{d^2u}{dz^2} \tag{2-275}$$

and

$$fu = \nu \frac{d^2v}{\partial z^2} \tag{2-276}$$

where $u = u(z)$, $v = v(z)$, and friction resulting from vertical shearing stresses is introduced through the kinematic viscosity coefficient, ν, assumed constant. For a thin layer of fluid, these equations specify a balance between the frictional stress acting on the upper surface of the layer, the frictional stress acting on the bottom surface on the layer, and the Coriolis force. The situation is illustrated in Figure 2-61 for the two topmost layers.

To obtain a solution to (2-275) and (2-276), we begin by differentiating the first of these, with respect to z, twice. Thus,

$$-f\frac{d^2v}{dz^2} = \nu \frac{d^4u}{dz^4}$$

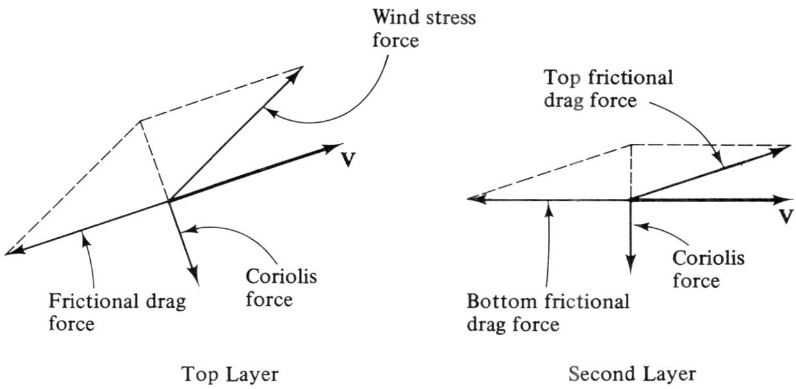

Fig. 2-61. Balance of forces acting on fluid for Ekman wind-drift currents.

which after substituting from (2-276) results in the differential equation whose solution we seek subject to appropriate boundary conditions, namely

$$\frac{d^4u}{dz^2} + \left(\frac{f}{\nu}\right)^2 u = 0 \tag{2-277}$$

Assuming a solution of the form

$$u = Ae^{kz} \tag{2-278}$$

substitution into (2-277) gives

$$\left(k^4 + \frac{f^2}{\nu^2}\right)Ae^{kz} = 0$$

from which it is seen that

$$k^4 + \left(\frac{f}{\nu}\right)^2 = 0$$

Solving for k, we obtain

$$k^2 = \pm i\frac{f}{\nu}$$

where $i \equiv \sqrt{-1}$. Also,

$$k = \pm(1 \pm i)\left(\frac{f}{2\nu}\right)^{1/2} = \pm\frac{1 \pm i}{H} \tag{2-279}$$

since

$$(-1)^{1/4} = \pm\frac{1 \pm i}{\sqrt{2}}$$

and where H is defined as

$$H \equiv \left(\frac{2\nu}{f}\right)^{1/2} \tag{2-280}$$

Substituting for k from (2-279) into the assumed solution of (2-278), we obtain the general solution

$$u = A_1 e^{(1+i)z/H} + A_2 e^{-(1+i)z/H} + A_3 e^{(1-i)z/H} + A_4 e^{-(1-i)z/H} \tag{2-281}$$

where the constants A_1, A_2, A_3, and A_4 can be evaluated from the boundary conditions.

Under our original assumptions, the required four boundary conditions are:

1. The wind stress, τ, on the sea surface is, from Newton's law of viscosity, equation (2-122), given by

$$\tau' = \nu \frac{du}{dz}\bigg|_{z=0} \tag{2-282}$$

where $\tau' \equiv \tau/\rho$ and

$$0 = \nu \frac{dv}{dz}\bigg|_{z=0} \tag{2-283}$$

where, without loss of generality, the x axis has been aligned in the direction of the wind.

2. At great depth the flow goes to zero such that

$$u\big|_{z=-\infty} = 0 \tag{2-284}$$

and

$$v\big|_{z=-\infty} = 0 \tag{2-285}$$

Now (2-282) and (2-284) can be applied directly to the general solution for u given by (2-281); while (2-283) and (2-285), which are conditions imposed on v, can be converted to conditions imposed on u through (2-275).

We will now summarize the boundary conditions. From (2-282),

$$\frac{du}{dz}\bigg|_{z=0} = \frac{\tau'}{\nu} \tag{2-286}$$

from (2-275) and (2-283),

$$\frac{d^3u}{dz^3}\bigg|_{z=0} = 0 \tag{2-287}$$

from (2-284),

$$u\big|_{z=-\infty} = 0 \tag{2-288}$$

and from (2-275) and (2-285),

$$\frac{d^2u}{dz^2}\bigg|_{z=-\infty} = 0 \tag{2-289}$$

Equation (2-288) makes it immediately apparent that our solution cannot contain negative, real exponents, and consequently $A_2 = A_4 = 0$, yielding the solution

$$u = A_1 e^{(1+i)z/H} + A_3 e^{(1-i)z/H} \tag{2-290}$$

Differentiating with respect to z, we obtain

$$\frac{du}{dz} = \frac{1+i}{H} A_1 e^{(1+i)z/H} + \frac{1-i}{H} A_3 e^{(1-i)z/H} \tag{2-291}$$

and evaluating this result at $z = 0$, we obtain

$$\left.\frac{du}{dz}\right|_{z=0} = \frac{\tau'}{\nu} = \frac{(1+i)A_1 + (1-i)A_3}{H} \tag{2-292}$$

[It is seen from (2-291) that the condition on the second derivative, as specified by the boundary condition of (2-289), is satisfied since no negative, real exponents exist.] The third derivative of u is given by

$$\frac{d^3u}{dz^3} = \left(\frac{1+i}{H}\right)^3 A_1 e^{(1+i)z/H} + \left(\frac{1-i}{H}\right)^3 A_3 e^{(1-i)z/H} \tag{2-293}$$

Recalling that

$$(1+i)^3 = -2(1-i)$$

and

$$(1-i)^3 = -2(1+i)$$

and evaluating (2-293) at $z = 0$, we find that

$$\frac{-2(1-i)A_1 - 2(1+i)A_3}{H^3} = 0$$

or

$$A_3 = -\frac{1-i}{1+i} A_1 = iA_1$$

Inserting this in (2-292) gives the result that

$$A_1 = \frac{\tau' H}{2\nu}\left(\frac{1-i}{2}\right)$$

Combining these last two equations with our solution for u as expressed by (2-290),

$$u = \frac{\tau' H}{2\nu}\left[\frac{1-i}{2} e^{(1+i)z/H} + \frac{1+i}{2} e^{(1-i)z/H}\right]$$

and rewriting,

$$u = \frac{\tau' H}{2\nu} e^{z/H}\left(\frac{e^{iz/H} + e^{-iz/H}}{2} + \frac{e^{iz/H} - e^{-iz/H}}{2i}\right)$$

or

$$u = \frac{\tau' H}{2\nu} e^{z/H}\left(\cos\frac{z}{H} + \sin\frac{z}{H}\right)$$

Finally,

$$u = \frac{\tau' H}{\sqrt{2}\nu} e^{z/H} \sin\left(\frac{z}{H} + \frac{\pi}{4}\right) \tag{2-294}$$

Application of (2-275) results in the equation for v,

$$v = -\frac{\tau' H}{\sqrt{2}\nu} e^{z/H} \cos\left(\frac{z}{H} + \frac{\pi}{4}\right) \tag{2-295}$$

These equations, (2-294) and (2-295), give the velocity components of a pure wind-drift current as a function of the depth. To begin an examination of the meaning of our results, consider the case for which $z = 0$. Thus,

$$u|_{z=0} \equiv u_0 = \frac{\tau' H}{\sqrt{2}\nu} \sin\frac{\pi}{4}$$

$$v|_{z=0} \equiv v_0 = -\frac{\tau' H}{\sqrt{2}\nu} \cos\frac{\pi}{4}$$

and

$$V|_{z=0} \equiv V_0 = (u_0^2 + v_0^2)^{1/2} = \frac{\tau' H}{\sqrt{2}\nu} \tag{2-296}$$

Consequently,

$$u_0 = V_0 \sin\frac{\pi}{4} = \frac{\sqrt{2}}{2} V_0$$

$$v_0 = -V_0 \cos\frac{\pi}{4} = -\frac{\sqrt{2}}{2} V_0$$

$$\mathbf{V}_0 = \frac{\sqrt{2}}{2} V_0 \mathbf{i} - \frac{\sqrt{2}}{2} V_0 \mathbf{j}$$

and therefore, as shown in Figure 2-62, the surface current vector, in the Northern Hemisphere, points in a direction 45° *cum sole* to the wind direction. Our solution, however, is only valid in the Northern Hemisphere, since in the

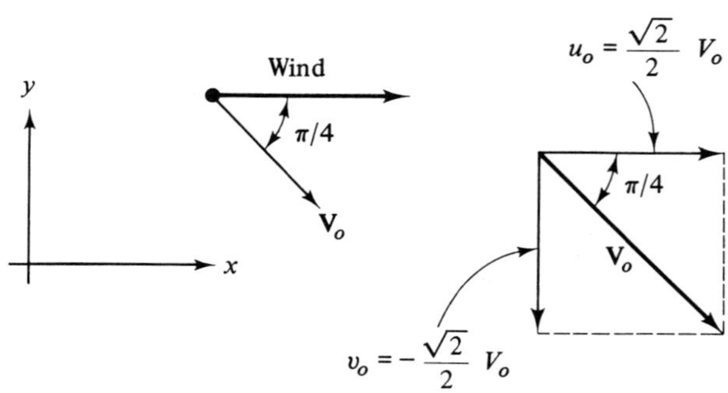

Fig. 2-62. Surface current in Northern Hemisphere for Ekman wind-drift.

Southern Hemisphere the latitude, θ, is taken as negative and, from (2-280), H is consequently imaginary. A solution valid for the Southern Hemisphere is obtained if the direction of the y axis of the coordinate system is reversed. This being done, we then find that, for the Southern Hemisphere, the surface water moves in a direction 45° to the left of the wind direction.

Next let us examine the case for which $z = -\pi H$. It follows that

$$u|_{z=-\pi H} = V_0 e^{-\pi} \sin \frac{3\pi}{4} = -V_0 e^{-\pi} \sin \frac{\pi}{4}$$

and

$$v|_{z=-\pi H} = -V_0 e^{-\pi} \cos \frac{3\pi}{4} = V_0 e^{-\pi} \cos \frac{\pi}{4}$$

which indicate that at this depth the current vector has decreased to $e^{-\pi}$ times the surface velocity ($\doteq \mathbf{V}_0/23$), but has also changed direction and is exactly opposite (Figure 2-63). In general, from (2-294) and (2-295), we conclude that the current vector between the surface and the deeper layers turns continuously *cum sole* while its magnitude decreases exponentially. These characteristics are shown in Figure 2-64, where the current vectors

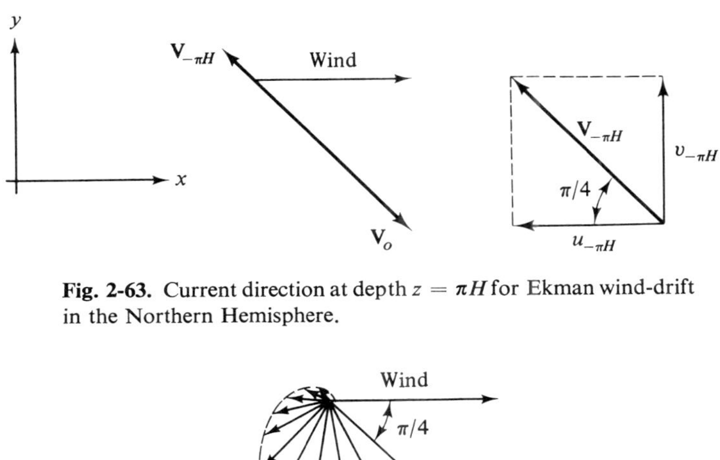

Fig. 2-63. Current direction at depth $z = \pi H$ for Ekman wind-drift in the Northern Hemisphere.

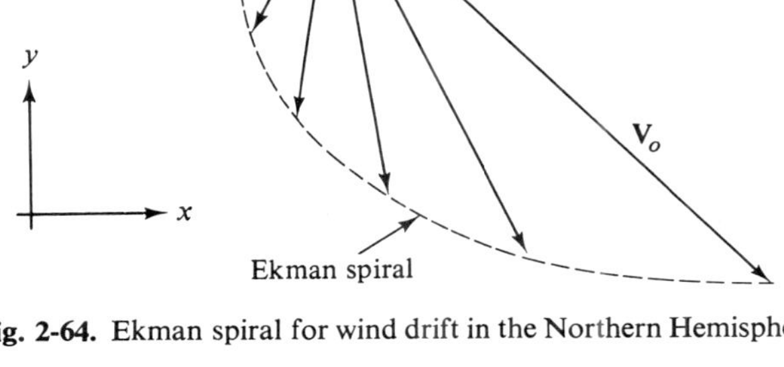

Fig. 2-64. Ekman spiral for wind drift in the Northern Hemisphere.

have been drawn for equal depth intervals. Projected on a horizontal plane the end points of the vectors form the logarithmic spiral known as the *Ekman spiral.* In our model, having taken an infinitely deep ocean, the velocity approaches zero asymptotically; however, by far the more significant currents are found above the depth $z = -\pi H$. This depth was called by Ekman the *depth of frictional resistance*, since for all practical purposes the fluid below this depth is unaffected by the surface wind. From the definition of (2-280), it should also be noted that H is independent of the wind stress but does increase with an increasing viscosity coefficient and with decreasing latitude. At the Equator, H approaches infinity. It also follows that the surface current is proportional to the wind stress.

To obtain some feeling for the numbers involved, consider a typical example for which $\tau = \rho\tau' = 1$ dyne/cm², $f = 0.7 \times 10^{-4}$ sec^{-1} ($\theta = 30°$N), and $\nu = 10$ cm²/sec. These give

$$H = \left(\frac{2\nu}{f}\right)^{1/2} = \left(\frac{2 \times 10}{0.7 \times 10^{-4}}\,\text{cm}^2\right)^{1/2} = 5.3\ \text{m}$$

$$V_0 = \frac{\tau' H}{\sqrt{2}\,\nu} = \frac{540}{10\sqrt{2}} = 38\ \text{cm/sec}$$

$$V_{-\pi H} = V_{-17.2\text{m}} = 1.5\ \text{cm/sec}$$

Finally, of particular significance to the oceanographer is the mass transport resulting from the Ekman current. We obtain the *mass transport* components S_x and S_y, having units of [mass]/([time][length]), by multiplying the velocity components u and v as given by (2-294) and (2-295) by the density ρ and integrating from $z = -\infty$ up to $z = 0$. Proceeding then,

$$S_x \equiv \int_{-\infty}^{0} \rho u\, dz = \frac{\rho\tau' H}{\sqrt{2}\,\nu} \int_{-\infty}^{0} e^{z/H} \sin\left(\frac{z}{H} + \frac{\pi}{4}\right) dz = 0$$

and

$$S_y \equiv \int_{-\infty}^{0} \rho v\, dz = -\frac{\rho\tau' H}{\sqrt{2}\,\nu} \int_{-\infty}^{0} e^{z/H} \cos\left(\frac{z}{H} + \frac{\pi}{4}\right) dz = -\frac{\tau H^2}{2\nu}$$

Since $S_x = 0$, the net mass transport in the Ekman layer is given solely by S_y and is thus directed 90° *cum sole* to the direction of the wind in the Northern Hemisphere.

2-32 Time-Averaged Form of the Momentum and Continuity Equations for Incompressible Flow

Most natural and engineering flows are of a turbulent character rather than being laminar. The *Glossary of Oceanographic Terms* (Baker et al., 1966) gives as the definition for laminar flow: "A flow in which the fluid moves smoothly in streamlines in parallel layers or sheets; a nonturbulent flow." Turbulent flow is then defined as: "A flow characterized by irregular, random velocity flucuations"; and for turbulence: "A state of fluid flow in which the instantaneous velocities exhibit irregular and apparently random fluctuations, so that in practice only statistical properties can be recognized and subjected to analysis. These fluctuations often constitute major deformations of the flow and are capable of transporting momentum, energy, and suspended matter at rates far in excess of the rate of transport by molecular diffusion and conduction in a nonturbulent or laminar flow." In geophysical fluid dynamics nonturbulent flow is the exception, not the rule. Waves, tides, ocean currents, the boundary layer of the earth's atmosphere, cumulus clouds, flow in rivers and canals, ship wakes, and aircraft wakes all represent fluid motion of a turbulent type. The list is almost endless.

It has been stated (Tennekes and Lumley, 1972) that a precise definition of turbulence is difficult to come by—that it is best described by its properties. For example:

Irregularity. All turbulent flows are random, thus making a deterministic approach impossible; statistical methods must be employed.

Continuum. The smallest turbulent scales are macrosopic rather than microscopic, and therefore turbulent flow can be treated as a continuum phenomenon subject to the equations of fluid mechanics.

Dissipation. In turbulent flow the viscous shear stresses result in the kinetic energy of the fluid being converted to internal energy by performing deformational work; bringing to mind the verse attributed to L. F. Richardson:

> Big whirls have little whirls that feed upon their velocity
> And little whirls have lesser whirls and so on to viscosity.

Diffusivity. The diffusity of turbulence causes rapid mixing together with greatly increased rates of transfer of momentum, heat, and mass. For example, turbulence greatly increases the transfer of energy between atmospheric winds and the ocean currents.

Large Reynolds Numbers. The Reynolds number is defined as the ratio of the inertial forces, $D\mathbf{V}/Dt$, to the viscous forces $[\nu\nabla^2\mathbf{V} + \frac{1}{3}\nu\nabla(\nabla \cdot \mathbf{V})]$. From a dimensional analysis of the Navier–Stokes equation, similar to that done earlier in the case of the hydrostatic equation, it is found that the Reynolds number $= LV/\nu$, where L is a characteristic length, V the magnitude of a characteristic velocity, and ν the kinematic viscosity. Turbulent flows always occur at large Reynolds numbers.

Turbulent Flows Are Flows. Unlike viscosity, which is a characteristic of fluids, turbulence is a feature of fluid flows in that it is not governed by the molecular properties of the fluid. The dynamics of turbulence are the same in all fluids.

Now in solving oceanographic problems through the use of the momentum equation (or of one of its forms, say, the Navier–Stokes equation), the results obtained can only be judged as to their validity by comparison with actual observations. Since, in general, we are concerned with fluid velocity and velocity changes, we can most directly verify our procedures by making velocity measurements. Since most naturally occurring flows are of a turbulent nature, the question arises as to exactly what velocity is to be measured. Clearly, we should measure the same velocity as that described by the equations we have used. If we have applied the Navier–Stokes equation, the velocity described is the one associated with small (macroscopic, yet not microscopic) fluid elements. Not only would we in practice encounter difficulty in making measurements of such small elements, but in the prevailing turbulent conditions, we usually are not (or cannot be) interested in all of the fine details of the motion. Rather, we are interested in the mean values of the parameters of the problem, the mean being taken over some time scale small with respect to the phenomenon of interest to us, but large with respect to the time scale of the turbulence. We want to use mean velocities (and pressures, temperatures, salinities, and densities) and therefore require that all of our defining equations be written in average form rather than in what we may call their *point property* form.

Before we proceed with this equation-averaging process, we will digress briefly to consider what is meant by mean values and fluctuations. Consider, as an example, the serial sequence of temperatures, T, as recorded by a rapidly responding sensing element at some fixed point in the ocean. Figure 2-65 illustrates the detailed record of T and the *running average* $\bar{T}$, which is a smoothed version of the temperature and depends on the time as well as the period of sampling or the averaging interval. (In general, of course,

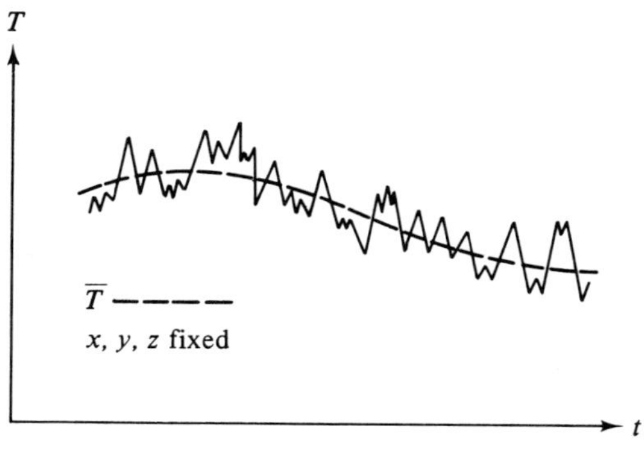

Fig. 2-65. Illustrating the relationship between the point value of a variable, T, and the mean value, $\bar{T}$.

both T and $\bar{T}$ are functions of the position coordinates x, y, and z.) Specifically,

$$\bar{T} \equiv \frac{1}{2\Delta t} \int_{t_0 - \Delta t}^{t_0 + \Delta t} T(t)\, dt \qquad (2\text{-}297)$$

represents a running average at some fixed point based on an averaging interval of $2\Delta t$. In general,

$$\bar{T} = \bar{T}(x, y, z, t_0, 2\Delta t)$$

Geometrically, (2-297) means that, having obtained a record of T at some fixed point over a period of time, the areas between the curve $T(t)$ and the time axis are measured over an interval of time $2\Delta t$, centered at $t = t_0$. This result is then used to construct average temperatures such that the areas above $T = \bar{T}$ are equal to the areas below this same line. Thus, it is possible to define the anomaly of the temperature from the running average as

$$T' \equiv T - \bar{T} \qquad (2\text{-}298)$$

A plot of T' would appear as in Figure 2-66, where the average value of T' itself will vanish (i.e., $\bar{T}' = 0$).

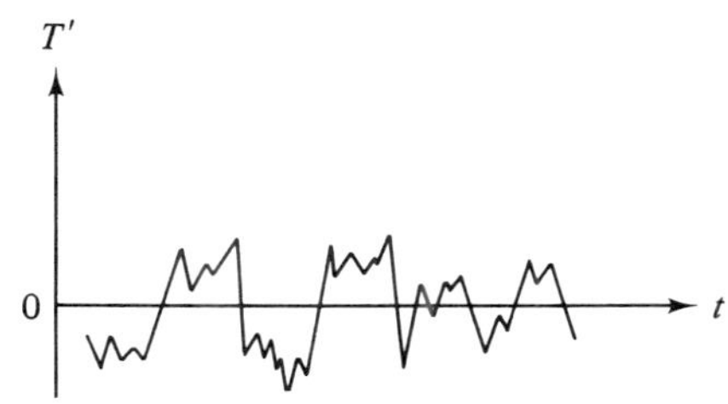

Fig. 2-66. Plot of the anomaly of T versus t.

Now, if the temperature $\bar{T}$ is to be characteristic of the temperature as a whole, the averaging interval $2\Delta t$ must be long enough to ensure that an adequate number of fluctuations is included; however, the adoption of a very long interval would quite likely remove important changes which are taking place in the temperature. All of the foregoing discussion, although applied specifically to the temperature, holds for any of the other field variables that may concern us, such as u, v, w, s, p, and ρ. For each of these the average of the anomaly from the running average is zero (e.g., $\bar{u}' = 0$), but it cannot be asserted a priori that mean values of the squares and products of anomalies are necessarily zero. For example, the product of T' and u' in general will not vanish (i.e., $\overline{T'u'} \neq 0$). For this particular example, it can be said that there will exist a net sensible heat transport associated with the turbulent fluctuations. If $\overline{T'u'} \neq 0$, T' and u' are said to be *correlated*. In the case of $\overline{T'u'} = 0$, the two variables are *uncorrelated*. Figure 2-67 illustrates

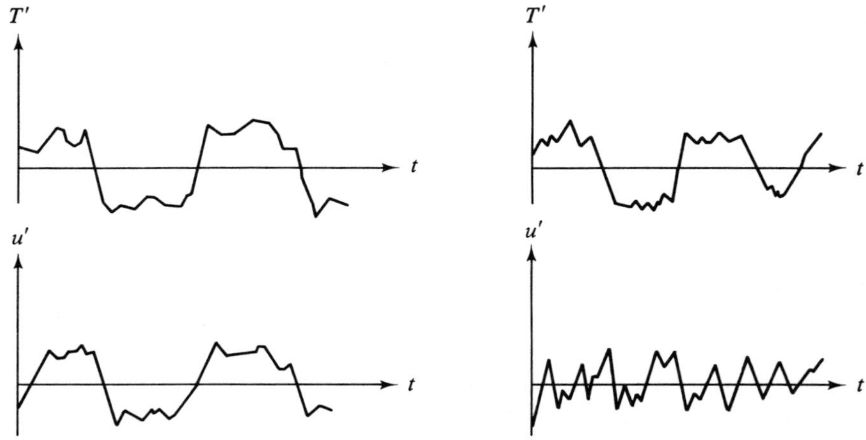

Fig. 2-67. Comparison between correlated and uncorrelated fluctuations.

this concept; the left pair of diagrams show the case where the fluctuating variables have the same sign for most of the time and T' and u' are correlated, while the right pair show the uncorrelated situation. A measure for the extent of correlation between the two variables T' and u' is the *correlation coefficient*, $r_{T,u}$, obtained by dividing $\overline{T'u'}$ (the covariance of T and u) by the square root of the product of $\overline{(T')^2}$ and $\overline{(u')^2}$ (the variances of T and u),

$$r_{T,u} \equiv \frac{\overline{T'u'}}{[\overline{(T')^2}\,\overline{(u')^2}]^{1/2}}$$

a dimensionless quantity that will always lie in the range

$$-1 \leq r_{T,u} \leq 1$$

Statistically, $r_{T,u}$ is a measure of the fraction of the variance of T (or of the fraction of the variance of u) that is related directly to that of $u(T)$. If $r_{T,u}^2 = 1$, the variables T and u are linearly dependent; if $r_{T,u} = 1$, there is a perfect positive correlation between T and u (i.e., an increase of one variable results in a linear increase of the other); if $r_{T,u} = -1$, there is a perfect negative correlation. However, $r_{T,u}$ is a useful measure of the strength of the relationship only in a linear sense. When $r_{T,u} = 0$, the two variables are said to be uncorrelated even though they may be closely related, but where the relationship is not linear. Also, no cause-and-effect relationship between the variables is implied.

In 1895 Reynolds formulated a set of rules of approximation in the calculation of mean values. Although they are not universally applicable they are good approximations for situations where the fluctuations are numerous and random:

1. Quantities that have already been averaged may be taken as constants in subsequent averaging,

$$\langle \bar{T} \rangle = \bar{T} \qquad \langle \bar{T}u \rangle = \bar{T}\bar{u} \tag{2-299}$$

where the angle brackets, $\langle \cdot \rangle$, have the identical meaning as the overbar as defined by (2-297).

2. Averaging obeys the distributive law,

$$\overline{T + u} = \bar{T} + \bar{u} \qquad \overline{\frac{\partial T}{\partial x}} = \frac{\partial \bar{T}}{\partial x} \tag{2-300}$$

Using these, then, we obtain

$$\bar{T} = \langle \bar{T} + T' \rangle = \langle \bar{T} \rangle + \langle T' \rangle = \bar{T} + \bar{T}'$$

from which is obtained that $\bar{T}' = 0$, as stated previously. Also,

$$\begin{aligned} \overline{Tu} &= \langle (\bar{T} + T')(\bar{u} + u') \rangle \\ &= \langle \bar{T}\bar{u} \rangle + \langle \bar{T}u' \rangle + \langle T'\bar{u} \rangle + \langle T'u' \rangle \\ &= \bar{T}\bar{u} + \overline{T'u'} \end{aligned} \tag{2-301}$$

If vectors, such as velocities, are multiplied and averaged, the term $\overline{V_i V_j}$ involves nine components, one of which typically is

$$\overline{V_1 V_3} = \bar{V}_1 \bar{V}_3 + \overline{V_1' V_3'}$$

the last quantity not necessarily being zero even if $\bar{V}_1$ and $\bar{V}_3$ are zero.

Time-Averaged Continuity Equation

Let each point property be represented, as in (2-298), as the sum of its average and a fluctuating or turbulent component,

$$\begin{aligned} u &\equiv \bar{u} + u' \qquad v \equiv \bar{v} + v' \qquad w \equiv \bar{w} + w' \\ \rho &\equiv \bar{\rho} + \rho' \qquad p \equiv \bar{p} + p' \qquad s \equiv \bar{s} + s' \end{aligned} \tag{2-302}$$

Substituting into the continuity equation,

$$\frac{\partial \rho}{\partial t} + \nabla \cdot \rho \mathbf{V} = 0$$

we obtain

$$\frac{\partial}{\partial t}(\bar{\rho} + \rho') + \frac{\partial}{\partial x}(\bar{\rho} + \rho')(\bar{u} + u') + \frac{\partial}{\partial y}(\bar{\rho} + \bar{\rho}')(\bar{v} + v') + \frac{\partial}{\partial z}(\bar{\rho} + \rho')(\bar{w} + w') = 0$$

Averaging, we obtain

$$\frac{1}{2\Delta t}\int_{t_0-\Delta t}^{t_0+\Delta t} \frac{\partial}{\partial t}(\bar{\rho} + \rho')\,dt + \frac{1}{2\Delta t}\int_{t_0-\Delta t}^{t_0+\Delta t} \frac{\partial}{\partial x}(\bar{\rho} + \rho')(\bar{u} + u')\,dt$$
$$+ \frac{1}{2\Delta t}\int_{t_0-\Delta t}^{t_0+\Delta t} \frac{\partial}{\partial y}(\bar{\rho} + \rho')(\bar{v} + v')\,dt + \frac{1}{2\Delta t}\int_{t_0-\Delta t}^{t_0+\Delta t} \frac{\partial}{\partial z}(\bar{\rho} + \rho')(\bar{w} + w')dt = 0$$

which yields, upon application of the rules of averaging,

$$\frac{\partial}{\partial t}(\langle\bar{\rho} + \rho'\rangle) + \frac{\partial}{\partial x}(\langle\bar{\rho}\bar{u}\rangle + \langle\bar{\rho}u'\rangle + \langle\rho'\bar{u}\rangle + \langle\rho'u'\rangle)$$
$$+ \frac{\partial}{\partial y}(\langle\bar{\rho}\bar{v}\rangle + \langle\bar{\rho}v'\rangle + \langle\rho'\bar{v}\rangle + \langle\rho'v'\rangle)$$
$$+ \frac{\partial}{\partial z}(\langle\bar{\rho}\bar{w}\rangle + \langle\bar{\rho}w'\rangle + \langle\rho'\bar{w}\rangle + \langle\rho'w'\rangle) = 0$$

or

$$\frac{\partial}{\partial t}(\bar{\rho} + \bar{\rho}') + \frac{\partial}{\partial x}(\bar{\rho}\bar{u} + \bar{\rho}\bar{u}' + \bar{\rho}'\bar{u} + \overline{\rho'u'}) + \frac{\partial}{\partial y}(\bar{\rho}\bar{v} + \bar{\rho}\bar{v}' + \bar{\rho}'\bar{v} + \overline{\rho'v'})$$
$$+ \frac{\partial}{\partial z}(\bar{\rho}\bar{w} + \bar{\rho}\bar{w}' + \bar{\rho}'\bar{w} + \overline{\rho'w'}) = 0$$

$$\frac{\partial\bar{\rho}}{\partial t} + \frac{\partial}{\partial x}(\bar{\rho}\bar{u} + \overline{\rho'u'}) + \frac{\partial}{\partial y}(\bar{\rho}\bar{v} + \overline{\rho'v'}) + \frac{\partial}{\partial z}(\bar{\rho}\bar{w} + \overline{\rho'w'}) = 0$$

In fluid dynamics the relative fluctuations of density, $\Delta\rho/\rho$, are usually negligible as compared with the relative fluctuations of the other variables (e.g., $\Delta u/u$, $\Delta v/v$, $\Delta w/w$), and the approximation is generally made that $\rho' \simeq 0$. The last equation above then reduces to

$$\frac{\partial\rho}{\partial t} + \frac{\partial\rho\bar{u}}{\partial x} + \frac{\partial\rho\bar{v}}{\partial y} + \frac{\partial\rho\bar{w}}{\partial z} = 0 \qquad (2\text{-}303)$$

or

$$\frac{\partial\rho}{\partial t} + \nabla \cdot \rho\bar{\mathbf{V}} = 0 \qquad (2\text{-}304)$$

which is the *averaged continuity equation*, having the same form as the point continuity equation except that u, v, and w are replaced by the averages $\bar{u}$, $\bar{v}$, and $\bar{w}$.

Time-Averaged Equation of Motion for Incompressible Flow

For incompressible flow the x-component equation of the fundamental form of the momentum equation of (2-196) can be written

$$\frac{\partial\rho u}{\partial t} + \frac{\partial}{\partial x}(\rho uu) + \frac{\partial}{\partial y}(\rho uv) + \frac{\partial}{\partial z}(\rho uw) - \rho fv$$
$$= -\frac{\partial p}{\partial x} + \mu\left(\frac{\partial^2 u}{\partial x^2} + \frac{\partial^2 u}{\partial y^2} + \frac{\partial^2 u}{\partial z^2}\right) \qquad (2\text{-}305)$$

where the additional Coriolis term $2\rho w\Omega \cos\theta$ has been dropped, as is usual, and where the viscous stress tensor has been replaced as in (2-207a). Again, let each point property be replaced by its average and a fluctuating component as given in (2-302), with the exception that $\rho' \simeq 0$, as was assumed in the derivation of the averaged continuity equation, after which the entire equation is averaged. Thus for the first term,

$$\begin{aligned}\frac{1}{2\Delta t}\int_{t_0-\Delta t}^{t_0+\Delta t}\frac{\partial \rho u}{\partial t}\,dt &= \frac{\partial}{\partial t}\frac{1}{2\Delta t}\int_{t_0-\Delta t}^{t_0+\Delta t}\rho(\bar{u}+u')\,dt \\ &= \frac{\partial}{\partial t}(\langle\rho\bar{u}\rangle + \langle\rho u'\rangle) = \frac{\partial}{\partial t}(\rho\bar{u})\end{aligned} \tag{2-306}$$

Applying this process to the other linear terms of (2-305), essentially similar results are obtained:

$$\frac{1}{2\Delta t}\int_{t_0-\Delta t}^{t_0+\Delta t}\rho f v\,dt = \frac{1}{2\Delta t}\int_{t_0-\Delta t}^{t_0+\Delta t}\rho f(\bar{v}+v')\,dt = \rho f\bar{v} \tag{2-307a}$$

$$\frac{1}{2\Delta t}\int_{t_0-\Delta t}^{t_0+\Delta t}\frac{\partial p}{\partial x}\,dt = \frac{\partial}{\partial x}\frac{1}{2\Delta t}\int_{t_0-\Delta t}^{t_0+\Delta t}(\bar{p}+p')\,dt = \frac{\partial \bar{p}}{\partial x} \tag{2-307b}$$

and

$$\begin{aligned}&\frac{1}{2\Delta t}\int_{t_0-\Delta t}^{t_0+\Delta t}\mu\left(\frac{\partial^2}{\partial x^2}+\frac{\partial^2}{\partial y^2}+\frac{\partial^2}{\partial z^2}\right)u\,dt \\ &\qquad = \mu\left(\frac{\partial^2}{\partial x^2}+\frac{\partial^2}{\partial y^2}+\frac{\partial^2}{\partial z^2}\right)\frac{1}{2\Delta t}\int_{t_0-\Delta t}^{t_0+\Delta t}(\bar{u}+u')\,dt \\ &\qquad = \mu\left(\frac{\partial^2\bar{u}}{\partial x^2}+\frac{\partial^2\bar{u}}{\partial y^2}+\frac{\partial^2\bar{u}}{\partial z^2}\right)\end{aligned} \tag{2-307c}$$

Averaging the nonlinear terms is only slightly more complicated. For example,

$$\begin{aligned}\frac{1}{2\Delta t}\int_{t_0-\Delta t}^{t_0+\Delta t}\frac{\partial}{\partial y}(\rho uv)\,dt &= \frac{\partial}{\partial y}\frac{1}{2\Delta t}\int_{t_0-\Delta t}^{t_0+\Delta t}\rho(\bar{u}+u')(\bar{v}+v')\,dt \\ &= \frac{\partial}{\partial y}(\langle\rho\bar{u}\bar{v}\rangle + \langle\rho\bar{u}v'\rangle + \langle\rho u'\bar{v}\rangle + \langle\rho u'v'\rangle) \\ &= \frac{\partial}{\partial y}(\rho\bar{u}\bar{v} + \rho\overline{u'v'})\end{aligned} \tag{2-307d}$$

Similar relations hold for the other two remaining nonlinear terms of (2-305). Putting all of this together, then, the averaged x-component momentum equation is

$$\begin{aligned}&\frac{\partial}{\partial t}(\rho\bar{u}) + \frac{\partial}{\partial x}(\rho\bar{u}\bar{u} + \rho\overline{u'u'}) + \frac{\partial}{\partial y}(\rho\bar{u}\bar{v} + \rho\overline{u'v'}) \\ &\qquad + \frac{\partial}{\partial z}(\rho\bar{u}\bar{w} + \rho\overline{u'w'}) - \rho f\bar{v} \\ &\qquad\qquad = -\frac{\partial\bar{p}}{\partial x} + \mu\left(\frac{\partial^2\bar{u}}{\partial x^2}+\frac{\partial^2\bar{u}}{\partial y^2}+\frac{\partial^2\bar{u}}{\partial z^2}\right)\end{aligned} \tag{2-308}$$

To put this result in the more frequently used form, the average form of the continuity equation, (2-303), is multiplied by $\bar{u}$,

$$\bar{u}\frac{\partial \rho}{\partial t} + \bar{u}\frac{\partial \rho\bar{u}}{\partial x} + \bar{u}\frac{\partial \rho\bar{v}}{\partial y} + \bar{u}\frac{\partial \rho\bar{w}}{\partial z} = 0$$

and subtracted from (2-308) to give

$$\begin{aligned}\frac{\partial \bar{u}}{\partial t} + \bar{u}\frac{\partial \bar{u}}{\partial x} + \bar{v}\frac{\partial \bar{u}}{\partial y} + \bar{w}\frac{\partial \bar{u}}{\partial z} - f\bar{v} &= -\frac{1}{\rho}\frac{\partial \bar{p}}{\partial x} + \frac{\mu}{\rho}\left(\frac{\partial^2 \bar{u}}{\partial x^2} + \frac{\partial^2 \bar{u}}{\partial y^2} + \frac{\partial^2 \bar{u}}{\partial z^2}\right) \\ &\quad - \frac{1}{\rho}\left[\frac{\partial}{\partial x}(\rho\overline{u'u'}) + \frac{\partial}{\partial y}(\rho\overline{u'v'}) + \frac{\partial}{\partial z}(\rho\overline{u'w'})\right]\end{aligned} \tag{2-309}$$

Averaging the y- and z-component point momentum equations in the same manner results in

$$\begin{aligned}\frac{\partial \bar{v}}{\partial t} + \bar{u}\frac{\partial \bar{v}}{\partial x} + \bar{v}\frac{\partial \bar{v}}{\partial y} + \bar{w}\frac{\partial \bar{v}}{\partial z} + f\bar{u} &= -\frac{1}{\rho}\frac{\partial \bar{p}}{\partial y} + \frac{\mu}{\rho}\left(\frac{\partial^2 \bar{v}}{\partial x^2} + \frac{\partial^2 \bar{v}}{\partial y^2} + \frac{\partial^2 \bar{v}}{\partial z^2}\right) \\ &\quad - \frac{1}{\rho}\left[\frac{\partial}{\partial x}(\rho\overline{u'v'}) + \frac{\partial}{\partial y}(\rho\overline{v'v'}) + \frac{\partial}{\partial z}(\rho\overline{v'w'})\right]\end{aligned} \tag{2-310}$$

and

$$\begin{aligned}\frac{\partial \bar{w}}{\partial t} + \bar{u}\frac{\partial \bar{w}}{\partial x} + \bar{v}\frac{\partial \bar{w}}{\partial y} + \bar{w}\frac{\partial \bar{w}}{\partial z} &= -\frac{1}{\rho}\frac{\partial \bar{p}}{\partial z} + \frac{\mu}{\rho}\left(\frac{\partial^2 \bar{w}}{\partial x^2} + \frac{\partial^2 \bar{w}}{\partial y^2} + \frac{\partial^2 \bar{w}}{\partial z^2}\right) - g \\ &\quad - \frac{1}{\rho}\left[\frac{\partial}{\partial x}(\rho\overline{u'w'}) + \frac{\partial}{\partial y}(\rho\overline{v'w'}) + \frac{\partial}{\partial z}(\rho\overline{w'w'})\right]\end{aligned} \tag{2-311}$$

Combining these three scalar equations into one vector equation yields

$$\frac{D\bar{V}_i}{Dt} + 2\epsilon_{ijk}\Omega_j\bar{V}_k = -\frac{1}{\rho}\frac{\partial \bar{p}}{\partial x_i} + \frac{\mu}{\rho}\frac{\partial^2 \bar{V}_i}{\partial x_j^2} - g\delta_{i3} - \frac{1}{\rho}\frac{\partial}{\partial x_j}(\rho\overline{V_i'V_j'}) \tag{2-312}$$

which is the averaged form of the equation of motion and is called the *Reynolds equation for incompressible, turbulent motion.* This result is seen to have the same form as the incompressible point equation except that average properties now appear in place of point properties and an additional term, called the *eddy* (or *turbulent* or *Reynolds*) *momentum term*, associated with the turbulent velocity fluctuations is added.

Time-Averaged Equations Expressing the Conservation of Salt and of Heat

If the same averaging techniques are applied to the point form of the conservation of salt equation as have just been used on the continuity and momentum equations, it should be possible to obtain a consistent result appropriate to application on a larger than continum scale. Returning then to (2-118),

$$\frac{\partial \rho s}{\partial t} + \nabla \cdot (\rho s \mathbf{V} + \mathbf{F}_s) = 0 \qquad (2\text{-}118)$$

where, as earlier, s is the salinity and $\mathbf{F}_s$ is the nonadvective flux of salt due to diffusion. Rewriting this to

$$\rho\left(\frac{\partial s}{\partial t} + \mathbf{V} \cdot \nabla s\right) + s\left(\frac{\partial \rho}{\partial t} + \nabla \cdot \rho \mathbf{V}\right) + \nabla \cdot \mathbf{F}_s = 0 \qquad (2\text{-}313)$$

the second term in parentheses is the continuity equation and as such is equal to zero. Further, if we assume that the molecular diffusion is negligible, an assumption that will be justified by later results, then (2-313) becomes merely

$$\frac{\partial s}{\partial t} + u\frac{\partial s}{\partial x} + v\frac{\partial s}{\partial y} + w\frac{\partial s}{\partial z} = 0 \qquad (2\text{-}314)$$

Substitution of the average and fluctuating quantities of (2-302) and time-averaging the entire equation gives for the first term

$$\frac{1}{2\Delta t}\int_{t_0-\Delta t}^{t_0+\Delta t} \frac{\partial s}{\partial t}\, dt = \frac{\partial}{\partial t}\frac{1}{2\Delta t}\int_{t_0-\Delta t}^{t_0+\Delta t} (\bar{s} + s')\, dt = \frac{\partial \bar{s}}{\partial t}$$

for the second term

$$\frac{1}{2\Delta t}\int_{t_0-\Delta t}^{t_0+\Delta t} u\frac{\partial s}{\partial x}\, dt = \frac{1}{2\Delta t}\int_{t_0-\Delta t}^{t_0+\Delta t} (\bar{u} + u')\frac{\partial}{\partial x}(\bar{s} + s')\, dt$$

$$= \frac{1}{2\Delta t}\int_{t_0-\Delta t}^{t_0+\Delta t} \left(\bar{u}\frac{\partial \bar{s}}{\partial x} + \bar{u}\frac{\partial s'}{\partial x} + u'\frac{\partial \bar{s}}{\partial x} + u'\frac{\partial s'}{\partial x}\right) dt$$

$$= \left\langle \bar{u}\frac{\partial \bar{s}}{\partial x}\right\rangle + \left\langle \bar{u}\frac{\partial s'}{\partial x}\right\rangle + \left\langle u'\frac{\partial \bar{s}}{\partial x}\right\rangle + \left\langle u'\frac{\partial s'}{\partial x}\right\rangle = \bar{u}\frac{\partial \bar{s}}{\partial x} + \overline{u'\frac{\partial s'}{\partial x}}$$

and relations similar to the last equation for the remaining terms. Putting it all together, we obtain

$$\left(\frac{\partial \bar{s}}{\partial t} + \bar{u}\frac{\partial \bar{s}}{\partial x} + \bar{v}\frac{\partial \bar{s}}{\partial y} + \bar{w}\frac{\partial \bar{s}}{\partial z}\right) + \left(\overline{u'\frac{\partial s'}{\partial x}} + \overline{v'\frac{\partial s'}{\partial y}} + \overline{w'\frac{\partial s'}{\partial z}}\right) = 0 \qquad (2\text{-}315)$$

Now, for incompressible flow

$$s'\left(\frac{\partial u}{\partial x} + \frac{\partial v}{\partial y} + \frac{\partial w}{\partial z}\right) = 0$$

or

$$\frac{1}{2\Delta t}\int_{t_0-\Delta t}^{t_0+\Delta t}\left[s'\frac{\partial}{\partial x}(\bar{u}+u')+s'\frac{\partial}{\partial y}(\bar{v}+v')+s'\frac{\partial}{\partial z}(\bar{w}+w')\right]dt=0$$

$$\left\langle s'\frac{\partial \bar{u}}{\partial x}\right\rangle+\left\langle s'\frac{\partial u'}{\partial x}\right\rangle+\left\langle s'\frac{\partial \bar{v}}{\partial y}\right\rangle+\left\langle s'\frac{\partial v'}{\partial y}\right\rangle+\left\langle s'\frac{\partial \bar{w}}{\partial z}\right\rangle+\left\langle s'\frac{\partial w'}{\partial z}\right\rangle=0$$

$$\overline{s'\frac{\partial u'}{\partial x}}+\overline{s'\frac{\partial v'}{\partial y}}+\overline{s'\frac{\partial w'}{\partial z}}=0$$

Adding this zero quantity to (2-315), it follows that for incompressible flow

$$\frac{\partial \bar{s}}{\partial t}+\bar{u}\frac{\partial \bar{s}}{\partial x}+\bar{v}\frac{\partial \bar{s}}{\partial y}+\bar{w}\frac{\partial \bar{s}}{\partial z}=-\left[\frac{\partial}{\partial x}(\overline{s'u'})+\frac{\partial}{\partial y}(\overline{s'v'})+\frac{\partial}{\partial z}(\overline{s'w'})\right] \tag{2-316}$$

and that again the average form is similar to the point form of an equation [compare (2-314) with (2-316)] except that the variables are expressed as average quantities and the additional fluctuation terms appear. These last three terms are the *eddy* (or *turbulent* or *Reynolds*) *diffusivity terms* and arise from the nonlinear advective quantities in the original point form of the equation.

Starting with the distribution of variables equation as applied to the temperature of the fluid, (2-215), and assuming that molecular conductivity is negligible, we have that

$$\frac{\partial T}{\partial t}+u\frac{\partial T}{\partial x}+v\frac{\partial T}{\partial y}+w\frac{\partial T}{\partial z}=0 \tag{2-317}$$

The similarity between (2-314) and (2-317) is obvious and permits us to write an average version from direct analogy to (2-316). Thus,

$$\frac{\partial \bar{T}}{\partial t}+\bar{u}\frac{\partial \bar{T}}{\partial x}+\bar{v}\frac{\partial \bar{T}}{\partial y}+\bar{w}\frac{\partial \bar{T}}{\partial z} =-\left[\frac{\partial}{\partial x}(\overline{T'u'})+\frac{\partial}{\partial y}(\overline{T'v'})+\frac{\partial}{\partial z}(\overline{T'w'})\right] \tag{2-318}$$

where the terms on the right of the equation are the *eddy* (or *turbulent* or *Reynolds*) *conductivity terms*. The equation as a whole merely expresses the conservation of heat for turbulent flow in the absence of radiation and when neglecting molecular conductivity.

2-33 Eddy Coefficients

Reyonolds Stress and the Eddy Viscosity

The last quantity in parentheses in the Reynolds equation of (2-312), $-\rho\overline{V_i'V_j'}$, is called the *Reynolds stress tensor* (in honor of the original developer of the theory) and will be deisgnated by

$$\tau_{ij}^r \equiv -\rho\overline{V_i'V_j'} \tag{2-319}$$

τ^r_{ij} is clearly a stress, having units of force per unit area, and is a symmetric tensor with a total of nine components, only six of which are independent,

$$\tau^r_{ij} = -\begin{bmatrix} \rho\overline{V'_1V'_1} & \rho\overline{V'_1V'_2} & \rho\overline{V'_1V'_3} \\ \rho\overline{V'_2V'_1} & \rho\overline{V'_2V'_2} & \rho\overline{V'_2V'_3} \\ \rho\overline{V'_3V'_1} & \rho\overline{V'_3V'_2} & \rho\overline{V'_3V'_3} \end{bmatrix}$$

Alternatively, in terms of an x, y, z coordinate system,

$$\tau^r_{ij} = -\begin{bmatrix} \rho\overline{u'u'} & \rho\overline{u'v'} & \rho\overline{u'w'} \\ \rho\overline{v'u'} & \rho\overline{v'v'} & \rho\overline{v'w'} \\ \rho\overline{w'u'} & \rho\overline{w'v'} & \rho\overline{w'w'} \end{bmatrix}$$

The three diagonal components of τ^r_{ij} are *normal stresses* and the off-diagonal ones are *shear stresses*. The Reynolds stresses can be considered as a component of a *total mean stress tensor*, S_{ij}. For laminar flow the viscous stress tensor τ_{ij} is given by (2-205), which for incompressible flow reduces to

$$\tau_{ij} = \mu\left(\frac{\partial V_i}{\partial x_j} + \frac{\partial V_j}{\partial x_i}\right)$$

After substitution of mean and fluctuating quantities and time averaging, τ_{ij} becomes the *mean viscous stress tensor*, $\bar{\tau}_{ij}$,

$$\bar{\tau}_{ij} = \mu\left(\frac{\partial \bar{V}_i}{\partial x_j} + \frac{\partial \bar{V}_j}{\partial x_i}\right) \tag{2-320}$$

The total mean stress can now be defined as the sum of the Reynolds stress and the mean viscous stress,

$$S_{ij} \equiv \bar{\tau}_{ij} + \tau^r_{ij} \tag{2-321}$$

or

$$S_{ij} = \mu\left(\frac{\partial \bar{V}_i}{\partial x_j} + \frac{\partial \bar{V}_j}{\partial x_i}\right) - \rho\overline{V'_iV'_j}$$

A better understanding of the physical meaning of the stress components τ^r_{ij} is obtained from a consideration of the simple flow situation shown in Figure 2-68. The stress component $\tau^r_{yx} = \tau^r_{xy} = -\rho\overline{u'v'}$ can be interpreted as the transport of x momentum through a surface normal to the y axis. In the case of

$$\bar{u} = \bar{u}(y) \qquad \bar{v} = \bar{w} = 0 \qquad \frac{d\bar{u}}{dy} > 0$$

it can be shown that $\overline{u'v'} \neq 0$; fluid particles with a turbulent fluctuation $v' > 0$ will travel upward and arrive at a layer $y + \Delta y$ from a region where a smaller mean velocity $\bar{u}$ prevails. In general, they will retain the velocity of their origin $\bar{u}$ and will thus give rise to a negative fluctuation component u' in the layer $y + \Delta y$. Conversely, particles of fluid with a turbulent fluctuation

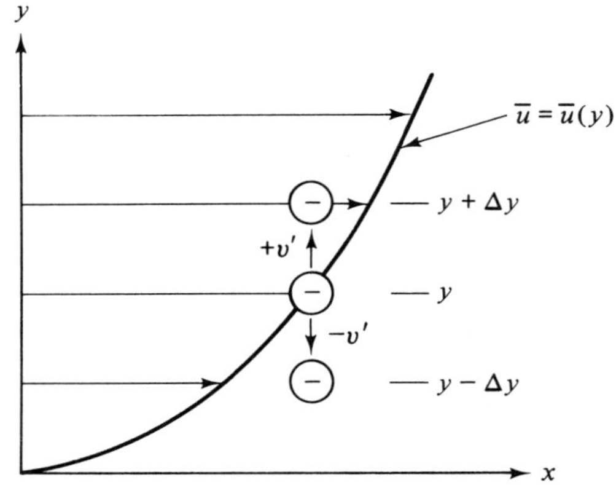

Fig. 2-68. Transport of x momentum in the y direction due to turbulent velocity fluctuations.

$v' < 0$ will travel downward and arrive at a layer $y - \Delta y$ and thus give rise to a positive fluctuation component u' in the layer $y - \Delta y$. On the average, then, a positive v' is usually associated with a negative u' and a negative v' is usually associated with a positive u'. Thus, it is to be expected that the time-averaged quantity $\overline{u'v'}$ will not only be different from zero but will, in fact, be negative. The Reynolds shearing stress $\tau^r_{yx} = -\rho\overline{u'v'}$ will be positive for this example, as is the appropriate laminar shearing stress $\mu(du/dy)$.

The introduction of the Reynolds stresses provides considerable insight into the general mechanism of turbulence. However, the mathematical difficulties still remain in that no universal, analytic method is known whereby these stresses can be expressed in terms of the mean velocities and their derivatives. Most of the progress made in relating theory with observation involves the introduction of certain plausible but empirical relations between the turbulent stresses and the mean flow. A rational theory does not exist. It is possible, in practice, to actually measure the velocity at a fixed point and to calculate the fluctuations from the mean and then determine the Reynolds stresses by time averaging. For example, Reichardt (1938) determined, in the laboratory, $-\overline{u'v'}$, S_{xy}/ρ, and $r_{u,v}$ (the correlation coefficient) for flow in a rectangular channel. Figure 2-69 shows the plot of $-\overline{u'v'}$, which is equal to τ^r_{xy}/ρ, the turbulent shearing stress divided by the density. Its value goes to zero at the center of the channel due to symmetry, whereas its maximum occurs near the wall, showing that turbulent friction has its largest value there. The dashed line S_{xy}/ρ shows the variation of the total shearing stress across the channel which was obtained from the measured pressure distribution and independently of the measured velocity. The difference between these two curves gives the ratio of the laminar stress to the density $\bar{\tau}_{xy}/\rho$. Over most of the section the two curves nearly coincide, showing that almost all of the shearing stress is due to turbulence. They diverge near

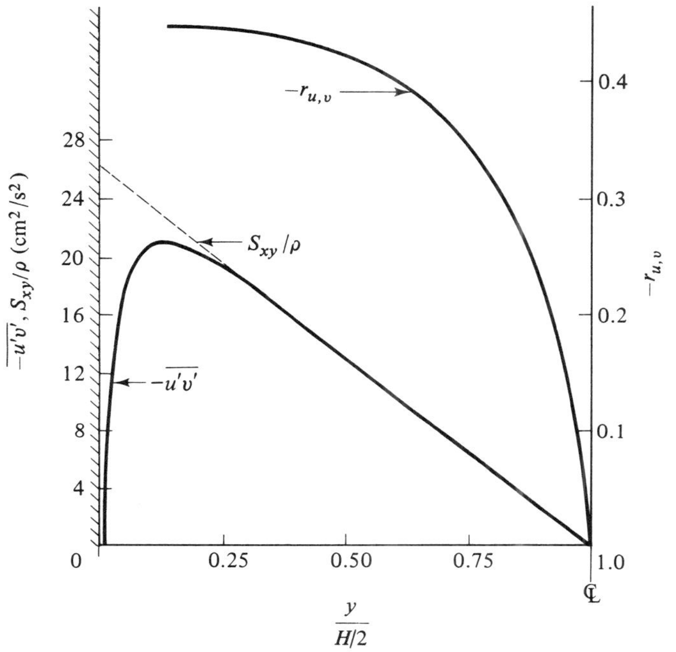

Fig. 2-69. Measurement of the Reynolds stress ($-\overline{u'v'}$), total stress (S_{xy}/ρ), and correlation coefficient ($r_{u,v}$) for flow in a rectangular channel, after Reichardt, 1938.

the boundary, $-\overline{u'v'}$ decreasing to zero, owing to fact that the turbulent fluctuations die out near the wall.

Now several semiempirical relations have been widely used to relate the turbulent stresses with the mean flow: Boussinesq's eddy viscosity theory, Prandtl's mixing-length theory, Taylor's vorticity-transport theory, von Kármán's similarity hypothesis. Only the first of these will concern us. In turbulence, two coefficients of viscosity can be defined (ignoring the bulk or second coefficient of viscosity, μ', since it does not appear in the incompressible form of the equations): one real and the other apparent. First, we have the molecular viscosity μ; second, we have the molar coefficient. By analogy with the form of Newton's viscosity relation, Boussinesq in 1877 introduced the concept of relating the turbulent stress to the shear strain rate. By analogy, the Reynolds stress becomes

$$\tau_{ij}^r \propto \frac{\partial \bar{V}_i}{\partial x_j}$$

or

$$\tau_{ij}^r = A \frac{\partial \bar{V}_i}{\partial x_j} \tag{2-322}$$

where A, the coefficient of proportionality, is called the *turbulent coefficient of viscosity* or the *eddy viscosity*, analogous to, but crucially different from, the molecular viscosity. The molecular viscosity is a property of the fluid and usually varies slowly with the state parameters of the fluid; the eddy viscosity is not a physical constant of the fluid but rather depends upon the type and scale of the motion considered, its degree of stratification, and interactions between them, and often varies considerably from one part of the fluid to another. It assumes different values for the same location at the same time when the time scale of the averaging process is changed. There is no known simple connection between the magnitude of A and any of these factors. The best that can be done is to make measurements for a number of different situations, estimate the value of the coefficient for these situations, and then hope that we can identify similar classes of motion to which we can attach an eddy momentum exchange coefficient that is at least of the proper order of magnitude. Application to a flow of such empirically determined results helps to get at important features of the motion and permits analytical and numerical solution of some of the forms of the equation of motion. This approach is, of course, merely a pragmatic one; it does not represent a real increase in our knowledge of the physical processes of turbulence. All that can be said with some certainty about the coefficient of eddy viscosity is that:

1. It's magnitude increases with the scale of the phenomenon under investigation,
2. It's magnitude is greatest where velocity shears are greatest.
3. It's magnitude is less for momentum exchanges in the vertical as compared with the horizontal due to the vertical stability of the ocean.

In recognition of these conclusions it is desirable to modify (2-322) and take different eddy exchange coefficients for each of the coordinate directions, this being especially true for the vertical. Thus, using (2-319), we assert that the eddy momentum term of (2-312) can be written as

$$\begin{aligned} -\frac{\partial}{\partial x_j}(\rho\overline{V_i'V_j'}) &= \frac{\partial}{\partial x_j}\left(A_{x_j}\frac{\partial\bar{V}_i}{\partial x_j}\right) \\ &= \frac{\partial}{\partial x_1}\left(A_{x_1}\frac{\partial\bar{V}_i}{\partial x_1}\right) + \frac{\partial}{\partial x_2}\left(A_{x_2}\frac{\partial\bar{V}_i}{\partial x_2}\right) + \frac{\partial}{\partial x_3}\left(A_{x_3}\frac{\partial\bar{V}_i}{\partial x_3}\right) \end{aligned} \tag{2-323}$$

or for situations in which we may wish only to distinguish between the vertical and the horizontal exchanges,

$$-\frac{\partial}{\partial x_j}(\rho\overline{V_i'V_j'}) = \frac{\partial}{\partial x_j}\left(A_H\frac{\partial\bar{V}_i}{\partial x_j}\right)_{j=1,2} + \frac{\partial}{\partial x_3}\left(A_V\frac{\partial V_i}{\partial x_3}\right) \tag{2-324}$$

it being understood that the x_3 axis is vertical.

As given in (2-322), the eddy viscosity coefficient, A, has units of [mass][length]$^{-1}$[time]$^{-1}$ or, more physically meaningful, of {[mass][length]/[time]} [length]$^{-2}$ (momentum per unit area), exactly those of the dynamic molecular viscosity, μ. Analogous to the kinematic molecular viscosity defined as $\nu = \mu/\rho$, a *kinematic eddy viscosity*,

$$K_{x_j} \equiv \frac{A_{x_j}}{\rho} \tag{2-325}$$

has been defined and called the *austauschkoeffizient* (*austausch* or *exchange coefficient*) by Wilhelm Schmidt (1917), a pioneer worker in this field. This form of the momentum exchange coefficient has the same units as ν, [length]^2 [time]$^{-1}$.

In sections that follow we will examine several different flow situations for which, by appropriate simplifications, we are able to calculate values of the viscosity coefficients. We will see that for geophysical scales, the coefficients may be of order 10^2 g/cm/sec for A_V and 10^6 g/cm/sec for A_H in oceanographic problems (see, for example, Sverdrup et al., 1942; Defant, 1961; Kamenkovich, 1977). These are to be compared with the order of magnitude of 10^{-2} g/cm/sec for the molecular viscosity. For turbulent flow then the Reynolds stresses and the turbulent friction terms will, in general, be orders of magnitude greater than the mean viscous stresses and their associated forces. For turbulent flow the viscous forces can be ignored with negligible error and the time-averaged equations of motion, equations (2-309) through (2-311), can be written after application of (2-324) as:

$$\begin{aligned}&\frac{\partial \bar{u}}{\partial t} + \bar{u}\frac{\partial \bar{u}}{\partial x} + \bar{v}\frac{\partial \bar{u}}{\partial y} + \bar{w}\frac{\partial \bar{u}}{\partial z} - f\bar{v}\\ &\quad = -\frac{1}{\rho}\frac{\partial \bar{p}}{\partial x} + \frac{1}{\rho}\left[\frac{\partial}{\partial x}\left(A_H\frac{\partial \bar{u}}{\partial x}\right) + \frac{\partial}{\partial y}\left(A_H\frac{\partial \bar{u}}{\partial y}\right) + \frac{\partial}{\partial z}\left(A_V\frac{\partial \bar{u}}{\partial z}\right)\right]\end{aligned} \tag{2-326}$$

$$\begin{aligned}&\frac{\partial \bar{v}}{\partial t} + \bar{u}\frac{\partial \bar{v}}{\partial x} + \bar{v}\frac{\partial \bar{v}}{\partial y} + \bar{w}\frac{\partial \bar{v}}{\partial z} + f\bar{u}\\ &\quad = -\frac{1}{\rho}\frac{\partial \bar{p}}{\partial y} + \frac{1}{\rho}\left[\frac{\partial}{\partial x}\left(A_H\frac{\partial \bar{v}}{\partial x}\right) + \frac{\partial}{\partial y}\left(A_H\frac{\partial \bar{v}}{\partial y}\right) + \frac{\partial}{\partial z}\left(A_V\frac{\partial \bar{v}}{\partial z}\right)\right]\end{aligned} \tag{2-327}$$

and

$$\begin{aligned}&\frac{\partial \bar{w}}{\partial t} + \bar{u}\frac{\partial \bar{w}}{\partial x} + \bar{v}\frac{\partial \bar{w}}{\partial y} + \bar{w}\frac{\partial \bar{w}}{\partial z}\\ &\quad = -\frac{1}{\rho}\frac{\partial \bar{p}}{\partial z} - g + \frac{1}{\rho}\left[\frac{\partial}{\partial x}\left(A_H\frac{\partial \bar{w}}{\partial x}\right) + \frac{\partial}{\partial y}\left(A_H\frac{\partial \bar{w}}{\partial y}\right) + \frac{\partial}{\partial z}\left(A_V\frac{\partial \bar{w}}{\partial z}\right)\right]\end{aligned} \tag{2-328}$$

Alternatively, combining these into one vector equation,

$$\frac{D\bar{V}_i}{Dt} + 2\epsilon_{ijk}\Omega_j\bar{V}_k = -\frac{1}{\rho}\frac{\partial \bar{p}}{\partial x_i} - g\delta_{i3} + \frac{1}{\rho}\frac{\partial}{\partial x_j}\left(A_{x_j}\frac{\partial \bar{V}_i}{\partial x_j}\right) \tag{2-329}$$

Eddy Diffusivity and Eddy Conductivity Coefficients for Incompressible Flow

The eddy diffusity terms of the time-averaged form of the conservation of salt equation for incompressible flow in (2-316) have their origin in the nonlinear advective terms of (2-314). Similar to the eddy viscosity quantities, they depend upon the type and scale of the flow and are not a physical constant of the fluid. Additionally, they depend upon the distribution of salinity within the flow. When multiplied by the density ρ, terms such as $-\overline{s'u'}$, $-\overline{s'v'}$, and $-\overline{s'w'}$ represent the flux of salt across a boundary just as the eddy momentum term of (2-312) represents a flux of momentum across a boundary. It is possible to obtain estimates of this salt flux due to turbulence through a time-series measurement of s and V_i; this has at times been done. As a matter of convenience, however, the same technique is frequently applied for the turbulent diffusion of salt as was applied for the turbulent diffusion of momentum, specifically, that of characterizing the process through the application of an eddy coefficient. Thus, it is assumed that

$$\overline{s'V_i'} \propto -\frac{\partial \bar{s}}{\partial x_i}$$

or

$$\overline{s'V_i'} = -K_s(x_i)\frac{\partial \bar{s}}{\partial x_i} \tag{2-330}$$

where $K_s(x_i)$ is the (kinematic) *eddy diffusivity coefficient for salt* having the components $K_s(x_1)$, $K_s(x_2)$, and $K_s(x_3)$ for eddy diffusion in the three coordinate directions with units of $[\text{length}]^2[\text{time}]^{-1}$. Similar to K_{x_i}, the austausch coefficient, $K_s(x_i)$, is a function of the type and scale of the motion as well as the distribution of s. Because of the inherent stability of oceanic stratification, it is usually only possible to distinguish between vertical and horizontal exchanges such that

$$\begin{aligned}
\overline{s'u'} &= -K_s(H)\frac{\partial \bar{s}}{\partial x}\\
\overline{s'v'} &= -K_s(H)\frac{\partial \bar{s}}{\partial y}\\
\overline{s'w'} &= -K_s(V)\frac{\partial \bar{s}}{\partial z}
\end{aligned} \tag{2-331}$$

Substituting these into (2-316), the time-averaged conservation of salt equation for incompressible flow becomes

$$\begin{aligned}
&\frac{\partial \bar{s}}{\partial t} + \bar{u}\frac{\partial \bar{s}}{\partial x} + \bar{v}\frac{\partial \bar{s}}{\partial y} + \bar{w}\frac{\partial \bar{s}}{\partial z}\\
&\qquad = \frac{\partial}{\partial x}\left(K_s(H)\frac{\partial \bar{s}}{\partial x}\right) + \frac{\partial}{\partial y}\left(K_s(H)\frac{\partial \bar{s}}{\partial y}\right) + \frac{\partial}{\partial z}\left(K_s(V)\frac{\partial \bar{s}}{\partial z}\right)
\end{aligned} \tag{2-332}$$

or, in indicial notation,

$$\frac{\partial \bar{s}}{\partial t} + \bar{V}_i \frac{\partial \bar{s}}{\partial x_i} = \frac{\partial}{\partial x_i}\left[K_s(x_i)\frac{\partial \bar{s}}{\partial x_i}\right] \tag{2-333}$$

Estimates of the saline eddy diffusion coefficient are fairly numerous (see, for example, Sverdrup et al, 1942; Defant, 1961; Kamenkovich, 1977), and unfortunately, fairly diverse. Lateral coefficients generally are of the order of 10^5 or 10^6 cm^2/sec, while vertical coefficients appear to be of order 10^{-1} cm^2/sec. On the average, then, for the open ocean, and on a geophysical scale, a ratio of $K_s(V)/K_s(H)$ of magnitude 10^{-6} to 10^{-7} appears to be a fair estimate. On the other hand, the molecular Fickian diffusion coefficient is of order 10^{-5} cm^2/sec, justifying the assumption of the negligibility of the molecular process in the derivation of (2-316).

Turning now to the time-averaged form of the distribution of variables equation when applied to the temperature, equation (2-318), we face the same problem of representing the eddy conductivity terms which we have just faced in the representation of the eddy momentum and eddy diffusion terms of (2-312) and (2-316), respectively. Again, these quantities depend upon the type and scale of the flow and the distribution of the variable (in this case the temperature). Quantities such as $-\overline{T'u'}$, $-\overline{T'v'}$, and $-\overline{T'w'}$, when multiplied by ρc, where c is the specific heat, will have units of energy flux (energy per unit area per unit time) and represent the flux of energy due to conductivity in turbulent flow. This is exactly analogous to quantities such as $\rho\overline{s'u'}$, $\rho\overline{s'v'}$, and $\rho\overline{s'w'}$, representing the turbulent flux of salt and $\rho\overline{u'u'}$, $\rho\overline{u'v'}$, $\rho\overline{u'w'}$, and so on, representing the turbulent flux of momentum. Again it is possible to obtain estimates of the heat flux through a time-series measurement of T and V_i, but, in general, resort is taken in the introduction of a plausible but empirical relation between the turbulent conductivity and the mean temperature gradient, namely

$$\overline{T'V_i'} \propto -\frac{\partial \bar{T}}{\partial x_i}$$

or

$$\overline{T'V_i'} = -K_T(x_i)\frac{\partial \bar{T}}{\partial x_i} \tag{2-334}$$

where $K_T(x_i)$ is the (kinematic) *eddy conductivity coefficient* having components $K_T(x_1)$, $K_T(x_2)$, and $K_T(x_3)$ in the three coordinate directions and, like the eddy diffusivity coefficient, having units of [length]^2[time]$^{-1}$. As in the case for the other eddy coefficients we have examined, it may be impossible to do more than to distinguish between the horizontal and the vertical exchanges, so that similar to (2-331),

$$\overline{T'u'} = -K_T(H)\frac{\partial \bar{T}}{\partial x}$$
$$\overline{T'v'} = -K_T(H)\frac{\partial \bar{T}}{\partial y} \qquad (2\text{-}335)$$
$$\overline{T'w'} = -K_T(V)\frac{\partial T}{\partial z}$$

Substitution of these into (2-318) yields a practical form for the time-averaged conservation of heat equation,

$$\frac{\partial \bar{T}}{\partial t} + \bar{u}\frac{\partial \bar{T}}{\partial x} + \bar{v}\frac{\partial \bar{T}}{\partial y} + \bar{w}\frac{\partial \bar{T}}{\partial z} = \frac{\partial}{\partial x}\left(K_T(H)\frac{\partial \bar{T}}{\partial x}\right) + \frac{\partial}{\partial y}\left(K_T(H)\frac{\partial \bar{T}}{\partial y}\right) + \frac{\partial}{\partial z}\left(K_T(V)\frac{\partial \bar{T}}{\partial z}\right) \qquad (2\text{-}336)$$

or, in indicial form,

$$\frac{\partial \bar{T}}{\partial t} + \bar{V}_i\frac{\partial \bar{T}}{\partial x_i} = \frac{\partial}{\partial x_i}\left[K_T(x_i)\frac{\partial \bar{T}}{\partial x_i}\right] \qquad (2\text{-}337)$$

The order of magnitude of the eddy conductivity coefficients is about the same as that for the eddy diffusivity coefficients, that is, $K_T(H)$ of order 10^5 to 10^6 cm^2/sec, $K_T(V)/K_T(H)$ of order 10^{-6} to 10^{-7} cm^2/sec (see, for example, Sverdrup et al, 1942; Defant, 1961; Kamenkovich, 1977). These can be compared to the molecular equivalent (laminar flow) of order 10^{-3} cm^2/sec and again we see the justification for dropping the molecular conductivity term in the derivation of (2-318).

2-34 Differences Between the Eddy Coefficients

It might appear that the eddy exchange coefficients for momentum, salt, and heat should be the same, since all of them are properties of the exchanged mass. Certainly, they have many things in common: they are all defined in a similar empirical fashion, they all are a function of the type and scale of flow, they all depend upon the distribution of the property, and they behave in the same way with respect to vertical versus horizontal transport due to the stability of oceanic stratification. We will see later that it may be useful to assume that two of them, salt diffusion and conductivity of heat, in special instances are equal. However, all three can be equal only in the case of the immediate and complete mixing of all the properties. It is possible through qualitative reasoning to show that, in general, momentum is the more readily transferrable of the three. When fluid elements change position, the change may be only temporary. For such a temporary change, the macroscopic element being displaced by the turbulence from its mean position will, because of collision, immediately exchange momentum with the fluid elements in its new surroundings. At the same time, some salt and heat exchange may occur, but these exchanges take place by the much slower process of molecular diffusion across the boundaries of the element. Thus, the larger elements rapidly exchange momentum in the new locality, but only slowly exchange salt or heat, since the surface area to mass ratio is small. A water parcel in random motion may exchange a large fraction of its momentum but only a small fraction of its salt or heat. This argument is borne out by actual flow measurements, which yield eddy viscosity coefficients larger than the coefficients of salt and heat.

2-35 Estimating the Eddy Coefficients

Putting the Reynolds stresses, the eddy diffusivity terms, and the eddy conductivity terms in a form analogous to that of the viscous stresses, as in (2-325), (2-330), and (2-334), respectively, permits (after additional simplifying assumptions) analytical solutions to be found for (2-329), (2-333), and (2-336). In fact, for the case where the coefficients are constant, the solutions are identical in form to those determined for viscous and/or laminar flow. To obtain numbers from these solutions, the magnitude of the various coefficients must be estimated through empirical methods. A general procedure for obtaining such estimates is: (1) reduce the complexity of (2-329), (2-333), or (2-336) so as to model some relatively simple process typical of a particular situation in the ocean; (2) solve the resulting equations assuming that the coefficients are either constants or of some mathematically tractible form; (3) measure the field variables, $\bar{V}_i$, $\bar{s}$, and $\bar{T}$, predicted by step 2; and (4) determine values of the eddy coefficients that will best fit the predicted values of the field variables to the observed values of the field variables. Naturally, such empirically derived coefficients may not be applicable under environmental conditions different from those under which the field measurements were made, but they will at least furnish a guide to the acceptable range of magnitude of the coefficients in dynamically similar situations.

In most of the remaining portion of this chapter we will examine a number of examples of the evaluation of eddy coefficients; the examples are not meant to be all-inclusive. In particular, attention should be directed to the way in which the magnitude of these coefficients depends upon the direction of the eddy flux (smaller across horizontal surfaces than vertical surfaces), upon the property exchanged (larger for momentum transfer than either salt or heat), and upon the scale of the motion (increase with scale) or the distribution of the variable.

Estimating the Eddy Viscosity for the Vertical Exchange of Momentum

One simple model that utilizes a vertical coefficient of eddy viscosity is that of Ekman wind drift. Earlier, in (2-280), we defined the depth of frictional influence, H, as

$$H \equiv \left(\frac{2\nu}{f}\right)^{1/2} \tag{2-280}$$

where ν is the kinematic molecular viscosity. The problem, as originally posed in Section 2-31, considered laminar flow. For the more general case of turbulent flow, the results are identical except that the vertical austausch coefficient, K_V, replaces ν wherever the latter appears. Thus, the depth of frictional influence is now given by

$$H \equiv \left(\frac{2K_V}{f}\right)^{1/2} = \left(\frac{K_V}{\Omega \sin \theta}\right)^{1/2} \tag{2-338}$$

Various investigators have empirically related H to the surface wind speed: for example, Thorade (1914) postulated

$$H(\text{meters}) = \frac{7.6\ W\,(\text{m/sec})}{\pi\sqrt{\sin \theta}}$$

for wind speeds, W, in excess of 6 m/sec and

$$H(\text{meters}) = \frac{3.67\ [W(\text{m/sec})]^{3/2}}{\pi\sqrt{\sin \theta}}$$

for winds speeds less than 6 m/sec. Assuming that these expressions are reasonably accurate, the austausch coefficient can be estimated by substitution of (2-338) to obtain $K_V(\text{cm}^2/\text{sec}) \simeq 4.3\ [W(\text{m/sec})]^2$ for $W > 6$ m/sec, $K_V(\text{cm}^2/\text{sec}) \simeq [W(\text{m/sec})]^3$ for $W < 6$ m/sec. Admittedly, these results can only be considered as crude and applicable only in the surface layers of the ocean, but they do provide a measure of K_V. For example, if $W = 3$ m/sec, $K_V \simeq 27$ cm²/sec; if $W = 10$ m/sec, $K_V \simeq 430$ cm²/sec. Durst (1924), based on wind speeds from 7.5 to 9.8 m/sec, made some calculations for various regions of the three oceans with results ranging from about 250 to 1300 cm²/sec for K_V.

Another approach is based on the assumptions that conditions are stationary in time but vary both horizontally and vertically, and that both lateral friction and the nonlinear terms of the acceleration are negligible. Consider the small parallelopiped of fluid bounded by two horizontal surfaces a distance Δz apart in the vertical, two vertical planes perpendicular to the current a distance apart Δx, and two vertical planes parallel to the current and a distance b from each other (Figure 2-70). The horizontal forces acting on the fluid inside the parallelopiped must balance since there is no acceleration.

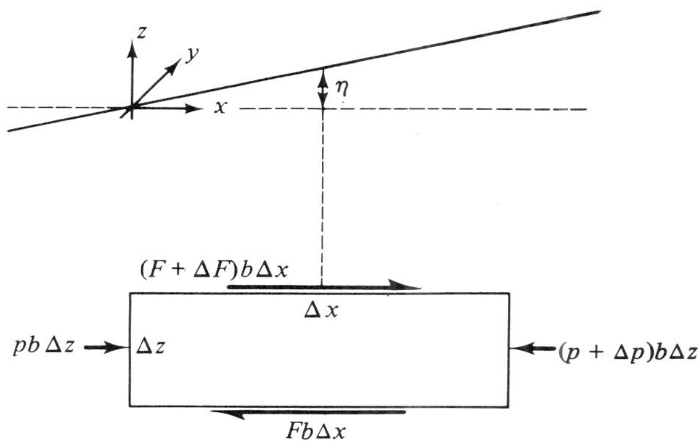

Fig. 2-70. Horizontal forces acting on fluid inside a parallelopiped.

Thus,

$$-b\,\Delta z\,\Delta p + b\,\Delta x\,\Delta F = 0$$

where ΔF is the difference in the horizontal stress between the top and bottom horizontal surfaces. In the limit,

$$-\frac{\partial p}{\partial x} + \frac{\partial F}{\partial z} = 0 \qquad (2\text{-}339)$$

The pressure at any depth z' is given by equation (2-268),

$$\frac{\partial p_{z'}}{\partial x} = \rho_0 g \frac{d\eta}{dx} + g \int_{z'}^{0} \frac{\partial \rho}{\partial x}\, dz \qquad (2\text{-}268)$$

where ρ_0 is the density at the surface. Substituting from (2-339), we obtain

$$-\frac{\partial F_{z'}}{\partial z'} + \rho_0 g \frac{d\eta}{\partial x} + g \int_{z'}^{0} \frac{\partial \rho}{\partial x} dz = 0$$

and integrating from $z' = z''$ up to $z' = 0$ yields

$$-\int_{z''}^{0} \frac{\partial F_{z'}}{\partial z'}\, dz' + \rho_0 g \frac{\partial \eta}{\partial x} \int_{z''}^{0} dz' + g \int_{z''}^{0} \int_{z'}^{0} \frac{\partial \rho}{\partial x}\, dz\, dz' = 0$$

$$-F_{z''} + F_0 + \rho_0 g \frac{d\eta}{dx} z'' - g \int_{0}^{z''} \int_{0}^{z'} \frac{\partial \rho}{\partial x}\, dz\, dz' = 0$$

where F_0 is the surface value of F_z and z'' is any depth. Taking $F_0 = 0$ and substituting the eddy stress for $F_{z''}$, we get, utilizing (2-322),

$$\left(\rho K_V \frac{\partial \bar{u}}{\partial z}\right)_{z''} = \rho_0 g \frac{d\eta}{\partial x} z'' - g \int_{0}^{z''} \int_{0}^{z'} \frac{\partial \rho}{\partial x}\, dz\, dz' \qquad (2\text{-}340)$$

Jacobsen (1913) applied this result to the mean conditions of the Kattegat to obtain the values of K_V shown in Table 2-1. The first, second, and fourth

TABLE 2-1. Example of the Estimation of the Eddy Viscosity for the Vertical Exchange of Momentum Via Equation (2-340)

z, z', z'' (m)	$\bar{u}$ (cm/sec)	$\frac{1}{\partial \bar{u}/\partial z}$ (sec)	$\frac{\partial \rho}{\partial x} \times 10^{10}$ (g/cm⁴)	$\frac{\partial \rho}{\partial x}\delta z \times 10^{8}$ (g/cm³)	$\int_0^{z'} \frac{\partial \rho}{\partial x} dz \times 10^{8}$ (g/cm³)	$\int_0^{z'} \frac{\partial \rho}{\partial x} dz\, \delta z' \times 10^{5}$ (g/cm²)	$\int_0^{z''}\int_0^{z'} \frac{\partial \rho}{\partial x} dz\, dz' \times 10^{5}$ (g/cm²)	$\rho \frac{\partial \eta}{\partial x} z'' \times 10^{5}$ (g/cm²)	$\frac{\rho}{g} K_V \frac{\partial \bar{u}}{\partial z} \times 10^{2}$ (g/cm²)	K_V (cm²/sec)
0.0			1.1	−2.8	0.0		0.0	0.0		
	7.4				−1.4	0.4				
−2.5		104	1.7	−4.2			0.4	2.8	2.4	2.5
	5.0				−5.6	1.4				
−5.0		60	2.2	−5.5			1.8	5.5	3.7	2.2
	0.8				−11.1	2.8				
−7.5		45	2.3	−5.8			4.6	8.2	3.6	1.6
	−4.8				−16.9	4.2				
−10.0		42	1.8	−4.5			8.8	11.0	2.2	0.9
	−10.7				−21.4	5.4				
−12.5		52	−1.7	4.2			14.2	13.8	−0.4	
	−15.5				−17.2	4.3				
−15.0		125	−4.7	11.8			18.5	16.5	−2.0	
	−17.5				−5.4	1.4				
−17.5		2500	−4.3	10.8			19.9	19.2	−0.7	
	−17.6				5.4	−1.4				
−20.0										

Source: Jacobsen (1913); in Proudman (1953)

columns were obtained from field observations. The value of $d\eta/dx$ was determined by assuming that $F_{z''} = 0$, where $\partial\bar{u}/\partial z = 0$. From (2-340), then,

$$\frac{d\eta}{dx} = \frac{1}{\rho_0 z''}\int_0^{z''}\int_0^{z'}\frac{\partial\rho}{\partial x}\,dz\,dz' \tag{2-341}$$

Then taking z'' approximately equal to -17.5 m and

$$\frac{1}{\rho_0}\int_0^{-17.5\text{ m}}\int_0^{z'}\frac{\partial\rho}{\partial x}\,dz\,dz' = 19.9 \times 10^{-5}\text{ cm}$$

it follows from (2-341) that

$$\frac{d\eta}{dx} = \frac{19.9 \times 10^{-5}\text{ cm}}{-17.5 \times 10^{2}\text{ cm}} = -1.1 \times 10^{-7}$$

approximately. The calculated values of K_V due to the special nature of the Kattegat are not all that significant in themselves, but the technique for the application of (2-340) is instructive.

Estimating the Eddy Viscosity for the Horizontal Exchange of Momentum

Assume that only lateral friction is important, that flow is purely zonal, that a longitudinal balance exists between the lateral friction force and the horizontal pressure gradient force, and that the eddy viscosity coefficient is constant. The horizontal time-averaged equations of motion, (2-326) and (2-327), reduce simply to

$$-\frac{d\bar{p}}{dx} + \frac{d}{dy}\left(A_H\frac{d\bar{u}}{dy}\right) = 0 \tag{2-342}$$

having taken the fluid acceleration and the Coriolis force as zero and $\bar{u} = \bar{u}(y)$ only. Integrating with respect to y,

$$\int d\left(A_H\frac{d\bar{u}}{dy}\right) = \int\frac{d\bar{p}}{dx}\,dy + C$$

or

$$A_H\frac{d\bar{u}}{dy} = \frac{d\bar{p}}{dx}y + C$$

Integrating again yields

$$\bar{u} = \frac{y^2}{2A_H}\frac{d\bar{p}}{dx} + Cy + C'$$

where C and C' are constants of integration. Now if b is the half-width of the current and taking that $\bar{u} = 0$ at $y = \pm b$, we get that

$$C = 0 \qquad C' = -\frac{b^2}{2A_H}\frac{d\bar{p}}{dx}$$

and therefore

$$\bar{u} = \frac{1}{2A_H}\frac{d\bar{p}}{dx}(y^2 - b^2) \tag{2-343}$$

In 1940, Montgomery and Palmén applied (2-343) to find a maximum estimate of the lateral eddy viscosity for the Equatorial Countercurrent of the Atlantic (see Figure 2-71). From hydrographic data, the slope of the sea was determined to be 3.8×10^{-8}, so $d\bar{p}/dx = -3.8 \times 10^{-5}$ g/cm²/sec², approximately, while from dynamic calculations $\bar{u}_0 = 65$ cm/sec. The width of the countercurrent was estimated as 300 km, so that $b = 1.5 \times 10^7$ cm. Substitution in (2-343) results in $A_H = 7 \times 10^7$ g/cm/sec, which is considered to be a very rough maximum estimate but, nonetheless, agrees well with other estimates.

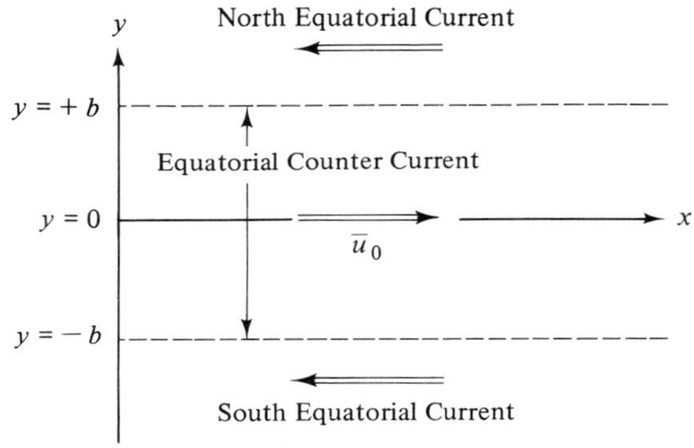

Fig. 2-71. Arrangement for estimating the lateral eddy viscosity via Montgomery and Palmen, 1940.

Estimating the Eddy Diffusivity for the Vertical Exchange of Salt

In the case of a stationary distribution of salinity, $\partial\bar{s}/\partial t = 0$. If the mean current is strictly zonal and mixing occurs mainly in the vertical (i.e., the horizontal advection of salt is balanced by the vertical diffusion of salt), the time-averaged conservation of salt equation for incompressible flow, (2-332), reduces to

$$\bar{u}\frac{\partial\bar{s}}{\partial x} = \frac{\partial}{\partial z}\left(K_{V,s}\frac{\partial\bar{s}}{\partial z}\right) \tag{2-344}$$

where for simplicity $K_{V,s}$ is used for the vertical kinematic eddy diffusion coefficient for salt rather than the previous $K_s(V)$. Integration over z between z_0 and z' gives

$$\int_{z_0}^{z'} \bar{u}\frac{\partial\bar{s}}{\partial x}\,dz = K_{V,s}\frac{\partial\bar{s}}{\partial z}\bigg|_{z_0}^{z'}$$

which if there is an extreme of $\bar{s}$ at the depth z_0, reduces to

$$(K_{V,s})_{z'} = \frac{1}{(\partial \bar{s}/\partial z)_{z'}} \int_{z_0}^{z'} \bar{u} \frac{\partial \bar{s}}{\partial x} \, dz \tag{2-345}$$

Thus, values of $K_{V,s}$ may be obtained at any depth z' from a numerical integration of the right side of (2-345) over the range from z_0 to z' given a salinity extreme at depth z_0, the current velocity as a function of z, and the salinity distribution as a function of x and z.

Table 2-2 illustrates the application of (2-345) as made by Jacobsen (1913) to observations at two stations, A and B, in the Kattegat, a distance of 47 km apart with $z_0 = 0$. Again, because of the shallow nature of the Kattegat, the actual values of K_V are not that significant as compared to the application of the technique.

Now if, in the region of interest, there is no extreme of $\bar{s}$, it is still possible to obtain an estimate of the average value of the vertical eddy diffusivity by assuming that it is not a function of z (i.e., by taking $\partial K_{V,s}/\partial z = 0$). Equation (2-344) then becomes

$$\bar{u} \frac{\partial \bar{s}}{\partial x} = K_{V,s} \frac{\partial^2 \bar{s}}{\partial z^2}$$

or

$$K_{V,s} = \bar{u} \frac{\partial \bar{s}/\partial x}{\partial^2 \bar{s}/\partial z^2} \tag{2-346}$$

As an example, Montgomery (1939) applied this result to the Equatorial Current of the Atlantic. In passing from longitude 55°W to 25°W, the salinity was found to decrease from 36.9‰ to 36.0‰. Thus,

$$\frac{\partial \bar{s}}{\partial x} = \frac{-0.9‰}{3.3 \times 10^8 \text{ cm}} = -2.7 \times 10^9 \text{ ‰/cm}$$

Observations yielded the estimates that

$$\frac{\partial^2 \bar{s}}{\partial z^2} = -2.3 \times 10^{-7} \frac{‰}{\text{cm}^2} \qquad \bar{u} = 30 \text{ cm/sec}$$

so that

$$K_{V,s} = (30 \text{ cm/sec}) \frac{-2.7 \times 10^{-9} \text{ /‰cm}}{-2.3 \times 10^{-7} \text{ ‰/cm}^2} = 0.35 \text{ cm}^2\text{/sec}$$

Proudman (1953) gives a simple graphical method for the estimation of $K_{V,s}/\bar{u}$ from a vertical salinity profile with or without an extreme. Suppose that Figure 2-72 represents the mean of the vertical distribution of salinity at two stations A and B with the line AB being in the direction of the current. Points A and B denote the salinities at a given depth at the respective stations; point P is the average of the salinities A and B. It is the midpoint of line AB. Two lines parallel to APB are drawn, one at a depth δz less than z and one

TABLE 2-2. Example of the Estimation of the Eddy Diffusivity Coefficient for the Vertical Exchange of Salt Via Equation (2-345)

		$\bar{s}$							
z, z' (m)	$\bar{u}$ (cm/sec)	Station A (‰)	Station B (‰)	$\delta_x \bar{s}$ (‰)	$\bar{u}\frac{\partial \bar{s}}{\partial x}\delta z \times 10^5$ (‰-cm/sec)	$\int_0^{z'} \bar{u}\frac{\partial \bar{s}}{\partial x} dz \times 10^5$ (‰-cm/sec)	$\delta_z \bar{s}$ (‰)	$\frac{\partial \bar{s}}{\partial z} \times 10^3$ (‰/cm)	$(K_{V,s})_{z'}$ (cm²/sec)
0.0		18.1	18.2			0.0		0.0	
	7.4			0.25	−9.8		0.35		
−2.5		18.3	18.7			−9.8		−1.4	0.070
	5.0			0.45	−12.0		0.35		
−5.0		18.6	19.1			−21.8		−2.0	0.109
	0.8			0.25	−1.1		0.65		
−7.5		19.5	19.5			−22.9		−2.8	0.082
	−4.8			−0.25	−6.4		0.75		
−10.0		20.5	20.0			−29.3		−8.0	0.037
	−10.7			−0.35	−19.9		3.25		
−12.5		23.6	23.4			−49.2		−13.0	0.038
	−15.5			−0.05	−4.1		3.25		
−15.0		26.7	26.8			−53.3		−10.4	0.051
	−17.5			0.45	41.9		1.95		
−17.5		28.3	29.1			−11.4			
	−17.6								
−20.0									

Source: Jacobsen (1913); in Proudman (1953).

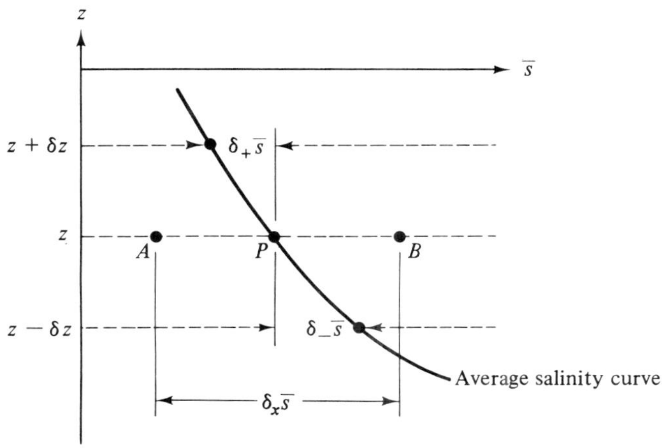

Fig. 2-72. Proudman's graphical method for estimating $K_{V,s}$ outside of a salinity extreme.

at a δz greater than z. Let $\delta_x\bar{s}$ be the difference in salinity between points A and B, $\delta_-\bar{s}$ the difference in the average salinity between depths $z - \delta z$ and z, $\delta_+\bar{s}$ the difference in the average salinity between depths $z + \delta z$ and z, δx the horizontal distance between the stations, and let all derivatives be evaluated at P.

From the average salinity curve,

$$\delta_+\bar{s} \doteq \frac{\partial \bar{s}}{\partial z}\delta z + \frac{\partial^2 \bar{s}}{\partial z^2}\frac{(\delta z)^2}{2} + \frac{\partial^3 \bar{s}}{\partial z^3}\frac{(\delta z)^3}{6}$$

$$\delta_-\bar{s} \doteq -\frac{\partial \bar{s}}{\partial z}\delta z + \frac{\partial^2 \bar{s}}{\partial z^2}\frac{(\delta z)^2}{2} - \frac{\partial^3 \bar{s}}{\partial z^3}\frac{(\delta z)^3}{6}$$

From the Taylor series and to the second order

$$\delta_+\bar{s} + \delta_-\bar{s} = (\delta z)^2\frac{\partial^2 \bar{s}}{\partial z^2}$$

Also, considering changes in the horizontal at depth z,

$$\delta_x\bar{s} \doteq \frac{\partial \bar{s}}{\partial x}\delta x$$

Substituting into (2-346) yields

$$\frac{K_{V,s}}{\bar{u}} = \frac{\partial \bar{s}/\partial x}{\partial^2 \bar{s}/\partial z^2} = \frac{\delta_x\bar{s}/\delta x}{(\delta_+\bar{s} + \delta_-\bar{s})/(\delta z)^2} \tag{2-347}$$

all of which can be determined without having to actually draw the average curve.

If point P corresponds to an extreme of salinity, either maximum or minimum, and if we are willing to take $\delta_-\bar{s} = \delta_+\bar{s}$ as an approximation,

(2-347) simplifies to

$$\frac{K_{V,s}}{\bar{u}} = \frac{1}{2} \frac{\delta_x \bar{s}}{\delta_+ \bar{s}} \frac{(\delta z)^2}{\delta x} \tag{2-348}$$

Application of (2-347) is illustrated by Defant (1936) for a salinity minimum in the Antarctic Intermediate Current. Table 2-3 details the procedure for the five stations A, B, C, D, and E, 1000 km apart, with $\delta z = 200$ m. With $\bar{u}$ approximately 2 cm/sec, this would yield values of $K_{V,s}$ ranging from 4 cm²/sec to 1.4 cm²/sec.

TABLE 2-3. Example of the Estimation of the Eddy Diffusivity Coefficient for the Vertical Exchange of Salt Via Equation (2-347)

	Salinity (‰) at Station:				
	A	*B*	*C*	*D*	*E*
z + 200 m	34.20	34.27	34.40	34.45	34.51
z	34.18	34.21	34.29	34.34	34.38
z − 200 m	34.22	34.28	34.44	34.49	34.54

$\delta_x \bar{s}$ (‰)	0.05	0.12	0.05	0.05
$\delta_+ \bar{s}$ (‰)	0.04	0.08	0.11	0.12
$\delta_- \bar{s}$ (‰)	0.06	0.11	0.15	0.16
$\dfrac{\delta_x \bar{s}}{\delta_+ \bar{s} + \delta_- \bar{s}}$	0.50	0.63	0.19	0.18
$(\delta z)^2/\delta x$ (cm)	4	4	4	4
$K_{V,s}/\bar{u}$ (cm)	2.00	2.50	0.76	0.72

Source: Defant (1936); in Proudman (1953).

Another simple graphical method for the estimation of $K_{V,s}/\bar{u}$ is given by Proudman (1953) using isohalines in a vertical section containing the current. Three isohalines for $\bar{s} - \frac{1}{2}\delta\bar{s}$, $\bar{s}$, and $\bar{s} + \frac{1}{2}\delta\bar{s}$ are shown in Figure 2-73. The points P_-, P, and P_+ mark the intersection of a horizontal line with the isohalines. P_-T, Q_+PQ_-, and RP_+ are vertical lines and TPR is a tangent to the isohaline $\bar{s}$ at point P. Let δx represent the distance P_-P_+, $\delta_+ z$ the depth interval Q_+P, $\delta_- z$ the depth interval PQ_-,

$$\delta z \equiv Q_+Q_- = \delta_+ z + \delta_- z$$

and

$$\delta' z \equiv P_-T + RP_+$$

Take all derivatives as being evaluated at P.

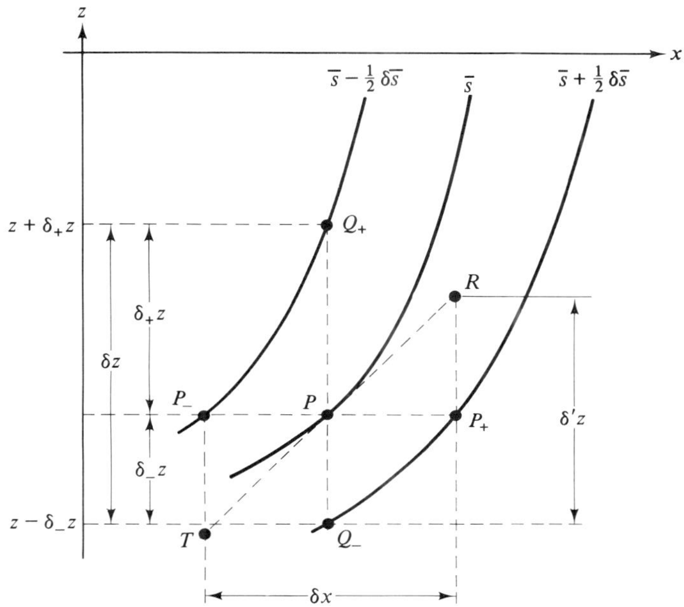

Fig. 2-73. Proudman's method for estimating $K_{V,s}$ from three isohalines.

At P the salinity along the tangent is not changing, so we may write approximately

$$\delta s \doteq \frac{\partial \bar{s}}{\partial x}\,\delta x + \frac{\partial \bar{s}}{\partial z}\,\delta' z = 0 \tag{2-349}$$

To the second order, salinity changes along Q_+PQ_- are approximately

$$\frac{1}{2}\,\delta\bar{s} \doteq \frac{\partial \bar{s}}{\partial z}\,\delta_+ z + \frac{1}{2}\,\frac{\partial^2 \bar{s}}{\partial z^2}(\delta_+ z)^2$$

and

$$-\frac{1}{2}\,\delta\bar{s} \doteq \frac{\partial \bar{s}}{\partial z}(-\delta_- z) + \frac{1}{2}\,\frac{\partial^2 \bar{s}}{\partial z^2}(-\delta_- z)^2$$

Adding,

$$\frac{\partial \bar{s}}{\partial z}(\delta_+ z - \delta_- z) + \frac{1}{2}\,\frac{\partial^2 \bar{s}}{\partial z^2}[(\delta_+ z)^2 + (\delta_- z)^2] = 0$$

If we accept that $\delta_+ z$ is about equal to $\delta_- z$, then

$$\tfrac{1}{2}[(\delta_+ z)^2 + (\delta_- z)^2] \doteq \left(\frac{\delta z}{2}\right)^2$$

and we see that

$$\frac{\partial \bar{s}/\partial z}{\partial^2 \bar{s}/\partial z^2} = \frac{(\delta z/2)^2}{\delta_- z - \delta_+ z} \tag{2-350}$$

Combining (2-346), (2-349), and (2-350) gives

$$\frac{K_{V,s}}{\bar{u}} = \frac{1}{4}\,\frac{\delta' z}{\delta_+ z - \delta_- z}\,\frac{(\delta z)^2}{\delta x} \tag{2-351}$$

This construction breaks down in the case of point P corresponding to a maximum or minimum of the vertical salinity profile since one of the points, Q_+ or Q_-, does not exist. A new construction is made with line Q_+PQ_- again vertical but with points Q_+ and Q_- located on the same isohaline (Figure 2-74). As before, δx represents the horizontal distance P_-P_+, $\delta_+ z$ the vertical distance Q_+P, $\delta_- z$ the vertical distance PQ_-, and δz the sum $(\delta_+ z + \delta_- z)$. Evaluating all derivatives at P yields

$$\delta \bar{s} \doteq \frac{\partial \bar{s}}{\partial x}\,\delta x \tag{2-352}$$

and

$$\begin{aligned} \frac{1}{2}\,\delta \bar{s} &\doteq \frac{1}{2}\,\frac{\partial^2 \bar{s}}{\partial z^2}(\delta_+ z)^2 \\ \frac{1}{2}\,\delta \bar{s} &\doteq \frac{1}{2}\,\frac{\partial^2 \bar{s}}{\partial z^2}(\delta_- z)^2 \end{aligned} \tag{2-353}$$

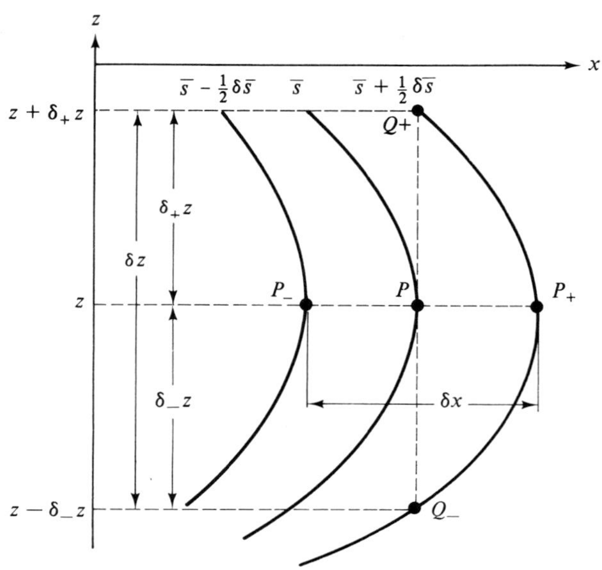

Fig. 2-74. Proudman's method for estimating $K_{V,s}$ from three isohalines with a low salinity tongue in the vertical section.

approximately. These last two equations give two slightly different approximations for $\frac{1}{2}\delta\bar{s}$, since $\delta_+ z$ and $\delta_- z$ are in general different. Taking an intermediate approximation,

$$\delta\bar{s} \doteq \frac{\partial^2\bar{s}}{\partial z^2}\left(\frac{\delta z}{2}\right)^2 \tag{2-354}$$

and substituting it and (2-352) in (2-346), we get

$$\frac{K_{V,s}}{\bar{u}} = \frac{1}{4}\frac{(\delta z)^2}{\delta x} \tag{3-355}$$

Estimating the Eddy Diffusivity for the Horizontal Exchange of Salt

Cases of stationary lateral mixing can be handled in a manner similar to that of the stationary vertical mixing discussed in the preceding section. The coordinate system can be oriented such that the mean current, designated as $\bar{u}$, is in the direction of the x axis and the assumption made that the only mixing that need be taken into account is horizontal and transverse to the current. Thus, (2-332) reduces to

$$\bar{u}\frac{\partial\bar{s}}{\partial x} = \frac{\partial}{\partial y}\left(K_{H,s}\frac{\partial\bar{s}}{\partial y}\right) \tag{2-356}$$

where for simplicity $K_{H,s}$ is used for the horizontal kinematic eddy diffusion coefficient for salt rather than the previous $K_s(H)$. Equation (2-356) may be written as

$$\bar{u}\frac{\partial\bar{s}}{\partial x} = K_{H,s}\frac{\partial^2\bar{s}}{\partial y^2} + \frac{\partial K_{H,s}}{\partial y}\frac{\partial\bar{s}}{\partial y}$$

Assuming that consideration be restricted to situations where either $K_{H,s}$ is constant ($\partial K_{H,s}/\partial y = 0$) or where the salinity along a line transverse to the current is either a maximum or a minimum ($\partial\bar{s}/\partial y = 0$), this result reduces to

$$K_{H,s} = \bar{u}\frac{\partial\bar{s}/\partial x}{\partial^2\bar{s}/\partial y^2} \tag{2-357}$$

and the methods discussed following (2-346) may be utilized by merely replacing z with y. For example, Proudman (1953) applied the graphical technique typified by (2-355) to the vertices of the mean isohalines of the Irish Sea, it being known that the salinity there is practically independent of depth. Writing (2-355) as

$$\frac{K_{H,s}}{\bar{u}} = \frac{1}{4}\frac{(\delta y)^2}{\delta x} \tag{2-358}$$

he obtained, after the substitution of values of δy and δx scaled directly from a horizontal isohaline chart, values of $K_{H,s}/\bar{u}$ ranging from 2.1×10^5 cm to 9.6×10^5 cm. Taking an average value of 5×10^5 cm and with

$\bar{u} = 0.5$ cm/sec, Proudman obtained an estimated value of 2.5×10^5 cm²/sec for $K_{H,s}$.

Estimating the Eddy Diffusivity from Temperature–Salinity Diagrams

Jacobsen (1927) suggested a method for using temperature–salinity (TS) diagrams in the calculation of eddy diffusion coefficients for the case of variable vertical mixing in the absence of a current. Proudman (1953) details three additional problems to which the same approach may be applied. It should be apparent from the previous sections that similar techniques could be employed using temperature data (in the absence of appreciable radiation) to estimate the eddy conductivity coefficient, K_T, as was described in using salinity data to estimate the eddy diffusion coefficient of salt. It is to be expected that the conductivity coefficient as determined from the temperature distribution will be, in general, different numerically from the salt diffusion coefficients as determined from the salinity distribution, even though the turbulence is the same. This is true since the eddy coefficients depend, not only upon the turbulence, but also upon the distribution of that which is being diffused. In certain specially simple circumstances it may be assumed that the mechanism for mixing is the same for both salinity and temperature, that the eddy coefficients depend only upon turbulence, and that K_s and K_T are therefore equal. Four of these simple cases have been examined by Jacobsen and Proudman.

Let us then assume that the coefficients of eddy diffusion are the same when determined from the salinity as when determined from the temperature. As a consequence, they will be the same when determined from any linear combination of $\bar{s}$ and $\bar{T}$. Let these coefficients be designated as

$$K_{V,s} = K_{V,T} \equiv K_V$$

and

$$K_{H,s} = K_{H,T} \equiv K_H$$

For a maximum or minimum in the TS diagram, four cases can be considered simultaneously. They are:

1. Stationary vertical mixing in the presence of a horizontal current (i.e., horizontal advection is balanced by vertical diffusion), for which the fundamental differential equations are, from (2-332) and (2-336),

$$\bar{u}\frac{\partial \bar{s}}{\partial x} = \frac{\partial}{\partial z}\left(K_V \frac{\partial \bar{s}}{\partial z}\right) \tag{2-359a}$$

and

$$\bar{u}\frac{\partial \bar{T}}{\partial x} = \frac{\partial}{\partial z}\left(K_V \frac{\partial \bar{T}}{\partial z}\right) \tag{2-359b}$$

2. Stationary lateral mixing in the presence of a horizontal current (horizontal advection balanced by horizontal diffusion), for which the fundamental equations are

$$\bar{u}\frac{\partial \bar{s}}{\partial x} = \frac{\partial}{\partial y}\left(K_H \frac{\partial \bar{s}}{\partial y}\right) \tag{2-360a}$$

and

$$\bar{u}\frac{\partial \bar{T}}{\partial x} = \frac{\partial}{\partial y}\left(K_H \frac{\partial \bar{T}}{\partial y}\right) \tag{2-360b}$$

3. Variable vertical mixing in the absence of a current (local rate of change is due to vertical diffusion), for which the fundamental equations are

$$\frac{\partial \bar{s}}{\partial t} = \frac{\partial}{\partial z}\left(K_V \frac{\partial \bar{s}}{\partial z}\right) \tag{2-361a}$$

and

$$\frac{\partial \bar{T}}{\partial t} = \frac{\partial}{\partial z}\left(K_V \frac{\partial \bar{T}}{\partial z}\right) \tag{2-361b}$$

4. Variable laterial mixing in the absence of a current (local rate of change is due to horizontal diffusion), for which the fundamental equations are

$$\frac{\partial \bar{s}}{\partial t} = \frac{\partial}{\partial y}\left(K_H \frac{\partial \bar{s}}{\partial y}\right) \tag{2-362a}$$

and

$$\frac{\partial \bar{T}}{\partial t} = \frac{\partial}{\partial y}\left(K_H \frac{\partial \bar{T}}{\partial y}\right) \tag{2-362b}$$

For each of these cases we require two TS curves:

a. For case 1, vertical TS diagrams through two stations, say A and B, a distance δx apart and such that the horizontal line AB connecting them is parallel to the mean current $\bar{u}$ (see Figure 2-75, case 1).

b. For case 2, horizontal TS diagrams at the same depth a distance δx apart, both transverse to the mean current $\bar{u}$, and passing through points A and B (see Figure 2-75, case 2).

c. For case 3, at the same location, vertical TS diagrams at two different times, say t_A and t_B, differing by a time interval δt (see Figure 2-75, case 3).

d. For case 4, at the same location, horizontal TS diagrams at two different times, t_A and t_B, differing by a time interval δt (see Figure 2-75, case 4).

The points of the TS diagrams correspond to points in the sea, and in cases 3 and 4, at one of two different times. For example, Figure 2-76 shows the TS plots for the two stations, A and B, and point P corresponds to a

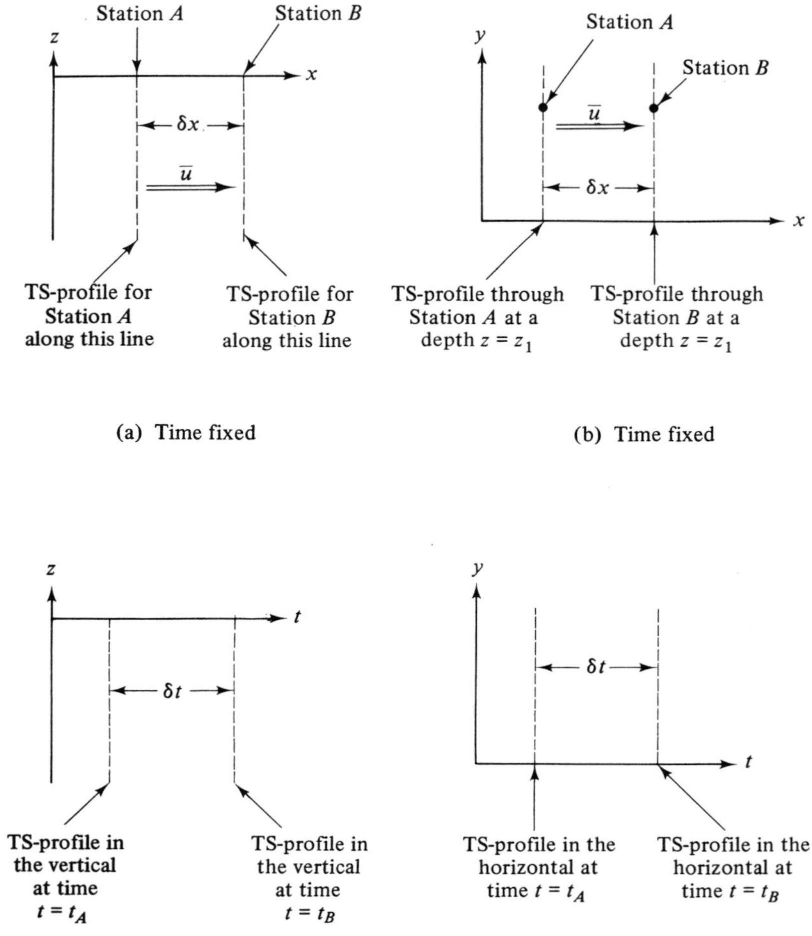

Fig. 2-75. Data required to obtain estimates of eddy diffusivities from TS diagrams.

specific depth at station A. In case 4, the point Q corresponds, on the other hand, to a specific point on the line in the sea at time $t = t_B$.

Assume that we want to find the diffusion coefficient at point P for case 1. With the aid of Figure 2-77 we define a new coordinate F such that the F axis is normal to point Q and goes through point P. The F distance between P and Q is designated δF. A line tangent to the B curve at point Q (perpendicular to the F axis) intersects the A curve at points Q_+ and Q_-. In case 1 (or case 3) let δz denote the difference in depths between the points

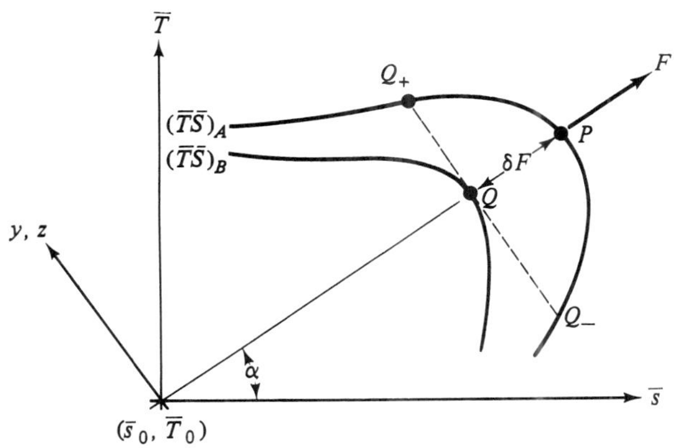

Fig. 2-76. Two TS diagrams and the F, z or F, y coordinate system.

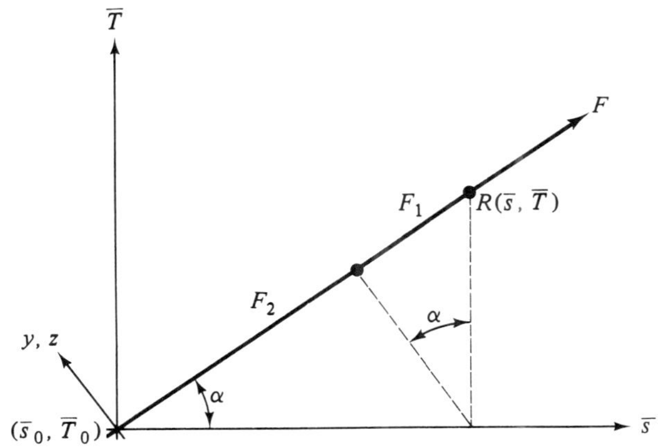

Fig. 2-77. F coordinate of any point R on the F axis.

in the sea corresponding to Q_+ and Q_-. (In cases 3 and 4, let δy denote the distance along the line in the sea between the points corresponding to Q_+ and Q_-.) Now the F coordinate of any point $R(\bar{s}, \bar{T})$ on the F axis is

$$F = F_1 + F_2 \tag{2-363}$$

where

$$\begin{aligned} F_1 &\equiv (\bar{T} - \bar{T}_0) \sin \alpha \\ F_2 &\equiv (\bar{s} - \bar{s}_0) \cos \alpha \end{aligned} \tag{2-364}$$

Multiplying (2-359a) by $\cos \alpha$, (2-359b) by $\sin \alpha$, and adding,

$$\bar{u}\left(\frac{\partial \bar{s}}{\partial x} \cos \alpha + \frac{\partial \bar{T}}{\partial x} \sin \alpha\right) = \frac{\partial}{\partial z}\left[K_V\left(\frac{\partial \bar{s}}{\partial z} \cos \alpha + \frac{\partial \bar{T}}{\partial z} \sin \alpha\right)\right] \tag{2-365}$$

The term in parentheses on the left side, from (2-363) and (2-364), is equal to $\partial F/\partial x$; the term in parentheses on the right side is equal to $\partial F/\partial z$. Thus,

$$\bar{u}\frac{\partial F}{\partial x} = \frac{\partial}{\partial z}\left(K_V \frac{\partial F}{\partial z}\right)$$

or

$$\bar{u}\frac{\partial F}{\partial x} = \frac{\partial K_V}{\partial z}\frac{\partial F}{\partial z} + K_V \frac{\partial^2 F}{\partial z^2} \tag{2-366}$$

Since the z axis and the F axis are orthogonal, $\partial F/\partial z = 0$, and (2-366) reduces to

$$K_V = \bar{u}\frac{\partial F/\partial x}{\partial^2 F/\partial z^2} \tag{2-367}$$

Proceeding in a manner similar to that used to establish (2-355) and with all derivatives referring to point P,

$$\delta F \doteq \frac{\partial F}{\partial x}\delta x \tag{2-368}$$

and

$$\delta_+ F \doteq \frac{1}{2}\frac{\partial^2 F}{\partial z^2}(\delta_+ z)^2$$

$$\delta_- F \doteq \frac{1}{2}\frac{\partial^2 F}{\partial z^2}(-\delta_- z)^2$$

$$\delta z \equiv \delta_+ z + \delta_- z$$

Taking $\delta_+ z$ as approximately equal to $\delta_- z$,

$$\left(\frac{\delta z}{2}\right)^2 \doteq \tfrac{1}{2}[(\delta_+ z)^2 + (\delta_- z)^2]$$

and since $\delta_+ F$ is equal to $\delta_- F$,

$$2\delta F \doteq \delta_+ F + \delta_- F \doteq \frac{1}{4}\frac{\partial^2 F}{\partial z^2}(\delta z)^2 \tag{2-369}$$

Substituting (2-368) and (2-369) into (2-367) yields

$$K_V = \frac{1}{8}\bar{u}\frac{(\delta z)^2}{\delta x} \tag{2-370}$$

for case 1. Equation (2-370) allows us to make an estimate of K_V from a knowledge of the average horizontal velocity between the two stations, the horizontal distance between the two stations, and the difference in depth between the two points Q_+ and Q_- determined from the pair of TS diagrams as in Figure 2-76.

If the same analysis is carried out for the remaining three cases, we would obtain: for case 2,

$$K_H = \frac{1}{8}\bar{u}\frac{(\delta y)^2}{\delta x} \tag{2-371}$$

for case 3,

$$K_V = \frac{1}{8}\frac{(\delta z)^2}{\delta t} \tag{2-372}$$

and for case 4,

$$K_H = \frac{1}{8}\frac{(\delta y)^2}{\delta t} \tag{2-373}$$

2-36 Elementary Examples of the Application of Eddy Coefficients

Lateral and Vertical Heat Conduction for a Stationary Temperature Distribution in an Ocean at Rest

Neumann and Pierson (1966) give an excellent discussion of Sverdrup's (1939) example of the elementary application of eddy coefficients to a problem involving the gross features of the temperature distribution in the ocean. Consider a meridional cross section extending from the North Pole to the South Pole of an ideal ocean covering the entire earth. Assuming that the ocean is at rest everywhere and that no changes occur in the zonal (x) direction, stationary temperature conditions are possible if, from the conservation of heat equation (2-336),

$$\frac{\partial}{\partial y}\left(K_{H,T}\frac{\partial \bar{T}}{\partial y}\right) + \frac{\partial}{\partial z}\left(K_{V,T}\frac{\partial \bar{T}}{\partial z}\right) = 0 \tag{2-374}$$

where $K_{H,T}$, and $K_{V,T}$ are the horizontal and vertical eddy conductivity coefficients, as before. Equation (2-374) merely states that under the assumed conditions the meridional thermal diffusion (or heat conduction) balances the vertical thermal diffusion. Taking the two conductivities as constant, we obtain the defining differential equation

$$K_{H,T}\frac{\partial^2 \bar{T}}{\partial y^2} + K_{V,T}\frac{\partial^2 \bar{T}}{\partial z^2} = 0 \tag{2-375}$$

Even though our ideal ocean was assumed to be at rest in relation to large-scale average flows ($\bar{u} = \bar{v} = \bar{w} = 0$), the eddy conductivity coefficients are assigned values of the same order of magnitude as are presumed to hold in the actual ocean. For boundary conditions, Neumann and Pierson wrote that

$$\bar{T} \equiv T_P + T_0 \cos\frac{\pi y}{2b} \qquad \text{at } z = 0 \tag{2-376}$$

where $\pm b$ is the meridional distance between the equator at $y = 0$ and either pole, $y = \pm b$, $z = 0$ is the sea surface, and T_P is the temperature at either pole taken as constant with depth. Equation (2-376) is in agreement with the general situation wherein the absorption of solar radiation by the surface layers of the ocean results in a heat gain in equatorial regions and a heat loss in the polar regions. No general solution of this boundary value problem was attempted by Sverdrup. Clearly, it is not fully posed, since only three boundary conditions (the temperature along the sea surface and along the two verticals at the poles) are specified, whereas four are in fact required (the fourth being the temperature distribution along the bottom). However, direct substitution readily verifies that a solution of (2-375) is

$$\bar{T}(y, z) = T_P + T_0 \frac{\cosh \alpha(z + D)}{\cosh \alpha D} \cos \frac{\pi y}{2b} \tag{2-377}$$

where $z = -D$ is the bottom and

$$\alpha \equiv \frac{\pi}{2b} \sqrt{K_{H,T}/K_{V,T}}$$

Notice that for this very simplified model, the stationary temperature distribution along a meridional section depends not upon the individual values of $K_{H,T}$ and $K_{V,T}$ but rather upon their ratio. Neumann and Pierson took this ratio as $K_{H,T}/K_{V,T} = 10^7$ and, with $b = 10{,}000$ km, arrived at a value of $\alpha = 0.497\ \text{km}^{-1}$. Figure 2-78 is the computed temperature section for an ocean of 6000 m depth when the temperatures were taken as $T_0 = 25°$C and $T_P = 0°$C. The solution is symmetrical about the equator.

Neumann and Pierson point out that the distribution arrived at shows similarities with some important observed facts. In both model and real ocean: (1) strong vertical temperature gradients develop in the upper levels of the ocean in tropical regions with the gradients decreasing in the lower levels, and (2) at any depth the vertical temperature gradient decreases

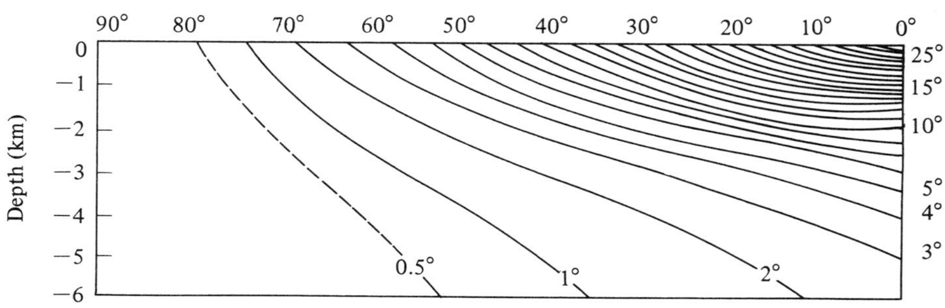

Fig. 2-78. Meridional temperature distribution due to eddy heat conductions in an idealized, motionless ocean. From Neumann and Pierson, *Principles of Physical Oceanography*, Prentice-Hall, Inc., 1966, p. 405. Reprinted by permission.

poleward with the fluid approaching homothermous conditions at higher latitudes. At any rate, the important result of Sverdrup's crude model is that it illustrates how some major stationary features of the gross temperature distribution of the ocean can be explained without recourse to fluid advection—that these features can be explained solely as the result of eddy heat conduction resulting from mixing.

Salinity Tongue in an Ocean at Rest

In a manner analogous to the previous example, Sverdrup et al. (1942) and Defant (1961) considered the case of an ocean at rest with no advection but with horizontal and vertical salt diffusion (two-dimensional). A stationary salinity distribution is possible if, from the conservation of salt equation (2-332),

$$\frac{\partial}{\partial x}\left(K_{H,s}\frac{\partial \bar{s}}{\partial x}\right) + \frac{\partial}{\partial z}\left(K_{V,s}\frac{\partial \bar{s}}{\partial z}\right) = 0 \tag{2-378}$$

where the horizontal diffusion has been assumed to be in the x direction only and $K_{H,s}$ and $K_{V,s}$ are the saline eddy diffusion coefficients. Taking the coefficients as constants,

$$K_{H,s}\frac{\partial^2 \bar{s}}{\partial x^2} + K_{V,s}\frac{\partial^2 \bar{s}}{\partial z^2} = 0 \tag{2-379}$$

At $x = 0$, let the boundary condition for the salt distribution in the vertical be

$$\bar{s} \equiv s_0 + \Delta s \cos\frac{\pi z}{2h} \tag{2-380}$$

where $2h$ is the thickness of the layer of fluid having an upper boundary at $z = +h$ and a lower boundary at $z = -h$. Take the boundary conditions at these upper and lower boundaries to be

$$\bar{s} \equiv \text{constant} \equiv s_0 \qquad \text{at } z = \pm h \tag{2-381}$$

A solution for this boundary value problem is

$$\bar{s} = s_0 + \Delta s e^{-\alpha x}\cos\frac{\pi z}{2h} \tag{2-382}$$

where now

$$\alpha \equiv \frac{\pi}{2h}\sqrt{K_{V,s}/K_{H,s}} \tag{2-383}$$

Again the salinity distribution depends upon the ratio of the diffusion coefficients and not on their individual values. Sverdrup, taking $K_{H,s}/K_{V,s} = 9 \times 10^7$ and $h = 300$ m, obtained the distribution shown in Figure 2-79.

Neumann and Pierson (1966) point out that this example shows that tongue-like distributions of salinity (or, for that matter, any conservative property) may be explained on the basis of diffusion effects alone, that no horizontal advection is required. It is a natural temptation for the oceano-

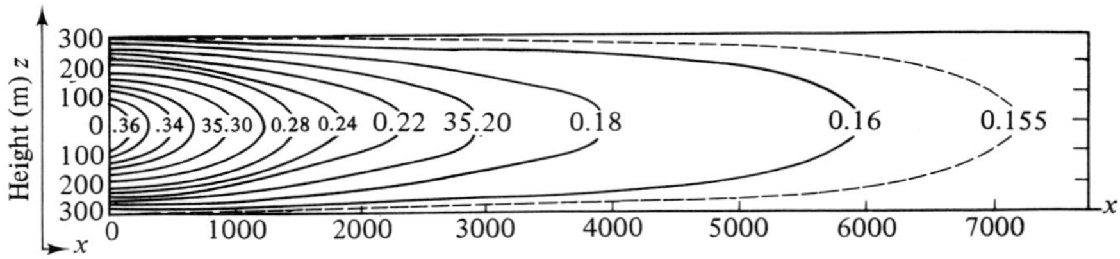

Fig. 2-79. Tongue-like distribution of salinity in a vertical section due to diffusion alone with boundary conditions of (2-380) and (2-381). From Sverdrup, Johnson, and Fleming, *The Oceans*, Prentice-Hall, Inc., 1942, 1970, p. 504. Reprinted by permission.

grapher to attribute such tongues to a current, with the resulting advection of relatively high or low salinity water into another body of water.

Tongue-Like Distributions of Salinity Resulting from Diffusion and Advection

Along with others, Defant (1929) and Thorade (1931) examined the distribution of a conservative property, $\bar{S}$, for several simple cases when both advection and diffusion occur. Assume again that the distribution of the property is stationary, that mixing occurs only in the vertical, and that the x axis of our coordinate system points in the direction of a current having a velocity $\bar{u}$ so that $\bar{v} = \bar{w} = 0$. Writing a conservation equation similar to (2-332) or (2-336) and simplifying for these conditions, we obtain that

$$\bar{u}\frac{\partial \bar{S}}{\partial x} = \frac{\partial}{\partial z}\left(K_{V,S}\frac{\partial \bar{S}}{\partial z}\right) \tag{2-384}$$

(horizontal advection is balanced by vertical diffusion), where $K_{V,S}$ is the vertical eddy diffusion coefficient of the particular property. Taking $K_{V,S}$ as constant, (2-384) becomes

$$\bar{u}\frac{\partial \bar{S}}{\partial x} = K_{V,S}\frac{\partial^2 \bar{S}}{\partial x^2} \tag{2-385}$$

First, take the boundary conditions such that the current has a uniform velocity $\bar{u} = u_0$ both in the horizontal and vertical between the upper and lower boundaries at $z = \pm h$. Assume that the distribution of $\bar{S}$ at $x = 0$ is given by

$$\bar{S} \equiv S_0 \cos\frac{\pi z}{2h} \tag{2-386}$$

so that $\bar{S} = 0$ at the boundaries $z = \pm h$. A solution to (2-385) is then

$$\bar{S} = S_0 e^{-\gamma x} \cos\frac{\pi z}{2h} \tag{2-387}$$

where

$$\gamma \equiv \frac{\pi^2 K_{V,S}}{4u_0 h^2} \tag{2-388}$$

Figure 2-80(a) illustrates the resulting tongue-like distribution of $\bar{S}$.

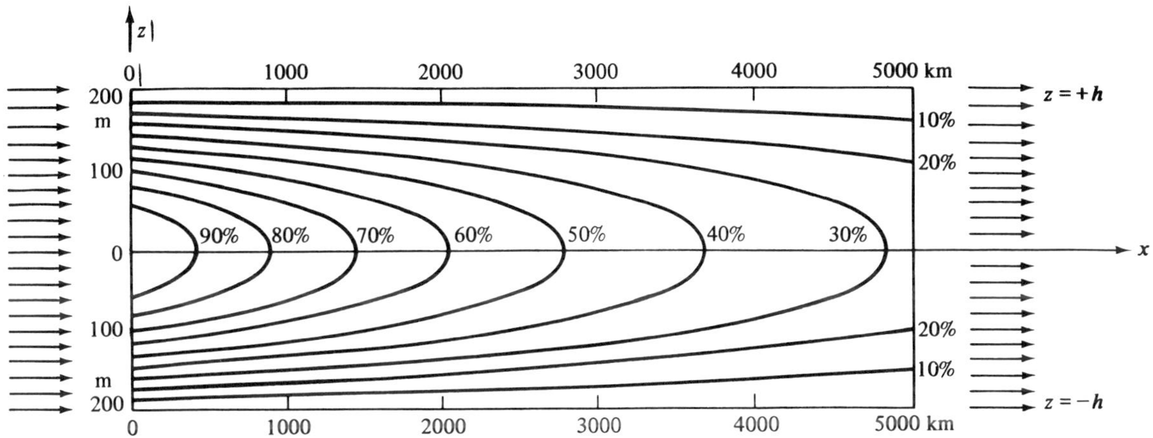

(a)

Fig. 2-80. Tongue-like distributions of a conservative property shown by isolines (Property concentrations expressed as percentages of the original 100% at source): (a) current is constant and Uniform, (b) symetric and constant current with a maximum along axis. From Neumann and Pierson, *Principles of Physical Oceanography*, Prentice-Hall, Inc., 1966, p. 407. Reprinted by permission.

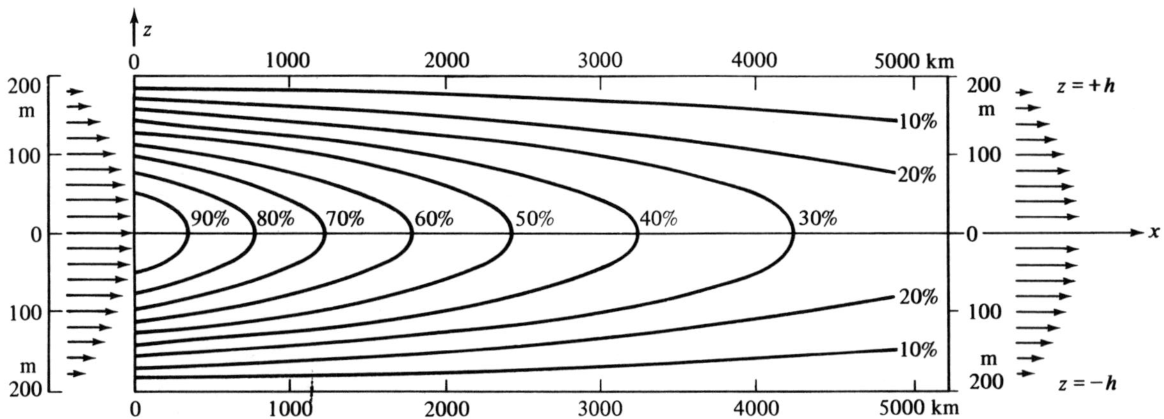

(b)

Fig.2-80. (Continued)

Suppose that the current, instead of being constant and uniform, was represented by the equation

$$\bar{u} \equiv u_0 \cos \frac{\pi z}{2h} \qquad (2\text{-}389)$$

so that $\bar{u} = u_0$ at $z = 0$ and $\bar{u} = 0$ at $z = \pm h$. The resulting distribution of $\bar{S}$ is then identical in form [see Figure 2-80(b)] to that shown in Figure 2-80(a); that is, another tongue-like distribution results.

Clearly, then, the appearance of the distribution of the property will be identical in form for the case represented by (2-382) (no current with two-dimensional diffusion in the horizontal and vertical) or for the cases of uniform flow with vertical diffusion, as given by (2-382), or with a symmetric current having an axis of maximum velocity; in all instances tongue-like distributions result and no definite conclusions can be drawn as to the currents.

THREE

Wind-Driven Ocean Circulation

3-1 Introduction

It is now our intention first to examine in detail three theoretical investigations [Sverdrup (1947), Stommel (1948), and Munk (1950)] that are concerned with only one particular aspect of the problem of the general circulation of the oceans: the relation of the permanent features of the horizontal circulation patterns to the distribution of mean atmospheric winds. These linear studies act as an introduction to the topic of bounded flow in a rotating fluid of enormous Reynolds number (the number resulting from dividing the terms involving inertial accelerations or forces by terms involving the viscous accelerations or forces). The practical application of these idealized mathematical studies to the circulation of real oceans lies in the fact that they act as building blocks to an understanding of more modern theories of dynamic oceanography involving nonlinear approaches and eventually, to attempts to combine wind-driven and thermohaline components in one model. They cannot be considered as ends within themselves, only as steps along a path whose end has not yet been reached.

Sverdrup (1947) demonstrates how an almost geostrophic flow over the major part of the oceans can be balanced locally by the stress imposed by the wind, but that such a flow cannot describe a closed circulation pattern. Stommel (1948) shows that the latitudinal variation of the vertical component of the earth's rotation vector is essential to complete the flow through a reproduction of the western boundary currents of the oceans, local regions of high relative vorticity. Munk (1950) succeeds in explaining many of the major, as well as some of the minor, details of the general circulation through the use of the more realistic wind stress data of Reid (1948) and by the inclusion of lateral (horizontal) frictional force of a Newtonian form.

Finally, in the last section of this chapter we look at, qualitatively, the effects that may be expected from an inclusion of the inertial terms of the equations of motion.

Before continuing, it is well to review the properties of geostrophic and Ekman wind drift currents (Section 2-31):

Geostrophic currents: a balance between the pressure gradient force and the Coriolis force, unaccelerated, the pressure force being due to the surface slope (barotrophic component) and the internal mass distribution (baroclinic component), barotropic (slope) current plus baroclinic (relative) current producing the total geostrophic current;

Ekman wind-drift currents: a balance between frictional forces and the Coriolis force, unaccelerated, depth of frictional influence typically 20 to 100 m, in the Northern Hemisphere the current vector between surface and deeper layers turns continuously clockwise while its magnitude decreases exponentially, net mass transport 90° to the right of the wind.

3-2 Sverdrup's Study of Wind-Driven Currents in a Baroclinic Ocean

We begin with the equations of motion in the form

$$\frac{\partial u}{\partial t} + u\frac{\partial u}{\partial x} + v\frac{\partial u}{\partial y} + w\frac{\partial u}{\partial z} - fv = -\frac{1}{\rho}\frac{\partial p}{\partial x} + \frac{1}{\rho}\left[\frac{\partial}{\partial x}\left(A_H\frac{\partial u}{\partial x}\right) + \frac{\partial}{\partial y}\left(A_H\frac{\partial u}{\partial y}\right) + \frac{\partial}{\partial z}\left(A_V\frac{\partial u}{\partial z}\right)\right] \tag{2-326}$$

$$\frac{\partial v}{\partial t} + u\frac{\partial v}{\partial x} + v\frac{\partial v}{\partial y} + w\frac{\partial v}{\partial z} + fu = -\frac{1}{\rho}\frac{\partial p}{\partial y} + \frac{1}{\rho}\left[\frac{\partial}{\partial x}\left(A_H\frac{\partial v}{\partial x}\right) + \frac{\partial}{\partial y}\left(A_H\frac{\partial v}{\partial y}\right) + \frac{\partial}{\partial z}\left(A_V\frac{\partial v}{\partial z}\right)\right] \tag{2-327}$$

and

$$\frac{\partial p}{\partial z} = -\rho g \tag{2-217}$$

where the overbars have been dropped for convenience and the hydrostatic approximation has been adopted in the vertical. Assume that:

1. The flow is stationary,

$$\frac{\partial u}{\partial t} = 0 \qquad \frac{\partial v}{\partial t} = 0$$

2. The nonlinear terms are small so that the field accelerations are given by

$$u\frac{\partial u}{\partial x} + v\frac{\partial u}{\partial y} + w\frac{\partial u}{\partial z} = 0$$

$$u\frac{\partial v}{\partial x} + v\frac{\partial v}{\partial y} + w\frac{\partial v}{\partial z} = 0$$

   Pragmatically, we are stating that the nonlinear terms are difficult to deal with mathematically.

3. Lateral friction is negligible in that the derivatives in the horizontal are much smaller than those in the vertical,

$$\frac{1}{\rho}\left[\frac{\partial}{\partial x}\left(A_H\frac{\partial u}{\partial x}\right) + \frac{\partial}{\partial y}\left(A_H\frac{\partial u}{\partial y}\right)\right] = 0$$

$$\frac{1}{\rho}\left[\frac{\partial}{\partial x}\left(A_H\frac{\partial v}{\partial x}\right) + \frac{\partial}{\partial y}\left(A_H\frac{\partial v}{\partial y}\right)\right] = 0$$

What we have said is that we require some friction, but that lateral friction is not significant as compared to vertical friction.

With these simplifications, (2-326) and (2-327) reduce to

$$\frac{\partial p}{\partial x} = \rho f v + \frac{\partial}{\partial z}\left(A_V\frac{\partial u}{\partial z}\right) \tag{3-1}$$

and

$$\frac{\partial p}{\partial y} = -\rho f u + \frac{\partial}{\partial z}\left(A_V\frac{\partial v}{\partial z}\right) \tag{3-2}$$

In these equations a balance between the left-hand term and the first term to the right of the equal sign represents geostrophic flow, while the two right-hand terms represent Ekman wind drift. Physically, the equations merely assert the the horizontal pressure gradient is balanced by the Coriolis force and the frictional stress exerted on horizontal surfaces.

Note that (3-1) and (3-2) involve horizontal derivatives of only the first order, and consequently place severe restrictions upon possible lateral boundary conditions. We cannot expect to satisfy more than two, one on x and one on y. Our model, then, will be that of a corner of the ocean. On the other hand, these equations are of second order in the vertical derivatives, and boundary conditions at the ocean surface and bottom can be imposed.

The next step is to integrate these equations in the vertical, thus making it possible to deal with the general case of a baroclinic ocean without a detailed specification of the vertical density distribution. Sverdrup assumed that at some moderate depth in the ocean, the horizontal pressure gradient vanished, that the barotopic and baroclinic pressure gradients balance each other, and that, therefore, the horizontal velocity goes to zero well before the bottom. The bottom frictional stress is thus zero and the ocean does not "know" anything about the sea-floor topography. When vertically integrating, this permits a depth $z = -d$ to be taken such that $u|_{z=-d}$ and $v|_{z=-d}$ are zero while d is kept small enough that the flow is uninfluenced by the bottom topography. Now define a new function P:

$$\frac{\partial P}{\partial x} \equiv \int_{-d}^{0} \frac{\partial p}{\partial x}\,dz \qquad \frac{\partial P}{\partial y} \equiv \int_{-d}^{0} \frac{\partial p}{\partial y}\,dz \tag{3-3}$$

Since the horizontal velocity vanishes at, and below, the depth $z = -d$, the integrals

$$S_x \equiv \int_{-d}^{0} \rho u\,dz \qquad S_y \equiv \int_{-d}^{0} \rho v\,dz \tag{3-4}$$

represent the components of the net mass transport by the currents. As horizontal boundary conditions, we specify that the lateral stress components at the surface, $z = 0$, and at the depth, $z = -d$, are

$$\begin{aligned} A_V \frac{\partial u}{\partial z}\bigg|_{z=0} &= \tau_x \qquad & A_V \frac{\partial u}{\partial z}\bigg|_{z=-d} &= 0 \\ A_V \frac{\partial v}{\partial z}\bigg|_{z=0} &= \tau_y \qquad & A_V \frac{\partial v}{\partial z}\bigg|_{z=-d} &= 0 \end{aligned} \tag{3-5}$$

where τ_x and τ_y are the x and y components of the wind stress. Integrating (3-1) and (3-2) from $z = -d$ up to $z = 0$ and introducing (3-3), (3-4), and (3-5) yields

$$\frac{\partial P}{\partial x} = fS_y + \tau_x \tag{3-6a}$$

$$\frac{\partial P}{\partial y} = -fS_x + \tau_y \tag{3-6b}$$

the terms of which are quite familiar. Omitting the stress components, these equations relate the mass transport to the density distribution; or, assuming homogeneous water in hydrostatic equilibirum ($\partial P/\partial x = \partial P/\partial y = 0$), they give the mass transport by pure wind-driven currents as determined by Ekman.

If equations (3-6) are cross-differentiated and subtracted, the pressure-related terms are eliminated and a *vorticity tendency equation* results. Thus, from (3-6a),

$$\frac{\partial^2 P}{\partial x\,\partial y} = \frac{\partial f}{\partial y} S_y + f\frac{\partial S_y}{\partial y} + \frac{\partial \tau_x}{\partial y}$$

and, from (3-6b),

$$\frac{\partial^2 P}{\partial x\,\partial y} = -f\frac{\partial S_x}{\partial x} + \frac{\partial \tau_y}{\partial x}$$

since $\partial f/\partial x = 0$ because of the choice of our coordinate system (x coordinate to the east, y coordinate to the north). Subtracting the second equation from the first, we obtain

$$\frac{\partial f}{\partial y}S_y + f\left(\frac{\partial S_x}{\partial x} + \frac{\partial S_y}{\partial y}\right) + \left(\frac{\partial \tau_x}{\partial y} - \frac{\partial \tau_y}{\partial x}\right) = 0 \qquad (3\text{-}7)$$

Physically, the first term accounts for changes in vorticity due to meridional flow, the second accounts for changes in vorticity from stretching due to the horizontal divergence of the mass transport, and the third accounts for the vorticity added by the wind torque.

Now for steady-state conditions there can be no net convergence or divergence of mass without flow over a variable depth; that is,

$$\frac{\partial S_x}{\partial x} + \frac{\partial S_y}{\partial y} = 0 \qquad (3\text{-}8)$$

for $\partial/\partial t = 0$ and $d =$ constant. Mathematically, this can be shown by starting with the continuity equation

$$\frac{\partial \rho}{\partial t} + \nabla \cdot \rho \mathbf{V} = 0$$

which for steady-state conditions reduces to

$$\frac{\partial \rho u}{\partial x} + \frac{\partial \rho v}{\partial y} + \frac{\partial \rho w}{\partial z} = 0$$

Integrating in the vertical from $z = -d$ up to $z = 0$,

$$\frac{\partial}{\partial x}\int_{-d}^{0} \rho u\,dz + \frac{\partial}{\partial y}\int_{-d}^{0} \rho v\,dz + \int_{-d}^{0} \frac{\partial \rho w}{\partial z}\,dz = 0$$

and applying the definitions of (3-4),

$$\frac{\partial S_x}{\partial x} + \frac{\partial S_y}{\partial y} + \rho w\Big|_{z=-d}^{z=0} = 0$$

Taking $w|_{z=0} = w|_{z=-d} = 0$, this result, called the *vertically integrated form of the continuity equation for steady-state conditions*, reduces to (3-8).

In view of this, (3-7) becomes

$$\frac{\partial f}{\partial y}S_y + \left(\frac{\partial \tau_x}{\partial y} - \frac{\partial \tau_y}{\partial x}\right) = 0 \qquad (3\text{-}9)$$

which is frequently referred to as *Sverdrup's curl equation* from the fact that

$$-\left(\frac{\partial \tau_x}{\partial y} - \frac{\partial \tau_y}{\partial x}\right) = \text{curl}_z\,\boldsymbol{\tau}$$

where curl_z is the z component of the curl operation. Thus, (3-9) is written alternatively as

$$\frac{\partial f}{\partial y} S_y = \text{curl}_z\, \boldsymbol{\tau} \tag{3-10}$$

and expresses a balance between the planetary vorticity due to motion in the meridional direction and the wind stress curl. Physically, since steady state is specified and since no frictional mechanism is provided to dissipate vorticity, water flowing to the north, for example, will acquire negative relative vorticity due to an increase of planetary vorticity to the north in the absence of any outside source of relative vorticity. Thus, it is apparent that no change in relative vorticity can occur at a point if we continually advect in water from the south, add positive relative vorticity from the wind stress, and continuously advect it out to the north.

For simplicity, assume that all the winds are zonal (i.e., $\tau_y = 0$). From (3-10),

$$S_y = -\frac{\partial \tau_x / \partial y}{df/dy} \tag{3-11}$$

where $\partial f / \partial y$ has been replaced by df/dy. From Figure 3-1,

$$dy = R_e\, d\theta$$

where R_e is the radius of the earth, and since $f \equiv 2\Omega \sin\theta$,

$$\frac{df}{dy} = \frac{df}{d\theta}\frac{d\theta}{dy} = \frac{2\Omega\cos\theta}{R_e} \tag{3-12}$$

Thus, (3-11) finally becomes

$$S_y = -\frac{\partial \tau_x}{\partial y}\left(\frac{R_e}{2\Omega\cos\theta}\right) \tag{3-13}$$

As an aside, note that if the zonal winds do not vary in the north-south direction, $\partial \tau_x / \partial y = 0$ and $S_y = 0$; that is, there is no net meridional mass transport; the Ekman transport is balanced by the geostrophic transport.

S_x, the net mass transport in the x direction, in the case of purely

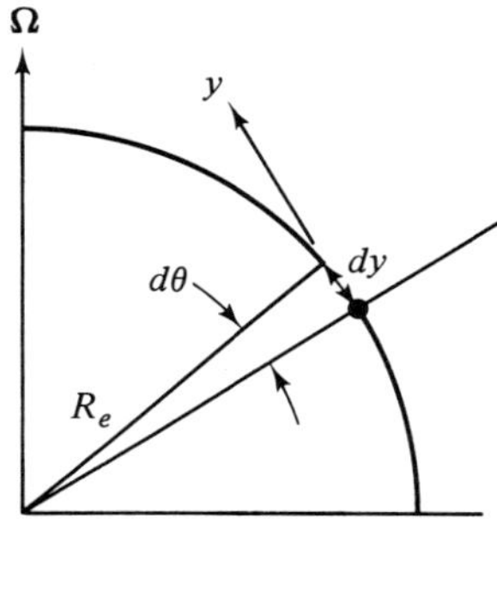

Fig. 3-1. Variation of the Coriolis parameter with latitude.

zonal winds is found from substituting (3-13) in the integrated continuity equation, (3-8),

$$\frac{\partial S_x}{\partial x} = \frac{\partial}{\partial y}\left[\frac{\partial \tau_x}{\partial y}\left(\frac{R_e}{2\Omega \cos\theta}\right)\right]$$

$$= \frac{R_e}{2\Omega}\left[\frac{\partial^2 \tau_x}{\partial y^2}\left(\frac{1}{\cos\theta}\right) + \frac{\partial \tau_x}{\partial y}\left(\frac{\sin\theta}{\cos^2\theta}\right)\frac{1}{R_e}\right]$$

or

$$\frac{\partial S_x}{\partial x} = \frac{1}{2\Omega \cos\theta}\left(\frac{\partial^2 \tau_x}{\partial y^2} R_e + \frac{\partial \tau_x}{\partial y} \tan\theta\right) \tag{3-14}$$

This result can satisfy boundary conditions at only one meridional boundary. Sverdrup introduced the kinematic condition that u at the eastern boundary is zero, $u|_{x=0} = 0$, and therefore $S_x|_{x=0} = 0$. Integrating (3-14) from $x = -L$ to $x = 0$ (Figure 3-2) yields

$$\int_{-L}^{0} \frac{\partial S_x}{\partial x}\,dx = \frac{1}{2\Omega \cos\theta}\left[R_e \int_{-L}^{0} \frac{\partial^2 \tau_x}{\partial y^2}\,dx + \tan\theta \int_{-L}^{0} \frac{\partial \tau_x}{\partial y}\,dx\right]$$

and then

$$S_x(x = -L) = -\frac{1}{2\Omega \cos\theta}\left[R_e \int_{-L}^{0} \frac{\partial^2 \tau_x}{\partial y^2}\,dx + \tan\theta \int_{-L}^{0} \frac{\partial \tau_x}{\partial y}\,dx\right] \tag{3-15}$$

Indicating average values by use of the overbar, (3-15) can be approximated as

$$S_x(x = -L) = -\frac{\Delta x}{2\Omega \cos\theta}\left[R_e \overline{\frac{\partial^2 \tau_x}{\partial y^2}} + \tan\theta\, \overline{\frac{\partial \tau_x}{\partial y}}\right] \tag{3-16}$$

where S_x is the zonal component of the net mass transport through a column of depth d and of unit width at a point with coordinate $x = -L$ and latitude θ; it has units of [mass]/([time][length]). It is apparent that S_x is related to the average value of the gradient and the curvature of the zonal wind stress in

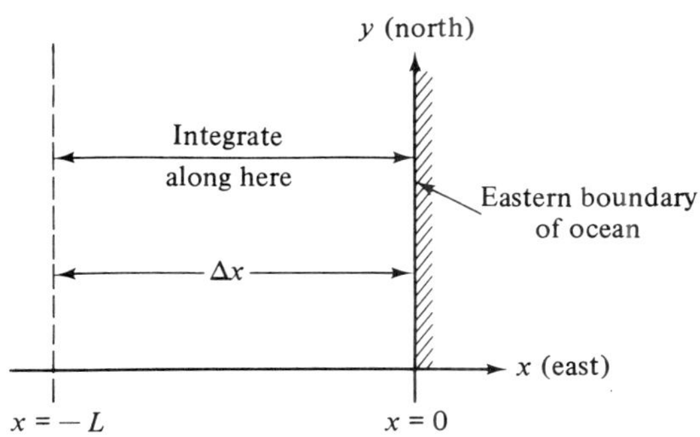

Fig. 3-2. Integration for the determination of net zonal mass transport, S_x.

the north-south direction between the point in question and the west coast. Sverdrup (1947) and Reid (1948) successfully applied these results in a study of the equatorial currents of the eastern Pacific Ocean. Comparison was made between transports determined using Sverdrup's theoretical approach and those calculated from actual oceanographic data obtained by the *Carnegie* in 1928–1929 and by the *Bushnell* in 1939. Both Sverdrup and Reid conclude that the predominant features of the distribution of mass and of mass transport can be accounted for semiquantitatively in terms of the wind stress alone. The details of the study are not within our present interest.

Reid points out that countercurrents (a current flowing adjacent to the main current but in the opposite direction) have long been recognized from ships' records in each of the equatorial oceans. Stockman (1945) has shown that a sufficient lateral variation in the velocity of a wind of constant direction can give rise to a countercurrent directed opposite to the prevailing wind. This situation is readily illustrated by a simple qualitative application of (3-16). Assume that the wind is zonal and constant at any latitude and that the profile is sinusoidal, as shown on the left of Figure 3-3. From a consideration of the "sign" of $\partial\tau_x/\partial y$ or $\partial^2\tau_x/\partial y^2$ at the latitudes for which one of them is zero, it is possible to obtain the "sign" of the east-west transport, S_x, qualitatively from (3-16). For example, along the latitude θ, corresponding to $y = y_1$, we see that $\partial\tau_x/\partial y < 0$ and $\partial^2\tau_x/\partial y^2 = 0$. From (3-16), $S_x > 0$ or the net zonal mass transport will be to the east. If this type of reasoning is repeated along the wind profile curve, S_x will appear schematically, as shown to the right in Figure 3-3. We see here then that the sinusoidal wind stress, always directed to the west, results in a region where the transport is directed to the east; that is, a countercurrent results.

Stommel (1957) has given a clear description of a very simple Sverdrup-type system: that of a *homogeneous* ocean, bounded by an eastern coast, and acted upon by a zonal wind stress. Figure 3-4 illustrates the resulting solution. The winds, represented by the heavy, shaded arrows above the sea surface, are zonal and exert a stress similar to the distribution of the Northern Hemisphere westerlies and trade winds over the actual ocean. The open arrows in the thin upper layer of frictional influence indicate horizontal transport in the Ekman layer (pure wind drift) and are everywhere proportional to and to the right of the wind. Between the maximum westerlies and the maximum trade winds there exists a downward motion of water resulting from a convergence of currents in the Ekman layer. To the north of the maximum westerlies and the south of the maximum trade winds, the vertical motion is upward, owing to the divergence imposed by the wind drift. From the bottom of the Ekman layer to the bottom of the ocean the vertical velocity component diminishes linearly to zero. At the latitudes of the maximum winds this vertical velocity is zero. A steady-state geostrophic current system can be constructed such that it matches by its field of divergence and convergence,

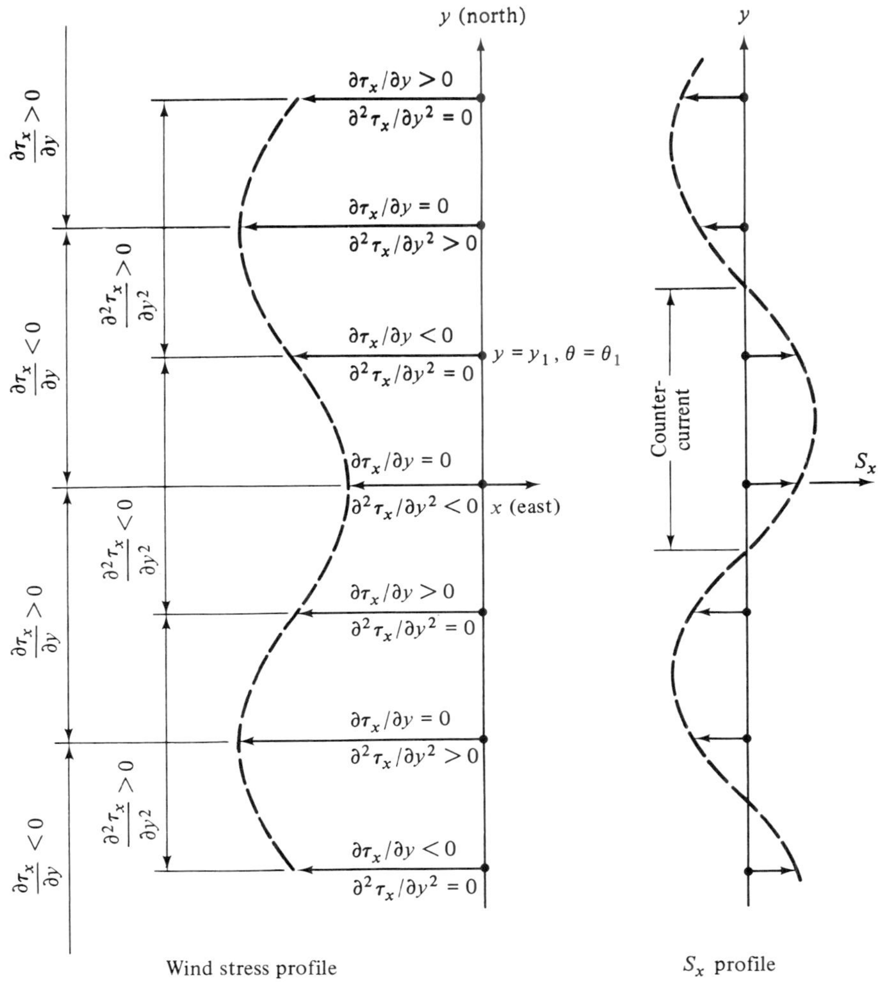

Fig. 3-3. A countercurrent resulting from a lateral variation of the wind stress.

because of the meridional variation of the Coriolis parameter, the field of convergence and divergence imposed by the Ekman layer, and that will satisfy the eastern coastal boundary condition. Thus, at midlatitudes (between the westerlies and the trades), the sea surface sloping downward to the east results in a geostrophic flow to the south; whereas north of the westerlies and south of the trades, the sea surface sloping up to the east results in a geostrophic flow to the north. The north-south slope of the sea surface (down to the north and to the south of midlatitudes) resulting from the mounding

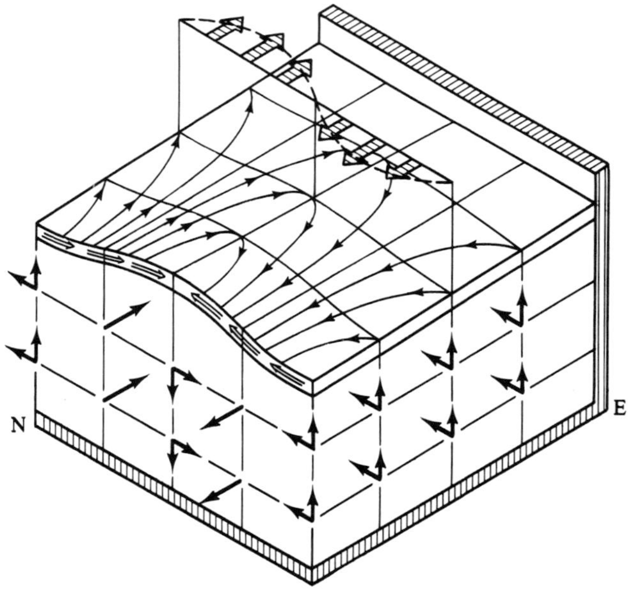

(a)

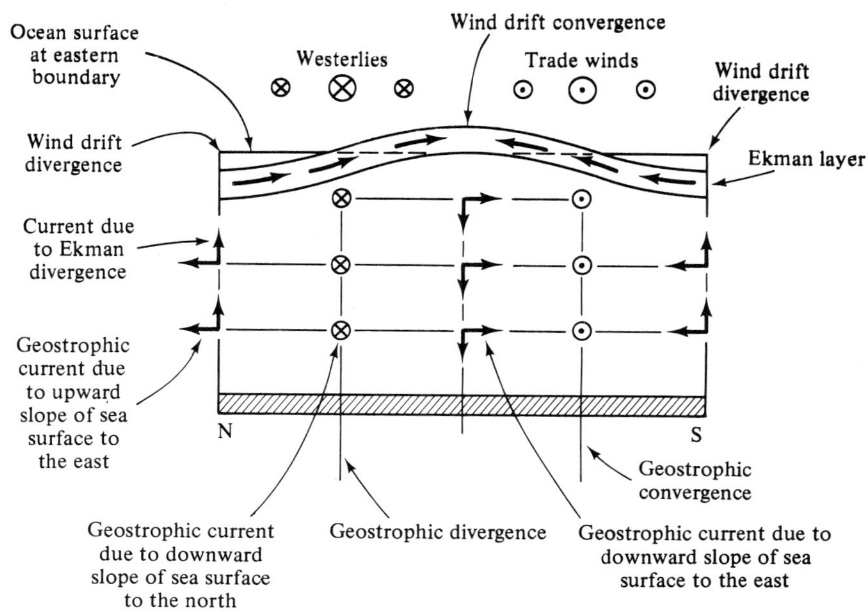

(b)

Fig. 3-4. Sverdrup-type solution in a homogeneous ocean. Stommel, 1957, from Neumann, *Ocean Currents,* Elsevier Publishing Co., 1968, p. 243.

due to wind-drift convergence and divergence, produces an eastward current along the latitude of maximum westerlies and a westward current along the latitude of maximum trades. Note that the north-south sea surface slope is zero at the eastern coastal boundary, so that the east-west geostrophic current at the boundary is zero, as required by the boundary condition. In Figure 3-4(a) the curved lines with the arrows on them are isobars; the arrowheads indicate the direction of the geostrophic flow (independent of depth).

3-3 Westward Intensification of Wind-Driven Ocean Currents

In the last section we saw that Sverdrup investigated wind-driven circulation in a central to eastern ocean region. The fluid was taken to be baroclinic and a solution was obtained for the mass transports. The basic physics of this model are such that it balances the local vorticity added by the wind the north-south transport of mass in the presence of a variable Coriolis parameter. We also noted several important deficiencies in this model, specifically:

1. Limited boundary conditions (the ocean could not be closed).
2. Intense horizontal gradients of flow cannot be represented, since all quantities involving such derivatives (i.e., $\partial u/\partial x$, $\partial u/\partial y$, etc.) were dropped on the way to obtaining the differential equations (3-1) and (3-2).

Now in viewing the general oceanic wind-driven circulation, one of the most striking features is the intensification of currents near the western borders. The Gulf Stream, the Kuroshio, and the Agulhas Current are examples. The physical reason for this bias in the circulation pattern remained obscure up until the recognition by Stommel (1948) that local regions of high relative vorticity must be present to complete the flow. It is now our intent to achieve an understanding of the physical mechanism behind western boundary currents through paralleling Stommel's investigation.

To physically represent the phenomenon of westward intensification, it will clearly be required to introduce at least one additional physical process into the basic equations of Sverdrup; it seems reasonable that additional friction be tried. Bottom friction is approximately zero, since a level of no motion above the topography of the bottom was assumed. The other possible type of friction is lateral friction, or drag, against the sides of the ocean. But before examining the analytic details, let us look at the problem qualitatively from the viewpoint of vorticity.

Suppose that we have the wind profile as shown in Figure 3-5. This distribution will cause the ocean to rotate clockwise, thereby producing

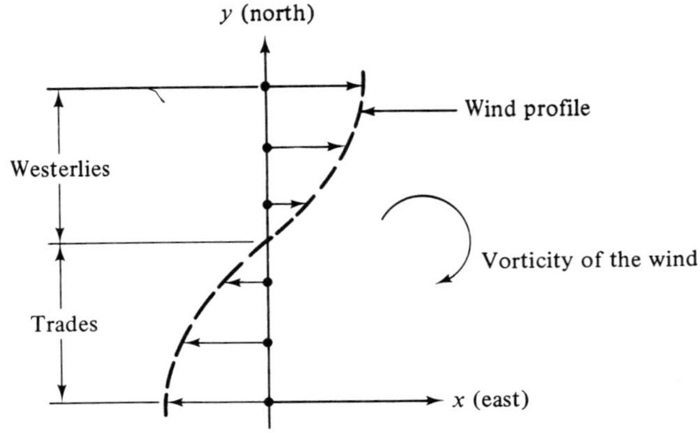

Fig. 3-5. Wind profile and associated vorticity.

negative or anticyclonic vorticity. On the other hand, relative vorticity changes induced by north-south fluid flow, due to the changes in the vertical component of the earth's rotation rate, are such as to conserve absolute vorticity if the fluid depth does not change; that is,

$$f + \zeta_z = \text{constant}$$

where f, the Coriolis parameter, is the vertical component of the planetary vorticity (positive in the Northern Hemisphere, increasing from the Equator to the Pole) and ζ_z is the relative or locally observed vorticity. The constant in this equation is the absolute vorticity. We have seen that a parcel of fluid in the Northern Hemisphere, moving from south to north, will tend to develop negative relative vorticity, while one moving from north to south will tend to develop positive vorticity. These two vorticity terms of relative and planetary character are the only ones used in Sverdrup's model.

Now, in a "whole" ocean circulation model, continuity requires that on the average as much water flows south as flows north. The net vorticity added by the wind cannot be balanced by a net meridional transport of fluid as in Sverdrup's model; we require some other mechanism which will act as a "sink" for the vorticity added by the wind. Let us see if lateral friction will alleviate this problem. Consider a simple rectangular ocean, as shown in Figure 3-6, with anticyclonic circulation. If lateral friction acts on the north-south boundaries, vorticity is introduced that is counter to the general circulation and consequently counter to the negative vorticity being introduced by the wind; we thus have our required sink.

Using the three types of terms—wind stress vorticity, frictional vorticity, and planetary vorticity—in an equation and demanding a steady-state condition (the relative vorticity at any point is a constant) permitted Stommel (1965) to make an estimate of their relative magnitude. It is to be

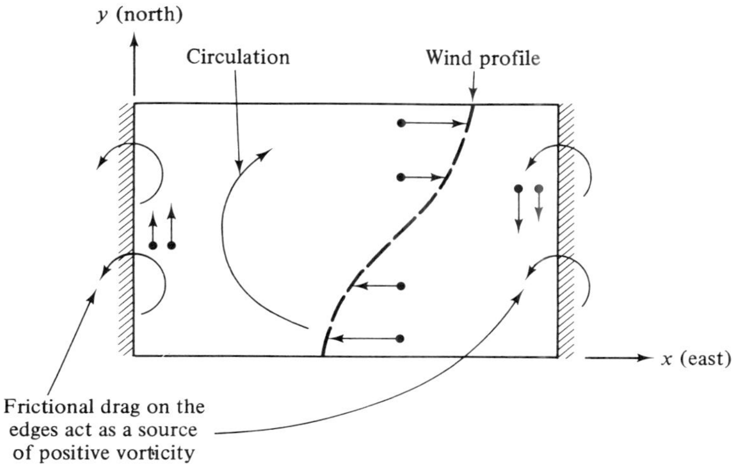

Fig. 3-6. Lateral friction acting as a sink for wind induced velocity.

pointed out that Stommel's discussion, which we parallel, is meant to indicate the sort of balance that must hold; it is a plausible demonstration rather than a rigorous one. The distribution of the wind stress vorticity tendency is regarded as fixed and is assigned a relative value of -1. If the circulation of the ocean is symmetric (i.e., a broad current system exists), the frictional vorticity tendency might be as much as one order of magnitude smaller, say $+0.1$, along either the east or west coast; and the planetary vorticity tendency would be perhaps of order -1 in the western part and $+1$ in the eastern part of the ocean. As shown in Table 3-1, an approximate balance of the vorticity tendencies exist in the eastern portion of the ocean, but the western part will not be near equilibrium; for such an ocean as this, a steady-state condition is not possible. On the other hand, if the current system is

TABLE 3-1. Vorticity Tendencies in an Ocean with a Symmetrical Circulation

Vorticity Tendency	*North-Flowing Currents in Western Side of Ocean*	*South-Flowing Currents in Eastern Side of Ocean*
Wind stress	−1.0	−1.0
Frictional	−0.1	+0.1
Planetary	−1.0	+1.0
Total	−1.9	+0.1

Source: Stommel (1965).

strongly asymmetrical, with streamlines crowding the western edge of the ocean, the near balance between the wind stress vorticity tendency and the planetary vorticity tendency in the eastern part of the ocean is not significantly altered and the frictional vorticity tendency remains small. However, on the western side, because of the crowding of the streamlines, the velocity and shear of the flow is great and the frictional and planetary vorticity tendencies are greatly enhanced, to perhaps an assumed order of magnitude greater than the wind stress tendency, say +10 and −10, respectively. A balance between the tendencies on the western and eastern sides of the ocean is thus obtained and a steady-state condition results. Table 3-2 illustrates the balance of terms in such an asymmetric circulation.

TABLE 3-2. Vorticity Tendencies in an Ocean with an Asymmetrical Circulation

Vorticity Tendency	*North-Flowing Currents in Western Side of Ocean*	*South-Flowing Currents in Eastern Side of Ocean*
Wind stress	−1.0	−1.0
Frictional	+10.0	+0.1
Planetary	−9.0	+0.9
Total	0.0	0.0

Source: Stommel (1965).

Stommel (1948) proposed a simple model of a wind-driven ocean that succeeded in showing the important effect of the variation of the Coriolis parameter with latitude on the current and at the same time closed Sverdrup's two-sided ocean model. Assume a rectangular ocean with the origin of a Cartesian coordinate system at the southwest corner, with the x axis to the east, the y axis to the north, and the z axis positive upward (Figure 3-7).

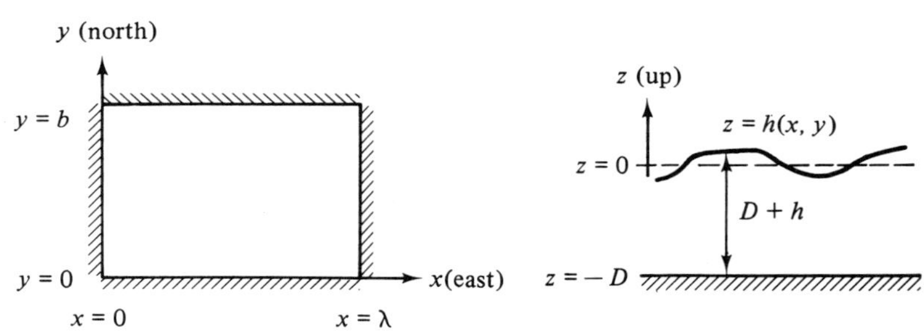

Fig. 3-7. Coordinates and boundaries of the rectangular ocean basin used by Stommel.

The lateral boundaries are at $x = 0$ and λ and $y = 0$ and b. The ocean is considered to be homogeneous and when at rest to be of constant depth D. In the presence of currents, the depth differs from D by the relatively small variable quantity $h(x, y)$. The total depth then is $(D + h)$.

The winds over the ocean are assumed to be zonal, the trades over the equatorial half and the prevailing westerlies over the poleward half of the basin, as shown in Figures 3-5 and 3-6. A simple functional form of the wind stress is taken as

$$\tau_x \equiv -F \cos \frac{\pi y}{b} \qquad \tau_y \equiv 0 \tag{3-17}$$

To keep the ocean from accelerating, frictional dissipative terms are required—let Ξ_x and Ξ_y be the lateral frictional components. With these restrictions, and assuming that the inertial terms are small (linearizing), the x and y components of the equation of motion are, similar to (3-1) and (3-2),

$$\frac{1}{\rho}\frac{\partial p}{\partial x} = fv + \frac{\partial}{\partial z}\left(K_V \frac{\partial u}{\partial z}\right) + \Xi_x \tag{3-18}$$

$$\frac{1}{\rho}\frac{\partial p}{\partial y} = -fu + \frac{\partial}{\partial z}\left(K_V \frac{\partial v}{\partial z}\right) + \Xi_y \tag{3-19}$$

where $K_V \equiv A_V/\rho$ is the austausch (kinematic) coefficient. The z component is the hydrostatic equation,

$$\frac{\partial p}{\partial z} = -\rho g \tag{2-217}$$

Since the water is homogeneous, the last equation can be integrated with respect to z,

$$p = -\rho g \int_h^z dz = \rho g(h - z) \tag{3-20}$$

and differentiated with respect to x and y to get

$$\frac{\partial p}{\partial x} = \rho g \frac{\partial h}{\partial x} \qquad \frac{\partial p}{\partial y} = \rho g \frac{\partial h}{\partial y} \tag{3-21}$$

which permits the substitution of the sea surface slope for the horizontal pressure gradients. Vertically integrating (3-18) and (3-19) and using (3-21) gives

$$g\int_{-D}^{h} \frac{\partial h}{\partial x}\, dz = \int_{-D}^{h} fv\, dz + \int_{-D}^{h} \frac{\partial}{\partial z}\left(K_V \frac{\partial u}{\partial z}\right) dz + \int_{-D}^{h} \Xi_x\, dz \tag{3-22}$$

and

$$g\int_{-D}^{h} \frac{\partial h}{\partial y}\, dz = -\int_{-D}^{h} fu\, dz + \int_{-D}^{h} \frac{\partial}{\partial z}\left(K_V \frac{\partial v}{\partial z}\right) dz + \int_{-D}^{h} \Xi_y\, dz \tag{3-23}$$

To simplify the analysis, Stommel took the quantities u and v to be independent of depth—a pragmatic assumption that does not seriously compromise the result. This simplification amounts to regarding the wind as a body force (as if the wind acts directly throughout the fluid volume) rather than exerting a surface stress. Writing that

$$K_V \frac{\partial u}{\partial z}\bigg|_{z=h} = \tau_x \qquad K_V \frac{\partial u}{\partial z}\bigg|_{z=-D} = 0$$

$$K_V \frac{\partial v}{\partial z}\bigg|_{z=h} = \tau_y \qquad K_V \frac{\partial v}{\partial z}\bigg|_{z=-D} = 0$$

and completing the integration of (3-22) and (3-23), we obtain

$$g(D+h)\frac{\partial h}{\partial x} = fv(D+h) + \tau_x + (D+h)\Xi_x \tag{3-24}$$

and

$$g(D+h)\frac{\partial h}{\partial y} = -fu(D+h) + \tau_y + (D+h)\Xi_y \tag{3-25}$$

For the lateral friction terms Ξ_x and Ξ_y, the simplest type of retarding force is used; that is, the retarding force is taken as being proportional to the fluid velocity,

$$(D+h)\Xi_x \propto -u \qquad (D+h)\Xi_y \propto -v$$

or

$$\begin{aligned}(D+h)\Xi_x &= -Ru\\ (D+h)\Xi_y &= -Rv\end{aligned} \tag{3-26}$$

where the proportionality constant R is a coefficient of friction with units of [length]/[time], whose magnitude will be determined at a later point in the analysis by "tuning" the solution so as to give reasonable fluid velocities.

To obtain a single equation expressing a vorticity balance, (3-17) and (3-26) are substituted into (3-24) and (3-25) and the former is differentiated with respect to y and the latter with respect to x,

$$\begin{aligned}&g(D+h)\frac{\partial^2 h}{\partial x\,\partial y} + g\frac{\partial h}{\partial y}\frac{\partial h}{\partial x}\\ &\quad = f\frac{\partial}{\partial y}(v(D+h)) + v(D+h)\frac{\partial f}{\partial y} + \left(\frac{\pi F}{b}\right)\sin\left(\frac{\pi y}{b}\right) - R\frac{\partial u}{\partial y}\end{aligned} \tag{3-27}$$

and

$$g(D+h)\frac{\partial^2 h}{\partial x\,\partial y} + g\frac{\partial h}{\partial x}\frac{\partial h}{\partial y} = -f\frac{\partial}{\partial x}[u(D+h)] - R\frac{\partial v}{\partial x} \tag{3-28}$$

Subtracting these two gives

$$\begin{aligned}&-f\left\{\frac{\partial}{\partial x}[u(D+h)] + \frac{\partial}{\partial y}[v(D+h)]\right\} - v(D+h)\frac{\partial f}{\partial y}\\ &\qquad = \left(\frac{\pi F}{b}\right)\sin\frac{\pi y}{b} + R\left(\frac{\partial v}{\partial x} - \frac{\partial u}{\partial y}\right)\end{aligned} \tag{3-29}$$

To this, the equation of continuity must be added. Recalling the vertically integrated, steady-state form of the continuity equation,

$$\frac{\partial S_x}{\partial x} + \frac{\partial S_y}{\partial y} = 0 \qquad (3\text{-}8)$$

we may write that

$$\frac{\partial}{\partial x}\int_{-D}^{h} \rho u\, dz + \frac{\partial}{\partial y}\int_{-D}^{h} \rho v\, dz = 0$$

which yields for $\rho =$ constant and u and v independent of depth,

$$\frac{\partial}{\partial x}[u(D + h)] + \frac{\partial}{\partial y}[v(D + h)] = 0 \qquad (3\text{-}30)$$

Thus, (3-29) reduces to

$$-v(D + h)\frac{\partial f}{\partial y} = \left(\frac{\pi F}{b}\right) \sin\frac{\pi y}{b} + R\left(\frac{\partial v}{\partial x} - \frac{\partial u}{\partial y}\right) \qquad (3\text{-}31)$$

The first two terms are those comprising Sverdrup's curl equation for zonal flow, (3-11), representing a balance between the meridional transport of vorticity and the curl of the wind stress. For the present case Stommel has introduced the last term, which expresses the additional vorticity induced due to lateral friction.

In the actual oceans, $h \ll D$ (h is on the order of centimeters, D on the order of meters), so that to a first degree of approximation (3-31) may be rewritten as

$$v\frac{D}{R}\frac{\partial f}{\partial y} + \left(\frac{\pi F}{bR}\right) \sin\frac{\pi y}{b} + \left(\frac{\partial v}{\partial x} - \frac{\partial u}{\partial y}\right) = 0 \qquad (3\text{-}32)$$

To the same degree of approximation the continuity equation (3-30) may be replaced by

$$\frac{\partial u}{\partial x} + \frac{\partial v}{\partial y} = 0 \qquad (3\text{-}33)$$

and now since the flow is two-dimensional and incompressible, it is possible to introduce a stream function ψ where

$$u = -\frac{\partial \psi}{\partial y} \qquad v = \frac{\partial \psi}{\partial x} \qquad (2\text{-}64)$$

Also, for brevity we define

$$\alpha \equiv \frac{D}{R}\frac{\partial f}{\partial y} \qquad \gamma \equiv -\frac{\pi F}{bR} \qquad (3\text{-}34)$$

and (3-32) is rewritten as

$$\left(\frac{\partial^2 \psi}{\partial x^2} + \frac{\partial^2 \psi}{\partial y^2}\right) + \alpha\frac{\partial \psi}{\partial x} = \gamma \sin\frac{\pi y}{b} \qquad (3\text{-}35)$$

This vorticity equation is the partial differential equation whose solution we seek, subject to the boundary conditions that there be no flow across the edges of our ocean. That is,

$$u(0, y) = u(\lambda, y) = v(x, 0) = v(x, b) = 0$$

which, expressed in terms of ψ,

$$\psi(0, y) = \psi(\lambda, y) = \psi(x, 0) = \psi(x, b) = 0 \qquad (3\text{-}36)$$

physically stating that the vertical boundaries of the ocean is a streamline.

In oceanography, the variation of f with latitude is called the *beta effect* and a tangent plane system of coordinates in which the Coriolis parameter is considered constant except where differentiated with respect to the meridional direction is called the *beta* (β) *plane*. If f is taken as a linear function of y,

$$f \equiv \beta y \qquad (3\text{-}37)$$

then, from (3-12),

$$\frac{\partial f}{\partial y} = \beta = \text{constant} = 2\Omega \cos \theta_0 / R_e$$

Thus, α is constant and (3-35) becomes a partial differential equation with constant coefficients.

The homogeneous equivalent of (3-35) with $\alpha =$ constant can be solved directly by the separation-of-variables technique. Adding this solution to a particular solution of (3-35) will give the general solution of (3-35); that is,

$$\psi \equiv \psi_1 + \psi_2 \qquad (3\text{-}38)$$

where ψ_2 is the general solution of the homogeneous equation and ψ_1 is the particular solution of the nonhomogeneous equation. If we assume that

$$\frac{\partial^2 \psi_1}{\partial y^2} = \gamma \sin \frac{\pi y}{b}$$

and integrate twice, we obtain

$$\psi_1 = -\left(\frac{\gamma b^2}{\pi^2}\right) \sin \frac{\pi y}{b} \qquad (3\text{-}39)$$

The homogeneous equation,

$$\frac{\partial^2 \psi_2}{\partial x^2} + \frac{\partial^2 \psi_2}{\partial y^2} + \alpha \frac{\partial \psi_2}{\partial x} = 0 \qquad (3\text{-}40)$$

in view of the boundary conditions of (3-36), may be treated so as to satisfy boundary conditions on ψ_2. For example,

$$\psi(0, y) = \psi_1(0, y) + \psi_2(0, y) = 0$$

and substituting for ψ_1 from (3-39) yields

$$\psi_2(0, y) = \left(\frac{\gamma b^2}{\pi^2}\right) \sin \frac{\pi y}{b} \qquad (3\text{-}41)$$

In a similar fashion we find that

$$\psi_2(\lambda, y) = \left(\frac{\gamma b^2}{\pi^2}\right) \sin \frac{\pi y}{b} = -\psi_1 \qquad (3\text{-}42)$$

and

$$\psi_2(x, 0) = \psi_2(x, b) = 0 \qquad (3\text{-}43)$$

To solve (3-40) by separation of variables, let

$$\psi_2 = X(x)\, Y(y)$$

where $X(x)$ is a function of x only and $Y(y)$ is a function of y only. Substituting this into (3-40), we obtain

$$X''Y + XY'' + \alpha X'Y = 0$$

and separating yields

$$\frac{X'' + \alpha X'}{X} = -\frac{Y''}{Y} \equiv n^2$$

where n is a constant. Solution of (3-40) is now reduced to finding the solutions to the two second-order, ordinary differential equations

$$X'' + \alpha X' - n^2 X = 0 \qquad (3\text{-}44)$$

and

$$Y'' + n^2 Y = 0 \qquad (3\text{-}45)$$

Since the characteristic equation of (3-44),

$$k^2 + \alpha k - n^2 = 0$$

has the real and unequal roots of

$$k_1 = -\frac{\alpha}{2} - \sqrt{\frac{\alpha^2}{4} + n^2} \qquad (3\text{-}46a)$$

and

$$k_2 = -\frac{\alpha}{2} + \sqrt{\frac{\alpha^4}{4} + n^2} \qquad (3\text{-}46b)$$

we get that

$$X(x) = C_1 e^{k_1 x} + C_2 e^{k_2 x} \qquad (3\text{-}47)$$

where C_1 and C_2 are constants. Equation (3-45), on the other hand, has complex conjugates as the roots of its characteristic equation and therefore results in the solution

$$Y(y) = C_3 \sin ny + C_4 \cos ny \qquad (3\text{-}48)$$

where C_3 and C_4 are also constants. As Stommel points out, these solutions are very general but reduce to simple closed forms when the boundary conditions are applied. First of all, from $\psi_2(x, 0) = 0$ we see that $C_4 = 0$, and from $\psi_2(x, b) = 0$ that $n = \pi/b$. Then for $\psi_2(0, y) = -\psi_1$,

$$\psi_2(0, y) = C_3 \sin\left(\frac{\pi y}{b}\right)(C_1 + C_2) = \left(\frac{\gamma b^2}{\pi^2}\right) \sin \frac{\pi y}{b}$$

which permits us to take

$$C_1 + C_2 = 1 \tag{3-49a}$$

if we identify C_3 as equal to $\gamma b^2/\pi^2$. Finally, from $\psi_2(\lambda, y) = -\psi_1$,

$$\psi_2(\lambda, y) = \left(\frac{\gamma b^2}{\pi^2}\right) \sin\left(\frac{\pi y}{b}\right)(C_1 e^{k_1\lambda} + C_2 e^{k_2\lambda}) = \left(\frac{\gamma b^2}{\pi^2}\right) \sin \frac{\pi y}{b}$$

or

$$C_1 e^{k_1\lambda} + C_2 e^{k_2\lambda} = 1 \tag{3-49b}$$

Solving the two equations (3-49) simultaneously, we obtain

$$C_1 = \frac{1 - e^{k_2\lambda}}{e^{k_1\lambda} - e^{k_2\lambda}} \tag{3-50a}$$

and

$$C_2 = \frac{e^{k_1\lambda} - 1}{e^{k_1\lambda} - e^{k_2\lambda}} \tag{3-50b}$$

Putting all the parts together yields

$$\psi = \left(\frac{\gamma b^2}{\pi^2}\right) \sin\left(\frac{\pi y}{b}\right)\left[\frac{(1 - e^{k_2\lambda})e^{k_1x} + (e^{k_1\lambda} - 1)e^{k_2x}}{e^{k_1\lambda} - e^{k_2\lambda}} - 1\right] \tag{3-51}$$

where k_1 and k_2 are given by (3-46). The lines of ψ = constant are streamlines of the horizontal flow. Using (2-64), the horizontal velocity components u and v are found by direct differentiation,

$$u = -\left(\frac{\gamma b}{\pi}\right) \cos\left(\frac{\pi y}{b}\right)(C_1 e^{k_1x} + C_2 e^{k_2x} - 1) \tag{3-52}$$

and

$$v = \left(\frac{\gamma b^2}{\pi^2}\right) \sin\left(\frac{\pi y}{b}\right)(C_1 k_1 e^{k_1x} + C_2 k_2 e^{k_2x}) \tag{3-53}$$

Stommel, to visualize the meaning of solution (3-51) and in order to show the role played by the different parameters, computed various examples. He examined three cases, in each of which the role of the Coriolis parameter is different but all involving the same effects of wind stress, friction, and the horizontal pressure gradients resulting from variations of the surface elevation, h. First, it is assumed that the Coriolis parameter is zero everywhere, the case of a nonrotating ocean; second, it is assumed that the Coriolis parameter is a nonzero constant everywhere, the case of a uniformly rotating ocean; and third, it is assumed that the Coriolis parameter is a linear function of latitude [as in (3-37)], this case being the one most nearly approximating the state in the real ocean.

The dimensions of the model ocean are taken as $\lambda = 10^9$ cm $= 10^4$ km, $b = 2\pi \times 10^8$ cm $= 2\pi \times 10^3$ km $\doteq 6300$ km, $D = 2 \times 10^4$ cm $=$ 200 m. In selecting D to be so small, Stommel is considering only the surface layers directly affected by the wind. The maximum wind stress, F, is taken as 1 dyne/cm^2. The only quantity for which a value remains to be specified is that of the friction coefficient R. A value of $R = 0.02$ was assumed since resulting fluid velocities approach those observed in the real ocean.

Case of a Nonrotating Ocean. For a nonrotating ocean the constants C_1 and C_2, from (3-34), (3-46), and (3-50), approximate (to within 1 percent according to Stommel)

$$C_1 = \frac{1 - e^{\pi\lambda/b}}{e^{-\pi\lambda/b} - e^{\pi\lambda/b}} \doteq 1$$

and

$$C_2 = \frac{e^{-\pi\lambda/b} - 1}{e^{-\pi\lambda/b} - e^{\pi\lambda/b}} \doteq e^{-\pi\lambda/b}$$

From (3-51), the equation for the streamlines is

$$\psi = \left(\frac{\gamma b^2}{\pi^2}\right) \sin\left(\frac{\pi y}{b}\right)[e^{(x-\lambda)\pi/b} + e^{-x\pi/b} - 1]$$

An actual plot of this result is shown in Figure 3-8; the streamlines are clearly symmetrical about a north-south and an east-west line and constitute a broad circulation with no tendency toward crowding in any region.

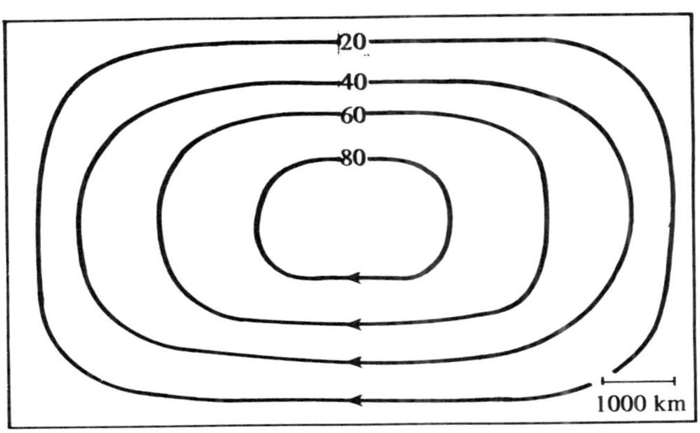

Fig. 3-8. Streamlines for a nonrotating or uniformly rotating ocean. From Stommel, *Trans. Amer. Geophy. Un.*, *Vol. 29*, American Geophysical Union, 1948, pp 202–206.

Case of a Uniformly Rotating Ocean. Taking the Coriolis parameter as a constant does not in any way change the constants C_1 and C_2 or the streamline pattern discussed above and shown in Figure 3-8.

Case of the Coriolis Parameter Being a Linear Function of Latitude. In the real ocean, the variation of the Coriolis parameter is nearly linear in low latitudes and β approximates 10^{-13} cm^{-1}/sec. Using this value, along with those specified for the other parameters, produces a complete lack of east-west symmetry (see Figure 3-9). Now the streamlines crowd the western border of the ocean while the circulation in the central and eastern portions remains broad and diffuse. The similarities between this result and the westward intensification in the real ocean is striking and Stommel concludes that the cause of both is the same—the variation of the Coriolis parameter with latitude.

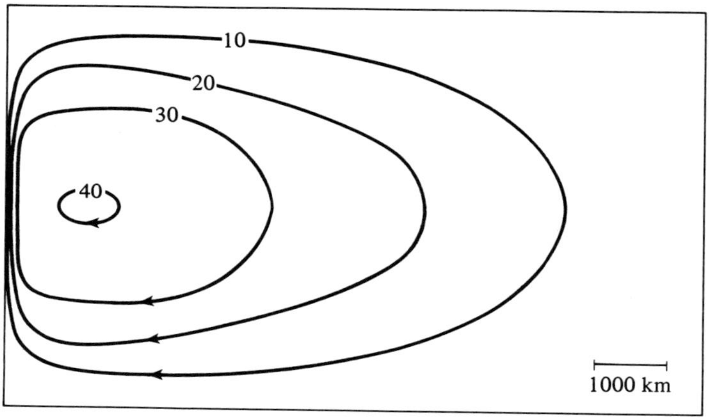

Fig. 3-9. Streamlines for an ocean in which the Coriolis parameter is a linear function of the latitude. From Stommel, *Trans. Amer. Geophy. Un., Vol. 29*, American Geophysical Union, 1948, pp. 202–206.

In all of the previous derivation, it has been assumed that the analysis is made in a Northern Hemisphere ocean. To extend the results and conclusions to the Southern Hemisphere, note that α is unaffected in crossing the Equator in that the latitude south of the Equator is taken as negative, $f = 2\Omega \sin \theta$ is negative, but β remains positive. If the wind stress is still given as

$$\tau_x = -F \cos \frac{\pi y}{b}$$

b will be less than zero, and γ simply changes sign. Figures 3-8 and 3-9 may be transformed to applying below the equator by simple reflection across the x axis, since from an examination of (3-52) and (3-53) it is evident that $u(x, y) = u(x, -y)$ and $v(x, y) = -v(x, -y)$. The crowding of streamlines, in the case of the Coriolis parameter being a linear function of the latitude, is therefore toward the western border of each ocean, regardless of hemisphere.

3-4 Munk's Theory of the Wind-Driven Ocean Circulation

In his book *The Gulf Stream*, Stommel in 1965 wrote: "By far the most distinguished theoretical investigation into the wind-driven ocean circulation in that conducted by Munk (1950), who succeeded in deducing many features of the mean ocean circulation from the wind-stress alone." In that which follows we will follow Munk's quantitative development of his theory in great detail, being of the opinion that in so doing, many useful techniques will be illustrated.

In Stommel's model for westward intensification, the ocean was assumed to be homogeneous, with the result that the horizontal currents extend to the bottom. This is in contrast with observations according to which the bulk of the mass transport in the main ocean currents takes place in the upper portions. This also lead to the mathematical complication that forced Stommel to resort to a rather arbitrary frictional force. To avoid this difficulty, Munk retained Sverdrup's integrated mass transport as the dependent variable, once again making it possible to examine the more general case of a baroclinic (nonhomogeneous) ocean without the necessity of having to specify the nature of the vertical distribution of density and current. As shown in Figure 3-10, the ocean is taken as having rectangular boundaries and with

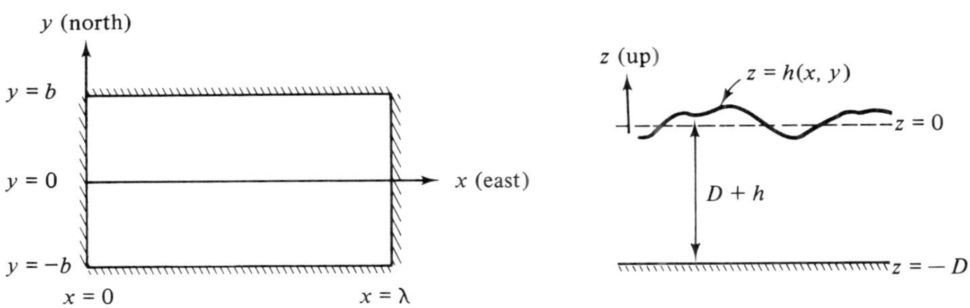

Fig. 3-10. Coordinates and boundaries of the rectangular ocean basin of Munk.

a plane surface when at rest. The x axis is to the east, the y axis to the north, and the z axis up; the plane $z = 0$ lies at the mean surface, the actual surface is at $z = h(x, y)$. The horizontal equations of motion for steady-state conditions, having taken the nonlinear terms as zero, are similar to those of (3-18) and (3-19):

$$-\rho f v = -\frac{\partial p}{\partial x} + \frac{\partial}{\partial x}\left(A_H \frac{\partial u}{\partial x}\right) + \frac{\partial}{\partial y}\left(A_H \frac{\partial u}{\partial y}\right) + \frac{\partial}{\partial z}\left(A_V \frac{\partial u}{\partial z}\right) \qquad (3\text{-}54)$$

and

$$\rho f u = -\frac{\partial p}{\partial y} + \frac{\partial}{\partial x}\left(A_H \frac{\partial v}{\partial x}\right) + \frac{\partial}{\partial y}\left(A_H \frac{\partial v}{\partial y}\right) + \frac{\partial}{\partial z}\left(A_V \frac{\partial v}{\partial z}\right) \qquad (3\text{-}55)$$

These equations have the same sort of terms as those of Stommel—Coriolis force, pressure gradient, lateral friction, and vertical friction—the difference being a more realistic form for the lateral friction terms. Thus, the solution hopefully will give a more realistic representation of what a frictionally dominated western boundary current looks like.

It is desired to vertically integrate (3-54) and (3-55) from some constant depth $z = -D$, at which depth the horizontal pressure gradients and the motion vanish (as in Sverdrup's model), up to the surface $z = h(x, y)$. Defining the vertically integrated pressure as

$$P \equiv \int_{-D}^{h} p\, dz \tag{3-56}$$

the integrals of the pressure gradients of (3-54) and (3-55) are

$$\int_{-D}^{h} \frac{\partial p}{\partial x}\, dz = \frac{\partial}{\partial x}\int_{-D}^{h} p\, dz - \frac{\partial h}{\partial x} p(z = h) - \frac{\partial D}{\partial x} p(z = -D)$$

and

$$\int_{-D}^{h} \frac{\partial p}{\partial x}\, dz = \frac{\partial}{\partial y}\int_{-D}^{h} p\, dz - \frac{\partial h}{\partial y} p(z = h) - \frac{\partial D}{\partial y} p(z = -D)$$

the last two terms of which are zero since $\partial D/\partial x = \partial D/\partial y = 0$ and $p(z = h) = 0$. Employing (3-56), then, yields

$$\begin{aligned} \int_{-D}^{h} \frac{\partial p}{\partial x}\, dz &= \frac{\partial P}{\partial x} \\ \int_{-D}^{h} \frac{\partial p}{\partial y}\, dz &= \frac{\partial P}{\partial y} \end{aligned} \tag{3-57}$$

Integration of the quantities in (3-54) and (3-55) involving the Coriolis parameter give the mass transport components S_x and S_y,

$$\begin{aligned} f\int_{-D}^{h} \rho u\, dz &= fS_x \\ f\int_{-D}^{h} \rho v\, dz &= fS_y \end{aligned} \tag{3-58}$$

and of those involving the vertical shearing stress terms, the x and y components of the surface wind stress, τ_x and τ_y,

$$\begin{aligned} \int_{-D}^{h} \frac{\partial}{\partial z}\left(A_V \frac{\partial u}{\partial z}\right) dz &= \tau_x \\ \int_{-D}^{h} \frac{\partial}{\partial z}\left(A_V \frac{\partial v}{\partial z}\right) dz &= \tau_y \end{aligned} \tag{3-59}$$

The horizontal shearing terms present a much more formidable task and certain approximations must be arbitrarily made if anything analytical is to be achieved. For example, taking both $A_H = \rho K_H$ and K_H, the kinematic eddy viscosity, as constant,

$$\begin{aligned}\int_{-D}^{h} \frac{\partial}{\partial x}\left(A_H \frac{\partial u}{\partial x}\right) dz &= K_H \int_{-D}^{h} \frac{\partial}{\partial x}\left(\rho \frac{\partial u}{\partial x}\right) dz \\ &= K_H\left[\frac{\partial}{\partial x}\int_{-D}^{h} \rho \frac{\partial u}{\partial x}\, dz - \frac{\partial h}{\partial x}\left(\rho \frac{\partial u}{\partial x}\right)_{z=h}\right] \\ &= K_H\left[\frac{\partial}{\partial x}\int_{-D}^{h} \frac{\partial \rho u}{\partial x}\, dz - \frac{\partial}{\partial x}\int_{-D}^{h} u \frac{\partial \rho}{\partial x}\, dz - \frac{\partial h}{\partial x}\left(\rho \frac{\partial u}{\partial x}\right)_{z=h}\right] \\ &= K_H\left\{\frac{\partial^2}{\partial x^2}\int_{-D}^{h} \rho u\, dz - \frac{\partial}{\partial x}\left[\frac{\partial h}{\partial x}(\rho u)_{z=h}\right.\right. \\ &\qquad \left.\left. + \int_{-D}^{h} u \frac{\partial \rho}{\partial x}\, dz\right] - \frac{\partial h}{\partial x}\left(\rho \frac{\partial u}{\partial x}\right)_{z=h}\right\}\end{aligned}$$

the retaining of all of which is clearly impossible. Munk assumed that the complete integral could be approximated as

$$\int_{-D}^{h} \frac{\partial}{\partial x}\left(A_H \frac{\partial u}{\partial x}\right) dz \doteq K_H \frac{\partial^2}{\partial x^2}\int_{-D}^{h} \rho u\, dz = K_H \frac{\partial^2 S_x}{\partial x^2}$$

having employed (3-58). The entire set of horizontal shearing terms then approximate, upon integration,

$$\int_{-D}^{h}\left[\frac{\partial}{\partial x}\left(A_H \frac{\partial u}{\partial x}\right) + \frac{\partial}{\partial y}\left(A_H \frac{\partial u}{\partial y}\right)\right] dz \doteq K_H\left(\frac{\partial^2}{\partial x^2} + \frac{\partial^2}{\partial y^2}\right) S_x \qquad (3\text{-}60a)$$

and

$$\int_{-D}^{h}\left[\frac{\partial}{\partial x}\left(A_H \frac{\partial v}{\partial x}\right) + \frac{\partial}{\partial y}\left(A_H \frac{\partial v}{\partial y}\right)\right] dz \doteq K_H\left(\frac{\partial^2}{\partial x^2} + \frac{\partial^2}{\partial y^2}\right) S_y \qquad (3\text{-}60b)$$

Putting it all together, from (3-57), (3-58), (3-59), and (3-60*a*, *b*), the vertically integrated equations of motion are

$$-fS_y = -\frac{\partial P}{\partial x} + K_H\left(\frac{\partial^2}{\partial x^2} + \frac{\partial^2}{\partial y^2}\right) S_x + \tau_x \qquad (3\text{-}61)$$

and

$$fS_x = -\frac{\partial P}{\partial y} + K_H\left(\frac{\partial^2}{\partial x^2} + \frac{\partial^2}{\partial y^2}\right) S_y + \tau_y \qquad (3\text{-}62)$$

Now since the variation of h with x and y is relatively small and a steady-state solution is sought, it is again possible, as in Sverdrup, to employ the vertically integrated continuity equation in the form

$$\frac{\partial S_x}{\partial x} + \frac{\partial S_y}{\partial y} = 0 \qquad (3\text{-}8)$$

and thus assert the existence of a mass *transport stream function* $\Pi = \Pi(x, y)$

such that

$$S_x \equiv -\frac{\partial \Pi}{\partial y} \qquad S_y \equiv \frac{\partial \Pi}{\partial x} \tag{3-63}$$

having units of [mass]/[time]. This is analogous to the existence of the volumn transport steam function $\psi(x, y)$ of (2-64), which exists if the flow is two-dimensional and divergentless (horizontal divergence of the velocity is zero); in this instance a mass transport steam function is said to exist if the flow is two-dimensional and the horizontal divergence of the mass transport is zero [as in (3-8)]. Substituting (3-63) into (3-61) and (3-62), we obtain

$$-f\frac{\partial \Pi}{\partial x} = -\frac{\partial P}{\partial x} - K_H\left(\frac{\partial^3 \Pi}{\partial x^2\, \partial y} + \frac{\partial^3 \Pi}{\partial y^3}\right) + \tau_x$$

and

$$-f\frac{\partial \Pi}{\partial y} = -\frac{\partial P}{\partial y} + K_H\left(\frac{\partial^3 \Pi}{\partial x^3} + \frac{\partial^3 \Pi}{\partial x\, \partial y^2}\right) + \tau_y$$

cross-differentiating,

$$-f\frac{\partial^2 \Pi}{\partial x\, \partial y} - \beta\frac{\partial \Pi}{\partial x} = -\frac{\partial^2 P}{\partial x\, \partial y} - K_H\left(\frac{\partial^4 \Pi}{\partial x^2\, \partial y^2} + \frac{\partial^4 \Pi}{\partial y^4}\right) + \frac{\partial \tau_x}{\partial y}$$

and

$$-f\frac{\partial^2 \Pi}{\partial x\, \partial y} = -\frac{\partial^2 P}{\partial x\, \partial y} + K_H\left(\frac{\partial^4 \Pi}{\partial x^4} + \frac{\partial^4 \Pi}{\partial x^2\, \partial y^2}\right) + \frac{\partial \tau_y}{\partial x}$$

and subtracting, we obtain

$$\left(K_H\nabla^4 - \beta\frac{\partial}{\partial x}\right)\Pi = -\text{curl}_z\, \boldsymbol{\tau} \tag{3-64}$$

where ∇^4 is the two-dimensional biharmonic operator,

$$\nabla^4 \equiv \frac{\partial^4}{\partial x^4} + 2\frac{\partial^4}{\partial x^2\, \partial y^2} + \frac{\partial^4}{\partial y^4}$$

As before, this results in a *vorticity tendency equation* and expresses a balance between three torques—that due to lateral friction, that due to planetary vorticity, and that due to the wind stress. The second and third terms of (3-64) constitute Sverdrup's curl equation, (3-10); this is what is left of (3-64) if one were to assume that higher-order derivatives were not important.

Eight boundary conditions, four on x and four on y are required for the integration of (3-64). We can specify that:

1. There is no flow through the boundaries,

$$\Pi = 0 \text{ at the boundaries} \tag{3-65}$$

which merely states that the boundaries constitute a streamline whose value is arbitrarily taken as zero.

2. There is no slippage (flow) along the boundary,

$$\frac{\partial \Pi}{\partial n} = 0 \text{ at the boundary} \tag{3-66}$$

where n is normal to the boundary. Equation (3-66) physically states that there is no mass transport along the boundary at the boundary. In terms of our rectangular ocean extending from $x = 0$ to $x = \lambda$ and from $y = -b$ to $y = +b$, these conditions become

$$\Pi(0, y) = \Pi(\lambda, y) = \frac{\partial \Pi(0, y)}{\partial x} = \frac{\partial \Pi(\lambda, y)}{\partial x} = 0 \tag{3-67}$$

and

$$\Pi(x, b) = \Pi(x, -b) = \frac{\partial \Pi(x, b)}{\partial y} = \frac{\partial \Pi(x, -b)}{\partial y} = 0 \tag{3-68}$$

Munk observes that it would be throught that the choice of arbitrary ocean boundaries, as above, would make it impossible to deal with actual conditions. However, he points out that the proper solution for the main features in the ocean circulation is expected to be fairly insensitive to the choice of boundaries, since these *main features* are similar in all oceans (see his Table 3), although the boundaries are not.

For a purely zonal wind circulation ($\tau_y = 0$), the stress on the sea surface can be represented as a sum of terms, each of the form

$$\tau_{x,n} \equiv a_n \cos \frac{n\pi y}{b} + d_n \sin \frac{n\pi y}{b} + c_n \tag{3-69}$$

where $n = 1, 2, 3, \ldots$. In essence, $n\pi/b$ can be considered to be a "wind wave number." For example, with $n = 1$,

$$\tau_{x,1} = a_1 \cos \frac{\pi y}{b} + d_1 \sin \frac{\pi y}{b} + c_1$$

and assuming for the moment that a_1, d_1, and c_1 are all positive, a plot of this component of the zonal wind would consist of the sum of the three right-hand terms. This is illustrated in Figure 3-11. The total zonal wind stress is thus representated as a sum of $\tau_{x,n}$'s, $\sum \tau_{x,n}$, and the curl$_z$ $\tau_{x,n}\mathbf{i}$ is a sum of terms like

$$\text{curl}_z\, \tau_{x,n}\mathbf{i} = -\frac{\partial \tau_{x,n}}{\partial y}$$

or

$$\text{curl}_z \tau_{x,n}\mathbf{i} = \frac{n\pi}{b}\left(a_n \sin \frac{n\pi y}{b} - d_n \cos \frac{n\pi y}{b}\right) \tag{3-70}$$

in which it is understood that the curl$_z$ $\tau_{x,n}\mathbf{i}$ means the z component of the curl of the n component of the zonal wind stress and that

$$\text{curl}_z\, \boldsymbol{\tau} = \sum \text{curl}_z\, \tau_{x,n}\mathbf{i}$$

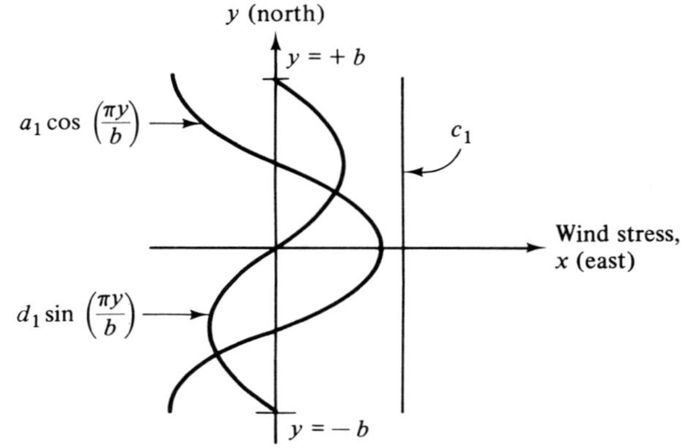

Fig. 3-11. Schematic plot of zonal wind stress terms for $n = 1$ in (3–69).

Since $\sum \tau_{x,n}$ is merely a linear combination of the $\tau_{x,n}$'s of (3-69), each $\tau_{x,n}$ can be used to solve the differential equation (3-64) independently and we may write that

$$\left(K_H \nabla^4 - \beta \frac{\partial}{\partial x}\right) \Pi_n = -\frac{n\Pi}{b}\left(a_n \sin \frac{n\pi y}{b} - d_n \cos \frac{n\pi y}{b}\right) \qquad (3\text{-}71)$$

A general solution of (3-71) will, as usual, consist of the sum of a particular solution of the nonhomogeneous equation and a general solution of its homogeneous equivalent. The particular integral is readily found from integrating

$$\frac{\partial^4 (\Pi_n)_1}{\partial y^4} = \frac{1}{K_H}\left(-\frac{n\pi}{b}\right)\left(a_n \sin \frac{n\pi y}{b} - d_n \cos \frac{n\pi y}{b}\right)$$

giving

$$(\Pi_n)_1 = -\frac{1}{m^3 K_H}(a_n \sin my - d_n \cos my)$$

or

$$(\Pi_n)_1 = -\frac{1}{m^4 K_H} \operatorname{curl}_z \tau_{x,n} \mathbf{i} \qquad (3\text{-}72)$$

where the subscript 1 designates the particular integral and where for brevity $m \equiv n\pi/b$. The homogeneous form of (3-71) was written by Munk in the form

$$\left(\nabla^4 - k^3 \frac{\partial}{\partial x}\right)(\Pi_n)_2 = 0 \qquad (3\text{-}73)$$

the subscript 2 indicating the general integral and k defined as

$$k \equiv \sqrt[3]{\beta / K_H} \qquad (3\text{-}74)$$

called the *Coriolis-friction wave number*, assumed constant. Equation (3-74) serves the purpose of scaling the variation of the Coriolis force to the frictional force.

Since we require that

$$\Pi_n(x, y) \equiv (\Pi_n(y))_1 + (\Pi_n(x, y))_2$$

satisfy the boundary conditions of (3-67), equivalently

$$(\Pi_n(y))_1 + (\Pi_n(0, y))_2 = (\Pi_n(y))_1 + (\Pi_n(\lambda, y))_2 = 0$$

and therefore

$$(\Pi_n(0, y))_2 = -(\Pi_n(y))_1 \tag{3-75}$$

This suggested to Munk that a convenient form of the general integral of (3-73) be taken as

$$(\Pi_n)_2 = X_n(x)\, Y_n(y) - (\Pi_n(y))_1$$

where X_n and Y_n are functions only of x and y, respectively. In this way the boundary condition of (3-75) is satisfied. Further, if $Y_n(y)$ is assumed to be

$$Y_n(y) \equiv \lambda K_H m^4 \beta^{-1}(\Pi_n(y))_1$$

then

$$(\Pi_n)_2 = (\Pi_n(y))_1[\lambda K_H m^4 \beta^{-1} X_n(x) - 1] \tag{3-76}$$

This form was selected by Munk as being, at a later time, of particular convenience. It involves the parameters of the problem, the choice of the combination $\lambda K_H m^4 \beta^{-1}$ being unitless so that the function $X_n(x)$ will be unitless, implying that $X(x)$ will be exponential in the variable x. Also, the assumed functional dependency on y simplifies the form of the complete integral Π_n, which from (3-71) is

$$\Pi_n = (\Pi_n(y))_1 \lambda K_H m^4 \beta^{-1} X_n(x) \tag{3-77}$$

or, from (3-72),

$$\Pi_n = -\lambda \beta^{-1} X_n(x) \operatorname{curl}_z \tau_{x,n}\mathbf{i} \tag{3-78}$$

To find the function $X_n(x)$, (3-76) is substituted into the homogeneous differential equation (3-73). The substitution requires the derivatives

$$\frac{\partial^4(\Pi_n)_2}{\partial x^4} = (\Pi_n)_1(\lambda K_H m^4 \beta^{-1} X_n^{iv}) \tag{3-79a}$$

$$\frac{\partial^4(\Pi_n)_2}{\partial y^4} = m^4(\Pi_n)_1(\lambda K_H m^4 \beta^{-1} X_n - 1) \tag{3-79b}$$

$$\frac{\partial^4(\Pi_n)_2}{\partial x^2\, \partial y^2} = -m^2(\Pi_n)_1(\lambda K_H m^4 \beta^{-1} X'') \tag{3-80a}$$

and

$$\frac{\partial(\Pi_n)_2}{\partial x} = (\Pi_n)_1(\lambda K_H m^4 \beta^{-1} X_n') \tag{3-80b}$$

and yields the nonhomogeneous ordinary differential equation

$$X^{iv} - 2m^2X'' - k^3X' + m^4X = \frac{k^3}{\lambda} \tag{3-81}$$

where the superscript indicates differentiation with respect to x and where the subscript n has been dropped for simplicity. The solution to (3-81) is again in the form of the sum of a particular integral and a general integral,

$$X \equiv X_1 + X_2$$

where the first term on the right is the former and the second is the latter. From (3-81) it is clear that we may take

$$X_1 = \frac{k^3}{\lambda m^4} \tag{3-82}$$

and assuming that X_2 is of the form

$$X_2 \equiv ce^{\alpha kx} \tag{3-83}$$

where c and α are constants, substitution into (3-81) yields after some rearrangement

$$\alpha^4 - 2\gamma_n^2\alpha^2 - \alpha + \gamma_n^4 = 0 \tag{3-84}$$

where

$$\gamma_n \equiv \frac{m}{k} = \frac{n}{k}\left(\frac{\pi}{b}\right)$$

This quartic can be factored to

$$(\alpha^2 - \gamma_n^2)^2 = \alpha \tag{3-85}$$

Equations (3-84) or (3-85) will have four roots: $\alpha_1, \alpha_2, \alpha_3$, and α_4, two of which are real and two of which are complex conjugates. Let the latter two be representated as

$$\alpha_1 = \alpha_1' + i\alpha_1''$$

and

$$\alpha_2 = \alpha_2' + i\alpha_2''$$

with $i = \sqrt{-1}$, $\alpha_1' = \alpha_2'$, and $\alpha_1'' = -\alpha_2''$. Munk illustrated the behavior of the four roots $\alpha_1, \alpha_2, \alpha_3$, and α_4 as functions of the parameter γ_n from the repeated numerical solution of (3-85) and obtained the results shown in Figure 3-12. If the minimum zonal wavelength is taken as $(2\pi/m)_{min} = 1500$ km, then

$$m_{\max} = \frac{2\pi}{1500}\ \text{km}^{-1}$$

and

$$(\gamma_n)_{\max} = \frac{m_{\max}}{k} = \frac{2\pi/1500}{0.016} = 0.26$$

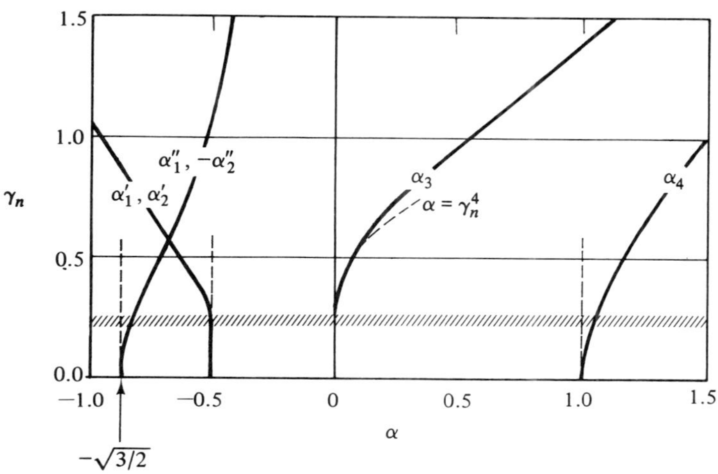

Fig. 3-12. Roots of the characteristic equation (3-85), dashed lines give the asymtotic solutions for $\gamma \ll 1$. From Munk, *Jour. of Meteor. Vol. 7, 79–93*, 1950.

where k has been estimated from (3-74),

$$k \doteq \left[\frac{2 \times 10^{-13} \text{ cm}^{-1}/\text{sec}}{5 \times 10^{7} \text{ cm}^2/\text{sec}}\right]^{1/3} = 0.016 \text{ km}^{-1}$$

From the figure it is apparent that for small values of γ_n the asymptotic solutions are reasonable approximations such that

$$\begin{aligned} \alpha_1 &= -\tfrac{1}{2}(1 + i\sqrt{3}) \\ \alpha_2 &= -\tfrac{1}{2}(1 - i\sqrt{3}) \\ \alpha_3 &= \gamma_n^4 \\ \alpha_4 &= 1 \end{aligned}$$

Substitution of these in (3-82) and (3-83) and summing gives

$$X = \frac{k^3}{\lambda m^4} + c_1 e^{-1/2(1+i\sqrt{3})kx} + c_2 e^{-1/2(1-i\sqrt{3})kx} + c_3 e^{\gamma_n^4 kx} + c_4 e^{kx}$$

or, rewriting,

$$\gamma_n^4 \lambda k x = 1 + c_1 e^{-1/2(1+i\sqrt{3})kx} + c_2 e^{-1/2(1-i\sqrt{3})kx} + c_3 e^{\gamma_n^4 kx} + c_4 e^{kx} \tag{3-86}$$

To evaluate the constants, the boundary conditions on x as given by (3-67) are applied through (3-77). Thus:

1. From $\Pi(0, y) = 0$, $X(0) = 0$, and

$$1 + c_1 + c_2 + c_3 + c_4 = 0 \tag{3-87}$$

2. From $\Pi(\lambda, y) = 0$, $X(\lambda) = 0$, and

$$1 + c_1 e^{-1/2(1+i\sqrt{3})k\lambda} + c_2 e^{-1/2(1-i\sqrt{3})k\lambda} + c_3 e^{\gamma_n{}^4 k\lambda} + c_4 e^{k\lambda} = 0 \tag{3-88}$$

3. From $\partial\Pi(0, y)/\partial x = 0$, $X'(0) = 0$, and

$$-\frac{1}{2}(1 + i\sqrt{3})kc_1 - \frac{1}{2}(1 - i\sqrt{3})kc_2 + \gamma_n^4 kc_3 + kc_4 = 0 \tag{3-89}$$

4. From $\partial\Pi(\lambda, y)/\partial x = 0$, $X'(\lambda) = 0$, and

$$-\frac{1}{2}(1 + i\sqrt{3})kc_1 e^{-1/2(1+i\sqrt{3})k\lambda} - \frac{1}{2}(1 - i\sqrt{3})kc_2 e^{-1/2(1-i\sqrt{3})k\lambda} + \gamma_n^4 kc_3 e^{\gamma_n{}^4 k\lambda} + kc_4 e^{k\lambda} = 0 \tag{3-90}$$

This rather formidable set of equations may be solved simultaneously, and under the assumptions that

$$\gamma_n^4 \ll 1 \qquad e^{-k\lambda} \ll 1 \qquad e^{-\gamma_n{}^4 k\lambda} \doteq 1 - \gamma_n^4 k\lambda \tag{3-91}$$

yield

$$\begin{aligned} c_1 &= -\frac{1}{2}\gamma_n^4 k\lambda\left(1 + i\frac{\sqrt{3}}{3}\right) \\ c_2 &= -\frac{1}{2}\gamma_n^4 k\lambda\left(1 - i\frac{\sqrt{3}}{3}\right) \\ c_3 &= \gamma_n^4 k\lambda - 1 \\ c_4 &= \gamma_n^4 k\lambda \end{aligned} \tag{3-92}$$

Subject to these results and to the approximations of (3-91), (3-86) becomes

$$X = -Be^{-\frac{1}{2}kx}\cos\left(\frac{\sqrt{3}}{2}kx + \frac{\sqrt{3}}{2k\lambda} - \frac{\pi}{6}\right) + 1 - \frac{1}{k\lambda}\left(kx - e^{-k(\lambda-x)} + 1\right) \tag{3-93}$$

West: $-Be^{-\frac{1}{2}kx}\cos\left(\frac{\sqrt{3}}{2}kx + \frac{\sqrt{3}}{2k\lambda} - \frac{\pi}{6}\right) + 1$

Central: $+1 - \frac{1}{k\lambda}(kx$

East: $+1 - \frac{1}{k\lambda}(kx - e^{-k(\lambda-x)} + 1$

where $B \equiv 2/\sqrt{3} - \sqrt{3}/k\lambda$. Also, differentiating with respect to x and expanding the resulting sine and cosine of the sum of angles,

$$\frac{X'}{k} = Be^{-\frac{1}{2}kx}\sin\left(\frac{\sqrt{3}}{2}kx + \frac{\sqrt{3}}{2k\lambda}\right) - \frac{1}{k\lambda}\left(1 - e^{-k(\lambda-x)}\right) \tag{3-94}$$

East: $Be^{-\frac{1}{2}kx}\sin\left(\frac{\sqrt{3}}{2}kx + \frac{\sqrt{3}}{2k\lambda}\right)$

Central: $-\frac{1}{k\lambda}(1$

West: $-\frac{1}{k\lambda}(1 - e^{-k(\lambda-x)}$

The meaning of the boxes will be made clear later. Notice that to the present degree of approximation, X does not depend upon m and consequently (3-78) may be written

$$\Pi = -\lambda\beta^{-1}X \operatorname{curl}_z \tau_{x,n}\mathbf{i} \qquad (3\text{-}95)$$

which can be computed directly from the observed zonal wind stress without expanding the stress in a Fourier series as suggested by (3-70), provided that the series converges so rapidly that for all significant terms the approximations of (3-91) are satisfied. Along any fixed latitude the mass transport stream function Π varies as $X(x)$ varies, and from the definition of (3-63), $X(x)$ can be interpreted as being proportional to the integrated northward transport of the current from $x = 0$ to some point $x = x_1$. The quantity $X'(x)$ is, in turn, proportional to the mass transport per unit x distance, S_y, at any point and its behavior along a line of constant latitude will be identical.

Figure 3-13 shows the mass transport streamlines $\Pi(x, y)$ in a rectangular ocean as computed from (3-95) for the mean annual zonal winds over the Pacific. To the left are plotted $\tau_x(y)$ and its curl $d\tau_x/dy$, at the bottom the function $X(x)$. These functions were combined graphically according to (3-95) to give the isolines of Π. The transport between adjacent solid lines is 10^7 m^3/sec. The zonal wind system is seen to divide the circulation into a number of gyres. The dividing lines between them occurring at $\theta = \theta_b$, where $d\tau_x/dy = 0$, corresponding to the latitudes of the maximum west winds of the northerly and southerly trades, and of the doldrums. The results compare favorably with those determined from oceanographic observations. The latitudinal axes of the gyres ($S_x = 0$) occur at latitudes θ_a, where $d\tau_x/dy$ has extreme values. The Sargasso Sea of the Atlantic Ocean would be associated with the inflection point of the mean wind stress between the westerlies and the trade winds. The inflection points between the doldrums and the northern and southern hemisphere trades determines the boundaries of the Equatorial Countercurrent.

Along any fixed latitude, Π varies only with $X(x)$ and we have noted that the functions $X(x)$ and $X'(x)$ can be physically interpreted as the north-south component of the integrated mass transport and the mass transport per unit length, repectively, drawn to an arbitrary scale. Consider sections at latitudes θ_a through the center of the gyres of Figure 3-13. When X and X' are calculated from (3-93) and (3-94) it is found that the equations fall naturally into three parts each one of which dominates in a given portion of the ocean corresponding to a western edge, a broad central section, and an eastern coastal strip (Figure 3-14).

Western Current Region. At the western edge of the ocean $x \ll \lambda$ and (3-93) can immediately be approximated as

$$X_w = -\frac{2}{\sqrt{3}}e^{-(1/2)kx}\cos\left(\frac{\sqrt{3}}{2}kx + \frac{\sqrt{3}}{2k\lambda} - \frac{\pi}{6}\right) + 1$$

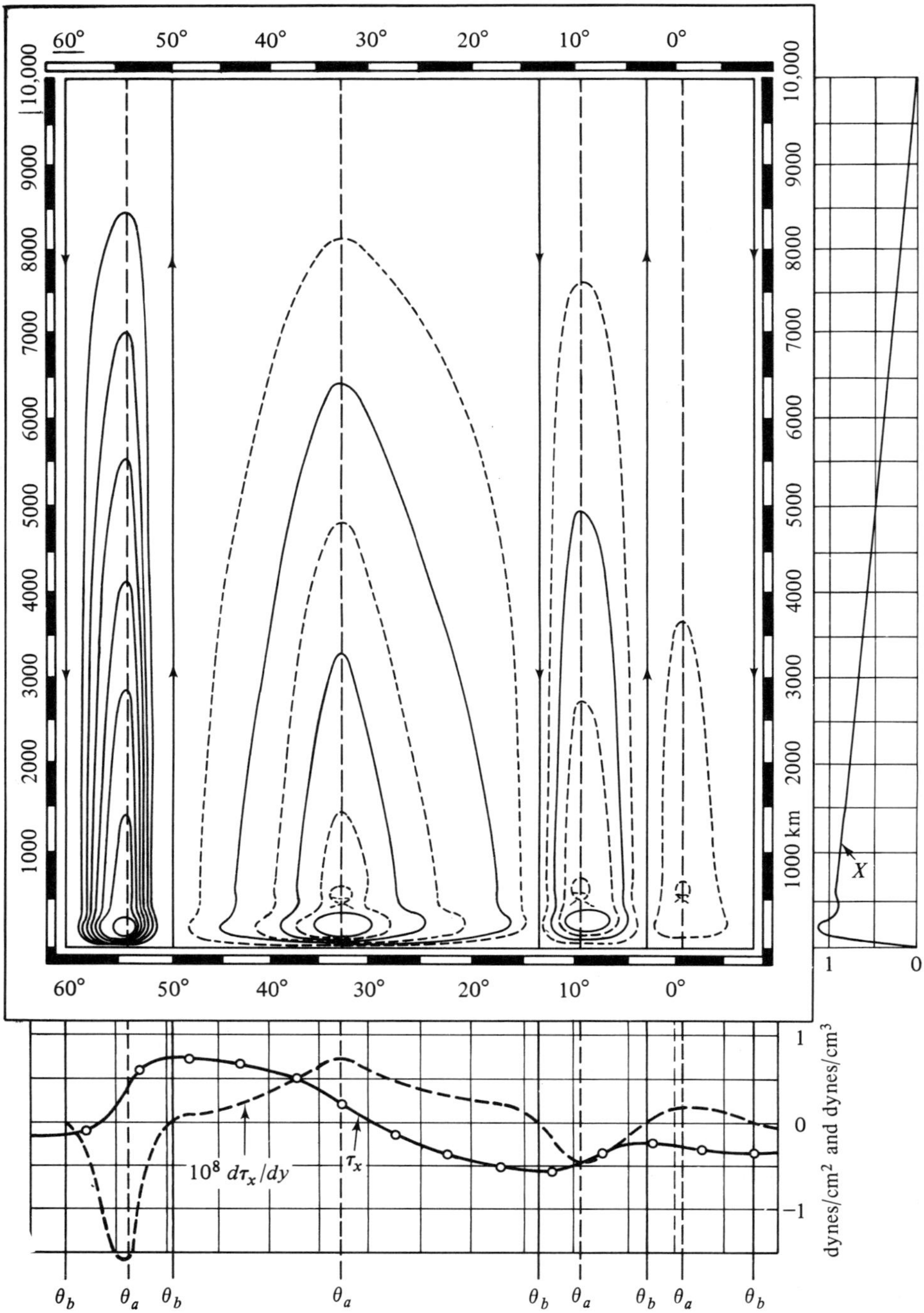

Fig. 3-13. Vertically integrated mass transport streamlines for the wind driven circulation in a rectangular ocean acted on by a zonal wind stress. From Munk, *Jour. of Meteor., Vol. 7, 79–93*, 1950.

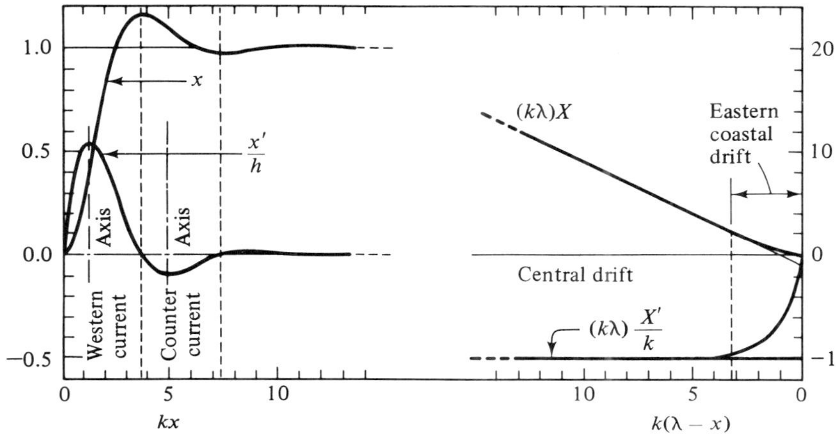

Fig. 3-14. Plot of east-west variation in mass transport ($\propto X$) and mass transport velocity ($\propto X'$); scale for central and eastern portions exaggerated relative to scale for western portion. From Munk, *Jour. of Meteor., Vol. 7, 79–93*, 1950.

having taken $\lambda = 6000$ km. Expanding the cosine,

$$X_w = -\frac{2}{\sqrt{3}} e^{-(1/2)kx}\left[\cos\left(\frac{\sqrt{3}}{2}kx - \frac{\pi}{6}\right)\cos\left(\frac{\sqrt{3}}{2k\lambda}\right) - \sin\left(\frac{\sqrt{3}}{2}kx - \frac{\pi}{6}\right)\sin\left(\frac{\sqrt{3}}{2k\lambda}\right)\right] + 1$$

and replacing

$$\cos\frac{\sqrt{3}}{2k\lambda} \longrightarrow 1 \qquad \text{and} \qquad \sin\frac{\sqrt{3}}{2k\lambda} \longrightarrow \frac{\sqrt{3}}{2k\lambda}$$

since $k\lambda \gg 1$, we get

$$X_w \doteq -\frac{2}{\sqrt{3}} e^{-(1/2)kx}\left[\cos\left(\frac{\sqrt{3}}{2}kx - \frac{\pi}{6}\right) - \frac{\sqrt{3}}{2k\lambda}\sin\left(\frac{\sqrt{3}}{2}kx - \frac{\pi}{6}\right)\right] + 1$$

which for $x \ll \lambda$ approximates

$$X_w \doteq -\frac{2}{\sqrt{3}} e^{-(1/2)kx}\cos\left(\frac{\sqrt{3}}{2}kx - \frac{\pi}{6}\right) + 1 \tag{3-96}$$

Differentiating,

$$\frac{X'_w}{k} = e^{-(1/2)kx}\left[\frac{1}{\sqrt{3}}\cos\left(\frac{\sqrt{3}}{2}kx - \frac{\pi}{6}\right) + \sin\left(\frac{\sqrt{3}}{2}kx - \frac{\pi}{6}\right)\right]$$

Expanding,

$$\frac{X'_w}{k} = e^{-(1/2)kx}\left[\frac{1}{\sqrt{3}}\left(\cos\frac{\sqrt{3}}{2}kx\cos\frac{\pi}{6} + \sin\frac{\sqrt{3}}{2}kx\sin\frac{\pi}{6}\right) + \sin\frac{\sqrt{3}}{2}kx\cos\frac{\pi}{6} - \cos\frac{\sqrt{3}}{2}kx\sin\frac{\pi}{6}\right]$$

$$= e^{-(1/2)kx}\left[\frac{1}{2}\cos\frac{\sqrt{3}}{2}kx + \frac{1}{2\sqrt{3}}\sin\frac{\sqrt{3}}{2}kx + \frac{\sqrt{3}}{2}\sin\frac{\sqrt{3}}{2}kx - \frac{1}{2}\cos\frac{\sqrt{3}}{2}kx\right]$$

and finally,

$$\frac{X'_w}{k} = \frac{2}{\sqrt{3}}e^{-(1/2)kx}\sin\frac{\sqrt{3}}{2}kx \tag{3-97}$$

Both (3-96) and (3-97) represent slightly underdamped oscillations, (3-96) being an exponentially decreasing oscillation around the number one while (3-97) is an exponentially decreasing oscillation around zero. The wave length of the oscillation is given by

$$L_W = \frac{4\pi}{\sqrt{3}\,k} = \frac{4\pi}{\sqrt{3}}\left(\frac{K_H}{\beta}\right)^{1/3} \tag{3-98}$$

These results correspond to the bracketed portions of (3-93) and (3-94) which are labeled as "west," and a plot of kx versus X_W and X'_W/k is shown to the left in Figure 3-14. A remarkable feature is the existence of a countercurrent to the east of the western boundary current. The locations and values of the first few extrema of X_W and X'_W are determined by Munk. For example, the extrema of X'_W are found by differentiating (3-97), setting the result equal to zero, and solving for kx. This gives the axes of the western current and of the countercurrent. Dividing by (3-98) gives the location of these axes as fractions of the western wave length. The extrema of X_w, representing the limits of the currents, is obtained by setting (3-96) equal to zero and solving for kx. The results of this calculation are summarized in Table 3-3. It is apparent that the countercurrent has a magnitude of about 17 percent of that of the western current ($\exp -\pi/\sqrt{3} \doteq 0.09/0.55 \doteq 0.17$). Analysis of actual data by Wüst (1936) and Iselin (1936) for the Gulf Stream indicate the existence of such a

TABLE 3-3. Extrema of X_W and X'_W According to (3–96) and (3–97), Respectively

	Location (see Figure 3-14)			
	Western Current Axis	*Western Current Limit*	*Countercurrent Axis*	*Countercurrent Limit*
x/L_W	$\frac{1}{6}$	$\frac{3}{6}$	$\frac{4}{6}$	1
X_W	0.45	1.17	1.09	0.97
X'_W/k	0.55	0.00	−0.09	0.00

Source: W. H. Munk, *J. Meteor.*, Vol. 7, pp. 79–93, © 1950 by the American Meteorological Society.

countercurrent with a maximum speed of about 19 percent of that of the main current (Wüst). Further, Munk points out the favorable comparison of his function X_W with the transport function Π across the Gulf Stream and the Kuroshio as computed from oceanographic data; the form of the curves of Figure 3-15 should be compared to the form of the function X on the left side of Figure 3-14.

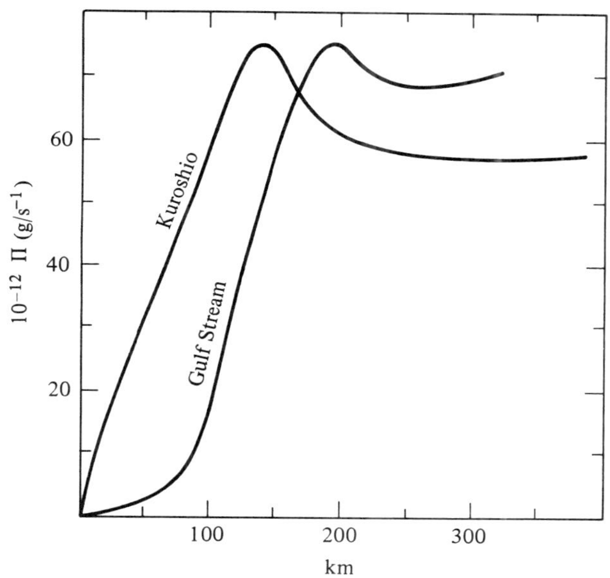

Fig. 3-15. The mass transport stream function Π across the Gulf Stream and the Kuroshio as computed from oceanographic data. From Munk, *Jour. of Meteor., Vol. 7, 79–93*, 1950.

A review of the then-current oceanographic literature suggested to Munk typical values for the widths of western currents and countercurrent of 200 km and 250 km, repectively. Taking L_W equal to 400 km and 500 km, the corresponding values of K_H from (3-98) are 3.4×10^7 cm²/sec and 6.6×10^7 cm²/sec. This increase of K_H with distance from the shore is in agreement with Prandtl's mixing-length theory. Montgomery (1939) from salt diffusion measurements and Montgomery and Palmén (1940) from dynamic considerations obtained values of 4×10^7 cm²/sec and 7×10^7 cm²/sec in the region of the Equatorial Countercurrent of the Atlantic; as was indicated earlier, Munk assumed a value of $K_H = 5 \times 10^7$ cm²/sec in order to obtain an estimate for k.

Stommel (1965, p. 98) points out that more recent studies of the Gulf Stream give a cross-sectional width of the main current as one-third to one-fourth of the value obtained by Munk and that for a purely frictional theory the coefficient of lateral eddy viscosity should be reduced to about 10^6 cm^2/sec. Such an adjustment of K_H is of course permissible since definitive values are lacking; however, the inertial terms are then no longer negligible, and therefore it is preferable to formulate the problem in nonlinear form.

Central Ocean Drift. At a distance from both east and west boundaries the exponential multiplier of the cosine and sine functions of (3-93) and (3-94) will have damped out the "wiggle" and the remaining parts approximate to

$$X_C = 1 - \frac{x}{\lambda} \qquad \frac{X'_C}{k} = -\frac{1}{k\lambda} \tag{3-99}$$

and, from (3-63) and (3-95),

$$S_y = \frac{\partial \Pi}{\partial x} = (-\lambda\beta^{-1} \operatorname{curl} \tau_x \mathbf{i})\left(-\frac{1}{\lambda}\right) = \beta^{-1} \operatorname{curl} \tau_x \mathbf{i} \tag{3-100}$$

These represent a broad constant drift as compared to the swift and relatively narrow western currents. Equation (3-100) is merely Sverdrup's curl equation (3-10), written for purely zonal winds, and represents a balance between geostrophic flow and pure wind drift. Its validity in midocean has been adequately demonstrated by the application of (3-100) to the equatorial currents of the eastern Pacific in the work of Sverdrup (1947) and Reid (1948). Figure 3-14 illustrates the behavior of (3-99). Writing these two equations in the form

$$k\lambda X_C = k(\lambda - x) \qquad k\lambda\left(\frac{X'_C}{k}\right) = -1 \tag{3-101}$$

it is apparent the first of these represents a straight line with a slope of unity when the left side is the ordinate and the right side is the abscissa; while the second, being independent of x, plots as a horizontal line.

Eastern Coastal Region. From (3-93) and (3-94) we have for the eastern solution

$$X_E = 1 - \frac{x}{\lambda} - \frac{1}{k\lambda}(1 - e^{-k(\lambda - x)}) \tag{3-102}$$

and

$$\frac{X'_E}{k} = -\frac{1}{k\lambda}(1 - e^{-k(\lambda - x)}) \tag{3-103}$$

Rewriting the first of these as

$$k\lambda X_E = k(\lambda - x) - (1 - e^{-k(\lambda - x)})$$

it is apparent that $k\lambda X_E$ can be represented as the sum of two curves, $(k\lambda X_E)_1$ and $(k\lambda X_E)_2$, where

$$(k\lambda X_E)_1 \equiv k(\lambda - x) - 1 \tag{3-104a}$$

and

$$(k\lambda X_E)_2 \equiv e^{-k(\lambda - x)} \tag{3-104b}$$

the first being a straight line of slope unity, the second an exponentially decreasing function when the abscissa is $k(\lambda - x)$; see Figure 3-16. On the other hand, the relations for X'_E as given by (3-103) plots as an exponentially decreasing function with an asymtote of -1 when $k(\lambda - x)$ is used as the independent variable. The results from both (3-102) and (3-103) are shown to the right in Figure 3-14. Munk refers to this eastern portion as "an exponential slippage zone" whose effective width, from Figure 3-16, of approximately $k(\lambda - x) = \pi$ corresponds to about 200 km for $K_H = 5 \times 10^7$ cm²/sec.

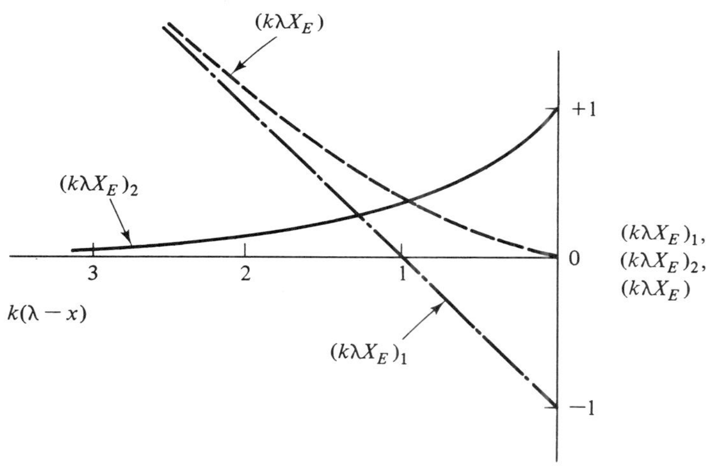

Fig. 3-16. Plot of $(k\lambda X_E)_1$, $(k\lambda X_E)_2$ and their sum, $(k\lambda X_E)$, for the eastern boundary condition as defined by (3-102) and (3-104).

Putting all the parts together, plots of x versus X and x versus X' along a gyre axis would appear as shown in Figure 3-17. Note that the sketches are not to scale and are intended only to illustrate the general form of the functions.

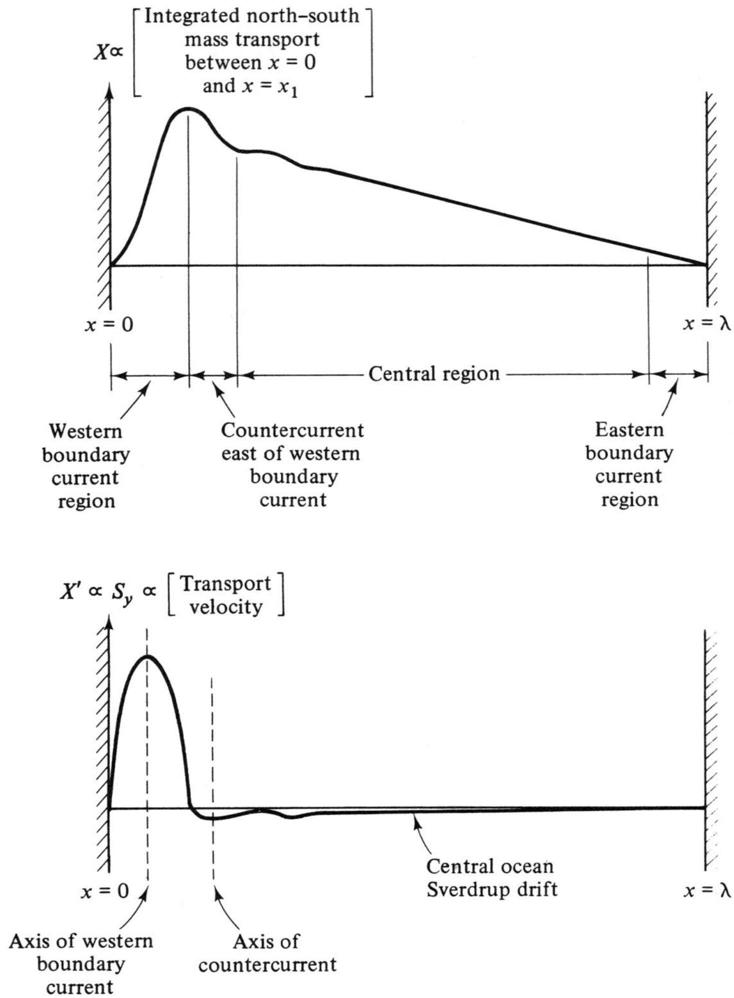

Fig. 3-17. Variation of X and X' [(3-93) and (3-94)] along the axis of a gyre (Fig. 3-13)-not to scale.

3-5 Qualitative Effects of the Nonlinear Terms in a Large-Scale Model of a Western Boundary Current

In the previous sections three models of the general circulation of the oceans have been examined in detail in the belief that the methods employed and the results obtained by three eminent investigators would be instructive; all three studies neglected the nonlinear terms in the equation of motion. Since 1950 various methods have been utilized to examine what the inclusion of nonlinear effects do to a model of the wind-driven ocean circulation. These include perturbation analyses, numerical approaches, and studies that take the interior flow as given but treat the boundary layer in a nonlinear manner. Veronis (1973) observes: "Perhaps the principal benefit from these investigations has been an increased understanding of the nonlinear process rather than an advance in our knowledge of the structure of the wind-driven gyres. The reason is that the models tend to be too idealized to admit incorporation of additional features; consequently, one cannot build an ordered hierarchy of successively more complicated models." Thus, to obtain a qualitative feeling for what nonlinear effects are going to do to a western boundary current region, we will restrict our discussion to Stommel's original model on the β-plane but include the nonlinear terms. Substituting (3-21) and (3-26) into (3-24) and (3-25), approximating $(D + h)$ by D, and adding the nonlinear terms, gives as the equations of motion

$$u\frac{\partial u}{\partial x} + v\frac{\partial u}{\partial y} - fv = -\frac{1}{\rho}\frac{\partial p}{\partial x} + \frac{\tau_x}{D} - Ru \tag{3-105}$$

$$u\frac{\partial v}{\partial x} + v\frac{\partial v}{\partial y} + fu = -\frac{1}{\rho}\frac{\partial p}{\partial y} + \frac{\tau_y}{D} - Rv \tag{3-106}$$

where R is Stommel's friction coefficient divided by D and where ρ can now be taken as a constant representing the vertically averaged density. Equations (3-24) and (3-25), and therefore (3-105) and (3-106), result from a vertical integration in which Stommel took u and v to be independent of depth; the fluid is barotropic. If the fluid is baroclinic, u and v are considered to represent vertical averages. This introduces additional approximations through the nonlinear terms, since in general the product of vertically averaged quantities [as in (3-105) and (3-106)] is not the same as the vertical average of the terms; thus, the resulting integrated model cannot be readily related to a Sverdrup model as was done earlier. To proceed, forming a vorticity equation by differentiating the first equation with respect to y and the second with respect to x, and subtracting the former from the latter, yields

$$u\frac{\partial}{\partial x}\left(\frac{\partial v}{\partial x} - \frac{\partial u}{\partial y}\right) + v\frac{\partial}{\partial y}\left(\frac{\partial v}{\partial x} - \frac{\partial u}{\partial y}\right) + \left(\frac{\partial u}{\partial x} + \frac{\partial v}{\partial y}\right)\left(\frac{\partial v}{\partial x} - \frac{\partial u}{\partial y}\right) + f\left(\frac{\partial u}{\partial x} + \frac{\partial v}{\partial y}\right) + \beta v = \frac{\partial}{\partial x}\left(\frac{\tau_y}{D}\right) - \frac{\partial}{\partial y}\left(\frac{\tau_x}{D}\right) - R\left(\frac{\partial v}{\partial x} - \frac{\partial u}{\partial y}\right) \tag{3-107}$$

In view of the integrated continuity equation, (3-33), the third and fourth terms on the left are zero and (3-107) reduces to

$$u\frac{\partial}{\partial x}\left(\frac{\partial v}{\partial x}-\frac{\partial u}{\partial y}\right)+v\frac{\partial}{\partial y}\left(\frac{\partial v}{\partial x}-\frac{\partial u}{\partial y}\right)+\beta v=\frac{\partial}{\partial x}\left(\frac{\tau_y}{D}\right)-\frac{\partial}{\partial y}\left(\frac{\tau_x}{D}\right)-R\left(\frac{\partial v}{\partial x}-\frac{\partial u}{\partial y}\right)$$

which in turn can be written in terms of the vertical component of the relative vorticity, ζ_z,

$$u\frac{\partial \zeta_z}{\partial x}+v\frac{\partial}{\partial y}(\zeta_z+f)=\frac{1}{D}\left(\frac{\partial \tau_y}{\partial x}-\frac{\partial \tau_x}{\partial y}\right)-R\zeta_z \tag{3-108}$$

In vector form this is equivalent to

$$\mathbf{V}\cdot\nabla_H(\zeta_z+f)=\frac{1}{D}\mathbf{k}\cdot(\nabla\times\boldsymbol{\tau})-R\zeta_z \tag{3-109}$$

where

$$\nabla_H\equiv\frac{\partial}{\partial x}\mathbf{i}+\frac{\partial}{\partial y}\mathbf{j}$$

This result can be scaled so as to exhibit important balances for large-scale flows in the manner done earlier when arriving at the hydrostatic equation. We do this by assuming that the variables of (3-108) can be written as follows:

$$x\equiv Lx' \qquad y\equiv Ly' \qquad u\equiv Uu'$$

$$v\equiv Uv' \qquad \zeta_z\equiv\left(\frac{U}{L}\right)\zeta_z' \qquad f\equiv f_0f'$$

$$\tau_x\equiv(Uf_0D)\tau_x' \qquad \tau_y\equiv(Uf_0D)\tau_y'$$

where the prime quantities are nondimensional and of order unity and the scales L, U, f_0, and D are determined by focusing attention to the magnitudes that reflect the processes being examined. The vorticity equation (3-108) then becomes

$$\left(\frac{U}{L}\right)^2\left(u'\frac{\partial \xi_z'}{\partial x'}+v'\frac{\partial \zeta_z'}{\partial y'}\right)+\left(\frac{Uf_0}{L}\right)v'\frac{\partial f'}{\partial y'}$$
$$=\left(\frac{Uf_0}{L}\right)\left(\frac{\partial \tau_y'}{\partial x'}-\frac{\partial \tau_x'}{\partial y'}\right)-\left(\frac{RU}{L}\right)\zeta_z'$$

Dividing by (Uf_0/L), we obtain

$$R_0\left(u'\frac{\partial \zeta_z'}{\partial x'}+v'\frac{\partial \zeta_z'}{\partial y'}\right)+v'\beta'=\left(\frac{\partial \tau_y'}{\partial x'}-\frac{\partial \tau_x'}{\partial y'}\right)-\epsilon\zeta_z' \tag{3-110}$$

where

$$R_0\equiv\frac{U}{f_0L} \qquad \epsilon\equiv\frac{R}{f_0} \tag{3-111}$$

the first of these nondimensional ratios being the *Rossby number*, named after the Swedish meteorologist C. G. Rossby, and the second a frictional decay coefficient. The significance of the Rossby number may be demonstrated by writing it in the form

$$R_0 = \frac{U^2/L}{f_0 U} \tag{3-112}$$

Now the advective terms in the equations of motion (e.g., $u\,\partial u/\partial x$, $v\,\partial u/\partial y$, etc.) represent the effects of inertia in the fluid, they account for the transport of momentum by the motion or inertia of the fluid. Such terms are referred to as being *inertial accelerations*. By expressing them nondimensionally,

$$u\frac{\partial u}{\partial x} = \left(\frac{U^2}{L}\right) u' \frac{\partial u'}{\partial x'}, \quad \text{etc.}$$

we conclude that the inertial accelerations are proportional to U^2/L, the numerator of (3-112). In the same fashion the denominator $f_0 U$ represents the Coriolis acceleration. The Rossby number is then a convenient measure of the magnitude of the inertial acceleration (or force/mass) to the Coriolis acceleration (or force/mass). When $U/f_0 L \gg 1$, Coriolis forces are in general negligible and result in only minor modifications of the flow pattern. This occurs if the rotation rate is slow (small f_0) or if the dimensions of the system are not large (small L); therefore, for everyday small-scale phenomena, regardless of the rotation of the earth, the Coriolis force in unimportant. On the other hand, if $U/f_0 L \ll 1$, the rotation of the system exerts an important influence. In many oceanographic situations this will be the case due to the characteristic size of the system being large.

Returning to (3-110), this vorticity equation can be written in terms of the stream function as

$$R_0\left(\frac{\partial \psi}{\partial x}\nabla_H^2 \frac{\partial \psi}{\partial y} - \frac{\partial \psi}{\partial y}\nabla_H^2 \frac{\partial \psi}{\partial x}\right) + \beta\frac{\partial \psi}{\partial x} + \epsilon\nabla_H^2 \psi = \mathbf{k} \cdot \nabla_H \times \boldsymbol{\tau} \tag{3-113}$$

where the primes have been dropped for simplicity. Expressing (3-110) in terms of ψ permits, among other things, a comparison to be made with the work of Sverdrup and Stommel discussed previously. With $R_0 = 0$ and $\epsilon = 0$ we retrieve a form comparable to Sverdrup's balance for the interior of the ocean, equation (3-9); while for $R_0 = 0$ and $\epsilon \neq 0$ we have a form comparable to Stommel's model, equation (3-35), of a Sverdrup type of interior flow with a western boundary layer. In general, in the interior of the ocean R_0 is so small that the nonlinear effects are negligible unless the boundary current penetrates the region.

The qualitative effect of nonlinearity can be obtained by considering the western boundary current region only and seeing how inclusion of the inertial terms modify Stommel's model. If we assume that the wind stress $\boldsymbol{\tau}$ can be neglected in this region, then (3-110) can be rewritten, again neglecting the primes, as

$$R_0 \mathbf{V} \cdot \nabla_H \zeta_z + v\beta = -\epsilon\zeta_z$$

or

$$\mathbf{V} \cdot \nabla_H (R_0 \zeta_z + f) = -\epsilon \zeta_z \tag{3-114}$$

Now the linear solution ($R_0 = 0$) of Stommel is

$$\mathbf{V} \cdot \nabla_H f = -\epsilon \zeta_z \tag{3-115}$$

stating that the rate of dissipation of the negative relative vorticity in the boundary region is balanced by the northward advection of planetary vorticity. When the nonlinear effects are included as in (3-114), in the southern half of the western boundary layer the northward advection of negative relative vorticity, $R_0\zeta_z$, will tend to reduce the rate of dissipation. This reduction is possible if $|\partial v/\partial x|$ decreases, which implies that the boundary layer must thicken in the region of its formation. At some point north of midlatitude the relative vorticty must start to increase from its large negative value (since it must be close to zero in the interior) and the sign of $\mathbf{V} \cdot \nabla_H(R_0\zeta_z)$ will change from negative to positive. This requires that $-\epsilon\zeta_z$ (or, equivalently, $-\epsilon\, \partial v/\partial x$) increase, which can practically be accomplished by a narrowing of the boundary layer. The overall effect is an introduction of a north-south asymmetry in the flow, with more dissipation taking place in the northern half of the boundary layer.

If the Rossby number is increased, the north-south asymmetry is intensified, the region of maximum $|\zeta_z|$ is shifted north, and the dissipation of vorticity occurs closer to the northwest corner of the model. If R_0 is so large such that the flow cannot dissipate all of the vorticity before it reaches the northern boundary, the flow to the east continues as an intense boundary current. Since the variation of planetary vorticity is essentially zero, a balance must exist between the nonlinear advection and the dissipation,

$$\mathbf{V} \cdot \nabla_H(R_0\zeta_z) = -\epsilon\zeta_z$$

With dissipation the narrowness of the boundary current and the intense eastward flow cannot be maintained; the streamlines leave the northern boundary to return to the interior of the basin. In this manner the inertial effects penetrate into the interior regions.

The flow described above is discussed by Veronis (1966, 1973). An example of his results is shown in Figure 3-18; the streamline pattern having been derived from a numerical integration of (3-113) with $\epsilon = 0.05$, $R_0 = 0.01$, and $\mathbf{k} \cdot \nabla_H \times \boldsymbol{\tau} = -\sin x \sin y$.

If a lateral friction term of a Newtonian form is assumed [as was done by Munk (1950) in the equations of motion (3-54) and (3-55)], (3-105) and (3-106) will acquire, in place of the terms $-Ru$ and $-Rv$, terms like

$$\frac{\partial}{\partial x}\left(K_H \frac{\partial u}{\partial x}\right) + \frac{\partial}{\partial y}\left(K_H \frac{\partial u}{\partial y}\right)$$

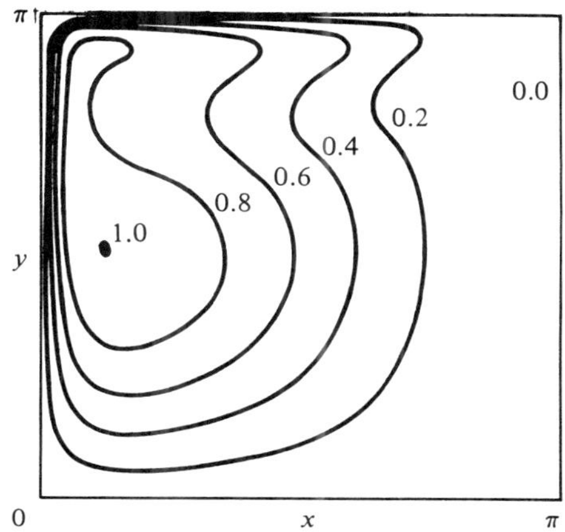

Fig. 3-18. Contours of the stream function for a nonlinear model of the wind-driven circulation of the ocean with $\epsilon = 0.05$, $R_0 = 0.01$, (3-113). Reprinted with permission from *Deep-Sea Res.*, Vol. 13, G. Veronis, "Wind-Driven Ocean Circulation—Part 2," Pergamon Press, 1966.

and

$$\frac{\partial}{\partial x}\left(K_H \frac{\partial v}{\partial x}\right) + \frac{\partial}{\partial y}\left(K_H \frac{\partial v}{\partial y}\right)$$

respectively. Cross-differentiation and subtraction results in the right side of (3-113) having the term $\epsilon'\nabla^4\psi$ ($\epsilon' \equiv K_H/f_0L$, ∇^4 being the two-dimensional biharmonic operator) in place of $\epsilon\nabla_H^2\psi$. These higher derivatives permit the imposition of four additional boundary conditions.

FOUR

The Hydrothermic Equations of a Binary Fluid

4-1 Introduction

It is our objective in this chapter to complete the systematic formulation of the fundamental equations governing the combined hydrodynamic and thermodynamic behavior of seawater treated as a binary fluid mixture of salt and water. These fundamental equations, two of which have already been developed in detail in Chapter 2, are the mathematical statements of four basic principles of physics governing the conservation of mass, the conservation of momentum, the conservation of energy, and the availability of energy. The first three lead to the four important equations concerning fluids—the continuity equation, the equation of motion, the total energy equation, and the first law of thermodynamics. The fourth principle results in the second law of thermodynamics and concerns itself with changes in entropy; it is not a conservation law but rather represents an inequality that imposes certain restrictions on such processes as diffusion, heat conduction, chemical reactions, and the dissipation of energy by viscous processes.

4-2 Generation and Dissipation of Mechanical Energy

The equation of motion, (2-195), can for our present purposes, be conveniently written as

$$\rho \frac{D\mathbf{V}}{Dt} = -2\rho\mathbf{\Omega} \times \mathbf{V} - \nabla p + \nabla \cdot \mathfrak{T} + \rho(\mathbf{g}' + \mathbf{g}'')$$

where now $\mathbf{g}'$ represents the combined Newtonian and centripetal acceleration of the earth and $\mathbf{g}''$ is the combined lunar and solar-tide-producing force. By forming the dot product of $\mathbf{V}$ on this equation we can obtain a description of the energy interconversions that take place in a flowing fluid. Thus, proceeding,

$$\frac{1}{2}\rho\frac{DV^2}{Dt} = -\mathbf{V} \cdot \nabla p + \mathbf{V} \cdot (\nabla \cdot \mathfrak{T}) + \rho\mathbf{V} \cdot (\mathbf{g}' + \mathbf{g}'') \tag{4-1}$$

The deflecting force $-2\rho\mathbf{\Omega} \times \mathbf{V}$ contributes nothing to (4-1) ($\mathbf{V} \cdot \mathbf{\Omega} \times \mathbf{V} = 0$) since it is directed normal to $\mathbf{V}$. That is, the deflecting force performs no work on the fluid system. If we multiply the continuity equation by $V^2/2$ and add the result to (4-1), split up the pressure and viscous contributions into two terms each through the use of the identities:

$$\mathbf{V} \cdot \nabla p = \nabla \cdot p\mathbf{V} - p\nabla \cdot \mathbf{V} \tag{4-2}$$

and

$$\mathbf{V} \cdot (\nabla \cdot \boldsymbol{\tau}) = \nabla \cdot (\mathfrak{T} \cdot \mathbf{V}) - \mathfrak{T} : \nabla\mathbf{V} \tag{4-3}$$

[this last identity is merely the equivalent of

$$V_j \frac{\partial \tau_{ij}}{\partial x_i} = \frac{\partial}{\partial x_i}(\tau_{ij}V_j) - \tau_{ij}\frac{\partial V_j}{\partial x_i}$$

see equation (1-17) for the definition of the double dot product], the terms of the resulting equation can be interpreted in terms of a stationary volume element through which the fluid flows. Hence,

$$\begin{aligned}\frac{\partial}{\partial t}\left(\frac{1}{2}\rho V^2\right) = &-\nabla \cdot \frac{1}{2}\rho V^2\mathbf{V} - \nabla \cdot p\mathbf{V} - p(-\nabla \cdot \mathbf{V}) \\ &- [-\nabla \cdot (\mathfrak{T} \cdot \mathbf{V})] - \mathfrak{T} : \nabla\mathbf{V} + \rho\mathbf{V} \cdot (\mathbf{g}' + \mathbf{g}'')\end{aligned} \tag{4-4}$$

where

1. $(\partial/\partial t)(\frac{1}{2}\rho V^2)$ is the local rate of increase of kinetic energy per unit volume of fluid.
2. $-\nabla \cdot \frac{1}{2}\rho V^2\mathbf{V}$ is the net rate of input of kinetic energy per unit volume of fluid due to advection; notice the similarity of this advection term to the advective terms in previously developed equations

—the equation of continuity, where $-\nabla \cdot \rho\mathbf{V}$ is the advection of mass, and the momentum equation, where $-\nabla \cdot \rho\mathbf{VV}$ is the advection of momentum.

3. $-\nabla \cdot p\mathbf{V}$ is the rate of work done per unit volume of fluid by the pressure of the surroundings on the volume element of the fluid.
4. $-p(-\nabla \cdot \mathbf{V})$ is the rate of *reversible* conversion to internal energy per unit volume of fluid (represents the reversible conversion from mechanical to thermal energy).
5. $-[-\nabla \cdot (\mathfrak{T} \cdot \mathbf{V})]$ is the rate of work done per unit volume of fluid by viscous forces on the volume element.
6. $-(\mathfrak{T} : \nabla\mathbf{V})$ is the rate of *irreversible* conversion to internal energy per unit volume of fluid (represents the irreversible conversion from mechanical to thermal energy).
7. $\rho\mathbf{V} \cdot (\mathbf{g}' + \mathbf{g}'')$ is the rate of work done per unit volume of fluid by the gravity and tidal forces on volume element of the fluid.

At this point it is not clear why we have ascribed the indicated physical significance to terms 4 and 6; their meaning cannot be properly appreciated until the total energy balance has been examined. This will be done later and it will be seen that these two terms enter the internal energy equation with opposite sign, with the result that the fluid may be heated or cooled internally.

Additional insight into the energy equation can be gained from an examination of its integrated form. Consider a finite volume of fluid whose boundaries move with the fluid velocity $\mathbf{V}$. The mass within the volume remains constant and the Reynolds transport theorem, (2-139), may be employed. Further, expressing $\mathbf{g}'$ in terms of the geopotential function χ' and taking the field of gravity relative to the earth as being sensibly invariant with respect to time, that is, $\partial\chi'/\partial t = 0$ (the time variant tidal components have not been included), we can write that

$$\frac{D\chi'}{Dt} = \mathbf{V} \cdot \nabla\chi' = -\mathbf{V} \cdot \mathbf{g}' \tag{4-5}$$

Also, from the continuity equation,

$$\nabla \cdot \mathbf{V} = \rho\frac{D\alpha}{Dt} \tag{4-6}$$

where α is the specific volume; finally, the first two terms of (4-4) are of course equivalent to the material derivative term of (4-1),

$$\frac{\partial}{\partial t}\left(\frac{1}{2}\rho V^2\right) + \nabla \cdot \frac{1}{2}\rho V^2\mathbf{V} = \frac{1}{2}\rho\frac{DV^2}{Dt} \tag{4-7}$$

Substituting these results into (4-4), integrating, and rewriting through the use of the Reynolds transport theorem, we obtain

$$\frac{d}{dt}\int_v (\tfrac{1}{2}\rho V^2 + \rho\chi')\,dv = -\int_v \nabla \cdot p\mathbf{V}\,dv + \int_v \rho p \frac{D\alpha}{Dt}\,dv + \int_v \nabla \cdot (\mathfrak{T} \cdot \mathbf{V})\,dv - \int_v (\mathfrak{T} : \nabla\mathbf{V})\,dv + \int_v \rho(\mathbf{V} \cdot \mathbf{g}'')\,dv \tag{4-8}$$

for the volume v moving with the fluid velocity $\mathbf{V}$. Defining the total kinetic energy E_K and the total geopotential energy E_P of the mass M as

$$E_k \equiv \int_v \tfrac{1}{2}\rho V^2\,dv \qquad E_p \equiv \int_v \rho\chi'\,dv \tag{4-9}$$

and using the divergence theorem to write

$$\int_v \nabla \cdot p\mathbf{V}\,dv = \int_A \mathbf{n} \cdot p\mathbf{V}\,dA$$

and

$$\int_v \nabla \cdot (\boldsymbol{\tau} \cdot \mathbf{V})\,dv = \int_A \mathbf{n} \cdot (\mathfrak{T} \cdot \mathbf{V})\,dA$$

where $\mathbf{n}$ is the outward unit normal at the point (x, y, z) on the surface A of the volume v, allows the writing of (4-8) as a sum of volume and surface processes,

$$\frac{d}{dt}(E_K + E_P) = \int_v \rho p \frac{D\alpha}{Dt}\,dv - \int_v (\mathfrak{T} : \nabla\mathbf{V})\,dv + \int_v \rho(\mathbf{V} \cdot \mathbf{g}'')\,dv - \int_A \mathbf{n} \cdot p\mathbf{V}\,dA + \int_A \mathbf{n} \cdot (\mathfrak{T} \cdot \mathbf{V})\,dA \tag{4-10}$$

There integrals may be physically interpreted as follows:

1. $G_p \equiv \int_v \rho p(D\alpha/Dt)\,dv$ represents the rate of generation of mechanical energy within the volume by the work due to the pressure associated with changes in the specific volume.

2. $G_g \equiv \int_v \rho(\mathbf{V} \cdot \mathbf{g}'')\,dv$ represents the rate of generation of mechanical energy within the volume by the work of the tide-producing forces.

3. $D_\tau \equiv \int_v (\mathfrak{T} : \nabla\mathbf{V})\,dv$ represents the rate of dissipation of mechanical energy within the volume by the action of the viscous stresses (this term is *always* positive and hence always contributes to a dissipation of mechanical energy).

4. $S_p \equiv -\int_A \mathbf{n} \cdot p\mathbf{V}\, dA$ represents the rate of generation of mechanical energy by the work of pressure at the boundaries of the volume and is associated with motion of the boundary relative to the earth.
5. $S_\tau \equiv \int_A \mathbf{n} \cdot (\mathfrak{T} \cdot \mathbf{V})\, dA$ represents the rate of generation of mechanical energy by the work of the viscous stresses at the boundaries.

In terms of the abbreviated symbols, (4-10) can be written

$$\frac{d}{dt}(E_K + E_P) = G_p + G_g - D_\tau + S_p + S_\mathfrak{T} \tag{4-11}$$

in which all of the terms can be positive or negative with the exception of $D_\mathfrak{T}$.

G_p as defined above represents a very general form; a familiar special case results when p is assumed uniform. For the pressure uniform,

$$\begin{aligned} G_p &= p \int_\upsilon \rho \frac{D\alpha}{Dt}\, d\upsilon \\ &= p \int_M \frac{D\alpha}{Dt}\, dM = p \frac{d}{dt} \int_M \alpha\, dM \\ &= p \frac{d\upsilon}{dt} \end{aligned}$$

which is the rate at which work is done by the pressure p in expanding the volume υ. However, if p is uniform within the volume and is continuous across its boundaries,

$$S_p = -p \int_A \mathbf{n} \cdot \mathbf{V}\, dA = -p \frac{d\upsilon}{dt}$$

Hence the sum $(G_p + S_p)$ is zero. This result is important in that it states that the pressure within and at the boundaries of volume can lead to a net production of mechanical energy only if the pressure is not alike throughout.

The generation quantity G_p can be expressed in terms of the sum of three physical processes—the bulk compression of the fluid associated with pressure changes, changes in fluid density due to mass diffusion (of salt and water), and changes in the fluid density due to the nonadvective flux of thermal energy. The first is important in the acoustics of fluids, while the other two represent the generation of energy by thermohaline processes and are presumed in the ocean to be of secondary importance compared with surface processes. In the atmosphere, however, the generation of mechanical energy by thermal processes is most important.

In the ocean, in so far as surface waves and the major current systems are concerned, S_p and $S_\mathfrak{T}$ are the major sources of mechanical energy. Gravity and capillary waves obtain energy from both differential surface pressures and tangential surface stresses. The energy supplied through the S_3 term is divided between wave and current systems, this energy source representing the primary agent for the maintainence of the quasi-steady currents in the face of the internal dissipation of energy by viscous effects.

If the volume υ to which the integrals of (4-10) apply is taken as representing that of all the oceans of the earth combined, S_p and S_3 apply to the total net effects over the sea surface and sea bottom and sides. The major contribution to S_p is that due to the coupling of differential pressures and motion at the air–sea interface since at the bottom and sides $\mathbf{n} \cdot \mathbf{V}$ is essentially zero. Two exceptions to this are at the mouths of rivers ($\mathbf{n} \cdot \mathbf{V} \neq 0$) and on infrequent occasions when the ocean bottom is disturbed by seismic activity or submarine volcanic activity. In both instances energy is added to the oceans; in the latter case it may at times be observed in the form of a tsunami.

The S_3 term may logically be divided into two parts,

$$S_3 \equiv (S_3)_{\text{sea surface}} + (S_3)_{\text{sea floor}}$$

The first of these surface integrals represents the supply of energy to waves and currents by wind stress; the second the supply of energy by bottom stresses. However, the latter contribution is predominantly negative, the net effect being that of a positive dissipation of energy rather than a positive supply. The negative of $(S_3)_{\text{sea floor}}$ physically represents the work done by the water in moving over the bottom in the presence of resisting tangential stresses. Implied here, then, is a slippage of the fluid along a solid interface. It is generally accepted that the fluid particles immediately adjacent to a solid surface adhere to it so that if the surface is stationary, the fluid velocity is zero at the surface. This, of course, is the no-slip boundary condition. In a thin laminar layer along the surface, however, the velocity gradient can be quite extreme, so that at the outer limit of a laminar layer, only a few millimeters removed from the boundary, the fluid velocity can attain a significant magnitude. For all practical purposes the surface viscous dissipation can be evaluated in terms of the velocity and stress at the effective boundary of the laminar skin of the fluid adjacent to the boundary. Thus, $(S_3)_{\text{sea floor}}$ effectively replaces that portion of the volume dissipation D_3 that applies to he laminar boundary layer. In any event, the boundary layer dissipation is important mainly in shallow seas, continental shelves, and tidal channels bordering the oceans proper. It is in these regions that the currents are significant near the bottom. It is suggested, in this connection, that the volume dissipative process D_3 is responsible for the majority of the energy dissipation in the oceans proper, while in bays and estuaries it appears that S_3 at the bottom and sides is the principal dissipative agent. In any case, kineti energy is dispersed from large-scale organized motion and eventually dissipated at the smallest scale by viscous processes, the loss of kinetic energy of macroscopic motion representing a gain of kinetic energy at the microscopic scale, which in turn is measurable only in terms of the thermal state of the fluid.

For the oceans as a whole, the total kinetic and potential energy is essentially the same from year to year. This implies that on the average the dissipation of energy balances the supply of energy at the sea surface along with the generation within the ocean by tidal and thermal processes. However, taking the supply at the surface to be nearly sufficient to maintain the currents and waves in the presence of dissipative processes, we may write that for all practical purposes

$$S_p + S_{\mathfrak{I}} \doteq D_\tau$$

this relation being valid only on the average for the oceans as a whole. Significant changes in the kinetic and potential energies can occur in certain regions of the sea, in which case the general relation (4-11) applies.

4-3 Conservation of Total Energy

The various forms of the mechanical energy equations [(4-4) and (4-10), for example] are all merely alternative forms of the same scalar relation derived directly from the equation of motion and hence from the principle of the conservation of momentum. A totally independent equation can be derived through the application of the principle of conservation of total (thermal plus chemical plus mechanical) energy. The formulation of this conservation principle requires the statement of the following postulates:

1. There is a quantity $\epsilon(x, y, z, t)$ that when added to the kinetic energy per unit mass gives the total energy per unit mass. This quantity is dependent upon the thermal state and chemical composition of the fluid and is called the *internal energy* per unit mass. It is asserted that the total energy of a system is the sum of the system's kinetic and internal energy. By kinetic energy we mean the macroscopic kinetic energy associated with the usual macroscopically observable velocity of the continuum. The kinetic energy of the molecules due to random thermal motions of the molecules, associated with temperature measurements instead of velocity measurements, is considered to be part of the internal energy. Also included in ϵ is the energy of interaction between molecules and any other forms of energy not explicity specified.

2. There is a vector $\mathbf{q}(x, y, z, t)$ that represents the *nonadvective flux of thermal energy* by conduction and radiation of heat.

3. There is a quantity $\tilde{\mu}$ such that $\tilde{\mu}\mathbf{F}_s$ ($\mathbf{F}_s$ is the nonadvective or diffusive mass flux vector for salt) represents the nonadvective flux of chemical energy due to diffusion. The quantity $\tilde{\mu}$ is called the *relative chemical potential*. It is "relative" in the sense that it corresponds to the chemical potential for

salt, $\tilde{\mu}_s$, minus the chemical potential for the water, $\tilde{\mu}_w$, in the mixture. That is,

$$\tilde{\mu} \equiv \tilde{\mu}_s - \tilde{\mu}_w$$

In general, the nonadvective flux of chemical energy for a mixture of *n different* constituents is

$$\sum_{j=1}^{n+1} \tilde{\mu}_j \mathbf{F}_j$$

For our binary system of salt and water, this becomes $(\tilde{\mu}_s \mathbf{F}_s + \tilde{\mu}_w \mathbf{F}_w)$ which reduces to $(\tilde{\mu}_s - \tilde{\mu}_w)\mathbf{F}_s$ or $\tilde{\mu}\mathbf{F}_s$ since from conservation considerations, $\mathbf{F}_s = -\mathbf{F}_w$. Now $\tilde{\mu}$ may be considered as a function of pressure, temperature, and salinity, being defined in that case as the rate of change of free energy with respect to salinity, while the pressure and temperature remain constant. The chemical potential is expressed in terms of energy per unit mass.

4. The total energy of a given finite mass of fluid moving with the fluid velocity can be altered by only four different processes:

a. By the efflux of thermal energy at its boundaries due to conduction and radiation.
b. By the efflux of chemical energy at its boundaries due to diffusion.
c. By the work done on the fluid inside the volume by the gravity and tide forces.
d. By the work done on the fluid due to surface stresses (including pressure) at the boundaries of the volume.

These postulates essentially imply that any other forms of energy conversion or transfer that might conceivably exist in the oceans are of minor importance.

From the foregoing it follows that for a volume υ whose boundaries move with the fluid velocity $\mathbf{V}$,

$$\frac{d}{dt}\int_{\upsilon} \rho\left(\frac{1}{2}V^2 + \epsilon\right) d\upsilon = -\int_A \mathbf{n} \cdot \mathbf{q}\, dA - \int_A \mathbf{n} \cdot \tilde{\mu}\mathbf{F}_s\, dA + \int_{\upsilon} \rho(\mathbf{g}' + \mathbf{g}'') \cdot \mathbf{V}\, d\upsilon + \int_A \mathbf{n} \cdot (\mathcal{P} \cdot \mathbf{V})\, dA \tag{4-12}$$

where A is the surface area of the volume υ and $\mathbf{n}$ is the outward unit normal. Application of the Reynolds transport theorem and the divergence theorem leads to

$$\int_{\upsilon} \Big[\rho \frac{D}{Dt}\left(\frac{1}{2}V^2 + \epsilon\right) + \nabla \cdot \mathbf{q} + \nabla \cdot \tilde{\mu}\mathbf{F}_s - \rho(\mathbf{g}' + \mathbf{g}'') \cdot \mathbf{V} - \nabla \cdot (\mathcal{P} \cdot \mathbf{V})\, d\upsilon = 0 \tag{4-13}$$

Finally, since the volume υ is of arbitrary size, (4-13) can be satisfied only if

$$\rho\frac{D}{Dt}\left(\frac{1}{2}V^2+\epsilon\right)=-\nabla\cdot\mathbf{q}-\nabla\cdot\tilde{\mu}\mathbf{F}_s+\rho(\mathbf{g}'+\mathbf{g}'')\cdot\mathbf{V}+\nabla\cdot(\mathcal{P}\cdot\mathbf{V}) \tag{4-14}$$

this being the fundamental differential equation expressing the conservation of total energy.

It is convenient to express this equation in a slightly different form by rewriting the left-hand side,

$$\frac{\partial}{\partial t}\rho\left(\frac{1}{2}V^2+\epsilon\right)-\left(\frac{1}{2}V^2+\epsilon\right)\frac{\partial\rho}{\partial t}+\nabla\cdot\left[\rho\mathbf{V}\left(\frac{1}{2}V^2+\epsilon\right)\right]-\left(\frac{1}{2}V^2+\epsilon\right)\nabla\cdot\rho\mathbf{V}=-\nabla\cdot\mathbf{q}-\nabla\cdot\tilde{\mu}\mathbf{F}_s+\rho(\mathbf{g}'+\mathbf{g}'')\cdot\mathbf{V}+\nabla\cdot(\mathcal{T}\cdot\mathbf{V})-\nabla\cdot p\mathbf{V} \tag{4-15}$$

and using the continuity equation to give

$$\frac{\partial}{\partial t}\rho\left(\frac{1}{2}V^2+\epsilon\right)=-\nabla\cdot\left[\rho\mathbf{V}\left(\frac{1}{2}V^2+\epsilon\right)\right]-\nabla\cdot\mathbf{q}-\nabla\cdot\tilde{\mu}\mathbf{F}_s+\rho(\mathbf{g}'+\mathbf{g}'')\cdot\mathbf{V}+\nabla\cdot(\mathcal{T}\cdot\mathbf{V})-\nabla\cdot p\mathbf{V} \tag{4-16}$$

whose terms we may physically interpret as:

1. $(\partial/\partial t)\rho(\frac{1}{2}V^2+\epsilon)$ is the local rate of gain of energy per unit volume.
2. $-\nabla\cdot[\rho\mathbf{V}(\frac{1}{2}V^2+\epsilon)]$ is the rate of energy input per unit volume due to advection (again note the similarity in form to the other advectional quantities examined previously).
3. $-\nabla\cdot\mathbf{q}$ is the rate of energy input per unit volume by thermal conduction and radiation.
4. $-\nabla\cdot\tilde{\mu}\mathbf{F}_s$ is the rate of energy input per unit volume by mass diffusion.
5. $\rho(\mathbf{g}'+\mathbf{g}'')\cdot\mathbf{V}$ is the rate of work done on the fluid per unit volume by gravitational and tidal forces.
6. $\nabla\cdot(\mathcal{T}\cdot\mathbf{V})$ is the rate of work done on the fluid per unit volume by viscous forces.
7. $-\nabla\cdot p\mathbf{V}$ is the rate of work done on the fluid per unit volume by pressure forces.

It is apparent that in (4-14) and (4-16), we have two forms of the same equation. They correspond to the two forms obtained for the continuity equation, the equation of motion, and the mechanical energy equation in that (4-14) describes the energy interchanges within a fluid from the viewpoint of an observer moving with the fluid, while (4-16) describes these same interchanges from the viewpoint of a stationary observer.

4-4 The First Law of Thermodynamics

Subtraction of the differential form of the mechanical energy equation (4-4) [after having substituted from (4-7)] from the differential form of the energy equation (4-14) [after replacing $\nabla \cdot (\mathcal{P} \cdot \mathbf{V})$ by $\nabla \cdot (\mathcal{T} \cdot \mathbf{V}) - \nabla \cdot p\mathbf{V}$] gives an equation for the change in the internal energy,

$$\rho \frac{D\epsilon}{Dt} = -\nabla \cdot \mathbf{q} - \nabla \cdot \tilde{\mu}\mathbf{F}_s - p(\nabla \cdot \mathbf{V}) + \mathcal{T} : \nabla\mathbf{V} \tag{4-17}$$

Or, if the continuity equation is used and indicial notation employed,

$$\rho\left(\frac{D\epsilon}{Dt} + p\frac{D\alpha}{Dt}\right) = -\frac{\partial q_i}{\partial x_i} - \frac{\partial \tilde{\mu} F_{s_i}}{\partial x_i} + \tau_{ij}\frac{\partial V_j}{\partial x_i} \tag{4-18}$$

Either of these last two equations represent a generalized form of the *first law of thermodynamics*. By analogy with the equation of mechanical energy, either may be called the *equation of thermal energy* and are in general found to be more convenient than the complete equation of energy. Physically, the terms of (4-17) can be interpreted as follows:

1. $\rho(D\epsilon/Dt)$ is the rate of gain of internal energy per unit volume following the fluid flow.
2. $-\nabla \cdot \mathbf{q}$ is the rate of internal energy input by conduction and radiation per unit volume.
3. $-\nabla \cdot \tilde{\mu}\mathbf{F}_s$ is the rate of internal energy increase per unit volume by mass diffusion.
4. $-p(\nabla \cdot \mathbf{V})$ is the *reversible* rate of internal energy increase per unit volume by compression.
5. $\mathcal{T} : \nabla\mathbf{V}$ is the *irreversible* rate of internal energy increase per unit volume by viscous dissipation.

In making a comparison of the mechanical energy equation with the first law, as expressed by (4-17), it is interesting to note that the terms $p(\nabla \cdot \mathbf{V})$ and $(\mathcal{T} : \nabla\mathbf{V})$, although common to both, appear with opposite signs. They in fact describe the interconversion of mechanical and thermal energy. The term $p(\nabla \cdot \mathbf{V})$ can be either positive or negative, depending upon whether the fluid is expanding or contracting, and therefore is a *reversible* mode of interchange. On the other hand, $(\mathcal{T} : \nabla\mathbf{V})$ can be shown to always be positive and hence represents an *irreversible* degradation of mechanical energy to thermal energy.

Finally, it is a relatively straightforward matter to reduce (4-18) to the special, more familiar form of the first law as quoted in introductory texts on thermodynamics. Assume that the following special conditions apply to a fluid—the pressure p is uniform throughout the volume υ moving with the fluid velocity $\mathbf{V}$, no viscous stresses are present, and the fluid consists of a single component. Under these restrictions the volume integral of (4-18), for a volume of fluid whose boundaries move with the fluid velocity, is

$$\int_{v} \rho\left(\frac{D\epsilon}{Dt} + p\frac{D\alpha}{Dt}\right) dv = -\int_{v} \nabla \cdot \mathbf{q}\, dv \tag{4-19}$$

Applying the transport and divergence theorems, we obtain

$$\frac{d}{dt}\int_{v} \rho\epsilon\, dv + p\frac{dv}{dt} = -\int_{A} \mathbf{n} \cdot \mathbf{q}\, dA \tag{4-20}$$

If E is the total internal energy of the volume and dQ/dt the net influx of heat, (4-20) becomes

$$dE + pdv = dQ \tag{4-21}$$

where

$$E \equiv \int_{v} \rho\epsilon\, dv \tag{4-22}$$

and

$$\frac{dQ}{dt} \equiv -\int_{A} \mathbf{n} \cdot \mathbf{q}\, dA \tag{4-23}$$

Equation (4-21) is recognized as the "common" form of the first law and is sometimes referred to as the first law for reversible processes, implying that diffusion and viscosity must be considered nonexistent or negligible.

4-5 Entropy and the Second Law of Thermodynamics

Experiments concerning the thermal behavior of fluids through measurement of pressure and specific volume are essentially of two types. In the first of these, the relationship between p and α can be established for a particular fluid while its temperature T is held constant ($dT = 0$), an isothermal process. Pairs of values of p and α are found to lie on a curve called an *isotherm*. In the second type of experiment the relationship between p and α can be determined when there is no exchange of heat Q with the surroundings ($dQ = 0$), an *adiabatic* process. Again pairs of values of p and α are found to lie on a curve, an adiabat. Thus, each fluid is characterized by two different reproducible families of curves in the p, α plane. (Actually, the adiabatic curves are reproducible if the experiment is carried out quasi-statically such that the process may be considered as essentially reversible. An adiabatic, reversible process is said to be *isentropic* and the established curves *isentropes*.) It is reasoned that since the isotherms are related to a unique thermal property—the temperature—the adiabats for a reversible process should be related to some other unique thermal prpperty. This property has been called *entropy*, and the isentropic curves are each characterized by a particular value of entropy. However, just as the isothermal curves on the p, α diagram do not uniquely define the temperature (but only its constancy), so also the isentropic curves do not uniquely define the entropy. To eliminate this ambiguity, an additional relationship between temperature and entropy is required.

Carl Eckart (1947) has observed that: "The wide-spread use of thermometers in the modern world has obscured the essential problems connected with temperature, and gives a somewhat misleading impression of common sense in regard to it. For better or worse, the reverse is true in regard to entropy. It is a concept that has not entered the realm of common sense in the same way as temperature, probably because we are not endowed with sensory organs that delude us into thinking that we perceive it directly. Yet the theoretical problems connected with the two are very similar indeed."

As a brief aside it seems appropriate to address the question: Who needs entropy? We observe that essentially materials contain only one kind of energy—internal energy (here "internal energy" is used in its broadest sense and is meant to include kinetic energy). They do not really contain heat and work as such. Energy may be transferred from a system in the form of heat or work, this energy coming from the internal energy of the material in the system. What is meant by the work or heat energy of a substance is in actuality the internal energy that could be transferred from the system in the form of work or heat. As an example, consider the isothermal, reversible expansion of a ideal gas from a high pressure to a lower pressure, with the resulting transfer of shaft work from the gas. During a closed, reversible, isothermal expansion, the work energy transferred from the gas is replaced by an equal amount of heat energy. The internal energy is constant. Clearly, the gas, if its lowest pressure is 1 atm, cannot practically expand further to produce additional shaft work even though its energy content is unchanged due to the isothermal process. Thus, at high pressure the internal energy of the gas can be converted to shaft work by an expansion; while at low pressure, although the internal energy is the same, no conversion of internal energy to shaft work by expansion can take place. Clearly, a major change has taken place but one that is not apparent from an examination of the internal energy. Since this change in the ability of the gas to do work is an important one, a state property that shows that a change has occurred is defined. This state property is entropy; and to answer the question posed above—we therefore need entropy.

In the formulation of the first law of thermodynamics as expressed by (4-21), the mechanical or shaft work done by a gas on the surroundings is $p\,dv$, the product of the pressure, an *intensive factor*, and the change in the volume, an *extensive factor*. (Intensive factors or variables are those which do not depend upon the mass; for example: pressure, specific volume, specific internal energy, etc. Extensive variables, such as volume, internal energy, etc., are mass dependent.) This form of relationship, the product of an extensive factor and an intensive factor, usually occurs in all expressions for work done—tension and length, surface tension and area, emf and charge, magnetic intensity and magnetization.

Kelvin first pointed out that it would not be unreasonable to assume that entropy is an extensive property of the fluid and that for a reversible isothermal process the change in entropy is directly proportional to the amount of heat added to the fluid at any stage in the process. Further, he suggested that the absolute thermodynamic temperature is an intensive property, like pressure, and that it is, in fact, the proportionality factor in the relation, that it is the same quantity which is constant along an isotherm (just as the entropy is the quantity constant along an adiabat), and that all of this holds regardless of the physical or chemical nature of the system. Thus, the heat added is a product of an intensive factor, the temperature, and an extensive factor, the change in entropy and therefore similar in form to all the work-done expressions. Extending this postulate to any reversible process, we can write that

$$dQ \equiv T\,dN \tag{4-24}$$

for a reversible process in a finite volume of fluid, where N is the total entropy of the volume and T the absolute thermodynamic temperature. Eckart (1947) and Reid (1954) have referred to all of the foregoing as the *Kelvin hypothesis*. It follows then that for an isentropic process, $dN = 0$, the equation for the family of isentropes is given by the entropy equaling a constant, similar to the equation for isotherms given by the temperature equaling a constant.

It is apparent that (4-24) has some rather severe limitations placed on it—it applies to a finite volume of fluid and not at a point in the fluid, and all processes must be reversible. That is, the system must be homogeneous with regard to temperature and have no diffusion or viscosity. All of these can result in the internal generation of entropy even though the heat added, dQ, is zero. To avoid these difficulties we begin by introducing the specific entropy, $\eta(x, y, z, t)$, having units of energy per degree absolute per unit mass, a function of position and time,

$$N \equiv \int_v \rho\eta\,dv \tag{4-25}$$

By writing (4-24) as

$$\frac{dQ}{dt} = T\frac{dN}{dt} \tag{4-26}$$

for a finite volume, then from (4-23), (4-25), and the transport theorem,

$$T\int_v \rho\frac{D\eta}{Dt}\,dv = -\int_A \mathbf{n}\cdot\mathbf{q}\,dA \tag{4-27}$$

Applying the divergence theorem, and since T must be uniform throughout the volume,

$$\int_v \left(\rho T \frac{D\eta}{Dt} + \nabla \cdot \mathbf{q}\right) dv = 0 \tag{4-28}$$

and the volume being of arbitrary size gives

$$\rho T \frac{D\eta}{Dt} + \nabla \cdot \mathbf{q} = 0 \tag{4-29}$$

for a reversible process. Comparing (4-19) with this last result, we see that

$$T\frac{D\eta}{Dt} = \frac{D\epsilon}{Dt} + p\frac{D\alpha}{Dt} \tag{4-30}$$

which can be adopted as a general form of the Kelvin hypothesis for a single component fluid.

We now introduce the following thermodynamic postulate:

> The state of any differential element of a fluid is determined uniquely and completely by $(n + 2)$ independent, intensive, thermal variables for a mixture containing n *different* constituents.

For a pure fluid, $n = 0$, and two intensive variables determine the state of a differential element. These might be T and p, T and α, α and p, T and η, and so on. It is convenient for the present to regard $\epsilon = \epsilon(\eta, \alpha)$. Taking the material derivative of $\epsilon(\eta, \alpha)$ yields

$$\frac{D\epsilon}{Dt} = \left(\frac{\partial \epsilon}{\partial \eta}\right)_\alpha \frac{D\eta}{Dt} + \left(\frac{\partial \epsilon}{\partial \alpha}\right)_\eta \frac{D\alpha}{Dt} \tag{4-31}$$

Comparing this result with (4-30), we see that

$$T = \left(\frac{\partial \epsilon}{\partial \eta}\right)_\alpha \tag{4-32a}$$

$$p = -\left(\frac{\partial \epsilon}{\partial \alpha}\right)_\eta \tag{4-32b}$$

where the subscript indicates the quantity to be held constant.

If we apply this postulate to seawater treated as a binary fluid mixture, then $n = 1$, $n + 2 = 3$, and ϵ is determined by three independent intensive variables, say η, α, and s, where s is the mass of the dissolved constituent (salt) per unit mass of fluid (seawater). Taking the material derivative of $\epsilon(\eta, \alpha, s)$ as in (4-31),

$$\frac{D\epsilon}{Dt} = \left(\frac{\partial \epsilon}{\partial \eta}\right)_{\alpha, s} \frac{D\eta}{Dt} + \left(\frac{\partial \epsilon}{\partial \alpha}\right)_{\eta, s} \frac{D\alpha}{Dt} + \left(\frac{\partial \epsilon}{\partial s}\right)_{\eta, \alpha} \frac{Ds}{Dt} \tag{4-33}$$

it is apparent that the extension of (4-32) to a binary fluid should be

$$T \equiv \left(\frac{\partial \epsilon}{\partial \eta}\right)_{\alpha, s} \tag{4-34a}$$

$$p \equiv -\left(\frac{\partial \epsilon}{\partial \alpha}\right)_{\eta, s} \tag{4-34b}$$

If more than one dissolved constituent is present, the s is replaced by s_1, $s_2, \ldots, s_n$. Equations (4-34) have been called (Reid, 1954) the *Gibbs extension of the Kelvin hypothesis*. They are not restricted in any way as far as thermal gradients, diffusion, or viscosity are concerned. They are not merely definitions, in that it is implied that all known empirical data with regard to any fluid are consistent with them.

Gibbs defined such quantities as

$$\tilde{\mu}_s \equiv \left(\frac{\partial E}{\partial m_s}\right)_{\eta, \alpha, m_w} \qquad \tilde{\mu}_w \equiv \left(\frac{\partial E}{\partial m_w}\right)_{\eta, \alpha, m_s}$$

where m_s, m_w are, for our purposes, the masses of salt and water in a mixture whose total mass $M = m_s + m_w$. If M is held fixed but, say, m_s is varied, then m_w must vary in the opposite sense (i.e., $dm_w = -dm_s$ for M constant). Thus for η, α, and M fixed,

$$\begin{aligned} dE &= \left(\frac{\partial E}{\partial m_s}\right)_{\eta, \alpha, m_w} dm_s + \left(\frac{\partial E}{\partial m_w}\right)_{\eta, \alpha, m_s} dm_w \\ &= \tilde{\mu}_s \, dm_s + \tilde{\mu}_w \, dm_w \\ &= (\tilde{\mu}_s - \tilde{\mu}_w) \, dm_s \\ &= \tilde{\mu} \, dm_s \end{aligned}$$

or

$$\tilde{\mu} = \left(\frac{\partial E}{\partial m_s}\right)_{\eta, \alpha, M} \equiv \left(\frac{\partial \epsilon}{\partial s}\right)_{\eta, \alpha} \tag{4-34c}$$

where

$$\tilde{\mu} \equiv (\tilde{\mu}_s - \tilde{\mu}_w)$$

$\tilde{\mu}$ is the relative chemical potential; it is the same quantity associated with the nonadvective flux of chemical energy due to diffusion, $\tilde{\mu}\mathbf{F}_s$, invoked previously in Section 4-3.

A few additional points can be made concerning chemical potentials, as introduced above. An important attribute of chemical potentials is that, like temperature, they must be the same for any two phases involving the same substance which are in equilibrium. Thus, for a salt solution in equilibrium with water vapor the chemical potential $\tilde{\mu}_w$ must be the same in the solution as in the vapor phase; that is,

$$\tilde{\mu}_w^l = \tilde{\mu}_w^v$$

If evaporation occurs, a nonequilibrium condition exists and then $\tilde{\mu}_w^l > \tilde{\mu}_w^v$; the reverse being true for condensation. The rate of production of entropy by evaporation is $R_w(\tilde{\mu}_w^l - \tilde{\mu}_w^v)$ per unit area, where R_w is the evaporative flux with units of [mass] [length]$^{-2}$ [time]$^{-1}$. For the equilibrium of salt crystals with the solution

$$\tilde{\mu}_s^s = \tilde{\mu}_s^l$$

Normally, such a saturation state does not occur for seawater. Similarly, for ice in equilibrium with seawater,

$$\tilde{\mu}_w^i = \tilde{\mu}_w^l$$

Continuing then, in view of the above, (4-33) can be written as

$$\frac{D\epsilon}{Dt} = T\frac{D\eta}{Dt} - p\frac{D\alpha}{Dt} + \tilde{\mu}\frac{Ds}{Dt} \tag{4-35}$$

where interestingly, each term on the right side is the product of an intensive parameter $(T, p, \tilde{\mu})$ and the specific representation of an extensive quantity (η, α, s).

If (4-35) is substituted into the generalized form of the first law as given by (4-18), the term involving α is eliminated as we obtain

$$\rho\left(T\frac{D\eta}{Dt} + \tilde{\mu}\frac{Ds}{Dt}\right) = \tau_{ij}\frac{\partial V_j}{\partial x_i} - \frac{\partial}{\partial x_i}(q_i + \tilde{\mu}F_{s_i}) \tag{4-36}$$

However, employing the conservation of salt equation, (2-118), written in the form

$$\rho\frac{Ds}{Dt} = -\nabla \cdot \mathbf{F}_s \tag{4-37}$$

and since

$$\nabla \cdot \left(\frac{\mathbf{q}}{T}\right) = \frac{1}{T}\nabla \cdot \mathbf{q} + \mathbf{q} \cdot \nabla\left(\frac{1}{T}\right) = \frac{1}{T}\nabla \cdot \mathbf{q} - \frac{\mathbf{q}}{T^2} \cdot \nabla T$$

we have that

$$\rho\frac{D\eta}{Dt} + \nabla \cdot \left(\frac{\mathbf{q}}{T}\right) = \frac{1}{T}\left(\mathfrak{T} : \nabla\mathbf{V} - \mathbf{F}_s \cdot \nabla\tilde{\mu} - \frac{\mathbf{q}}{T} \cdot \nabla T\right) \tag{4-38}$$

This represents an important alternative form of the first law for irreversible processes involving the rate of change of entropy. The terms on the right side are all of a common form—that of a nonadvective transfer quantity $(\mathfrak{T}, \mathbf{F}_s, \mathbf{q})$ multiplied by a gradient $(\nabla\mathbf{V}, \nabla\tilde{\mu}, \nabla T)$, which, as is known from empirical data, is related to the transfer quantity by which it is multiplied. That is, the viscous stresses are governed by the velocity gradients, the heat conduction in a material is proportional to the temperature gradient, and the diffusive flux is related directly to the gradient of the chemical potential. The quantities $\mathbf{V}$, $\tilde{\mu}$, and T all act as potential functions, potentials of $\mathfrak{T}$, $\mathbf{F}_s$, and $\mathbf{q}$, respectively.

Assuming for the time being that $\mathbf{q}$ represents the conductive flux only (radiation flux is assumed negligible or zero), and since Fourier's law of heat conduction states that the conductive flux is proportional to the temperature gradient, the last term in (4-38) gives that

$$-\frac{1}{T}\left(\mathbf{q}\cdot\nabla T\right) \propto \frac{(\nabla T)^2}{T} \geq 0 \qquad (4\text{-}39)$$

Clearly, the term $\mathfrak{T}:\nabla\mathbf{V}$, representing the rate of dissipation of mechanical energy per unit volume by the action of viscous stresses, must always be positive. In addition, since the gradient of the concentration is always opposite to the direction of the diffusive flux, and since the chemical potential increases with increasing concentration, the gradient of the chemical potential will always be opposite to the direction of the diffusive flux. $\mathbf{F}_s$ and $\nabla\tilde{\mu}$ will, therefore, be oppositely directed and the term $-\mathbf{F}_s\cdot\nabla\tilde{\mu}$ will always be positive. In view of this, it is thus asserted that under all conditions

$$\rho\frac{D\eta}{Dt} + \nabla\cdot\left(\frac{\mathbf{q}}{T}\right) \geq 0 \qquad (4\text{-}40)$$

which result is the mathematical statement of the *second law of thermodynamics*. It is a fundamental hypothesis—a hypothesis in that it represents an abstract generalization that agrees with, but cannot be proven by, experiment. Reid (1954) points out that the value of this law "lies not only in the collectivity of empirical fact but also in the expectation that it may predict relationships pertaining to the physical behavior of matter which might otherwise escape notice."

Finally, for a reversible process (no viscous dissipation, no mass diffusion, no temperature gradients), (4-38) reduces to

$$\rho\frac{D\eta}{Dt} + \frac{\nabla\cdot\mathbf{q}}{T} = 0$$

which in turn can be written as

$$d\eta = \frac{dQ_m}{T}$$

where dQ_m is the heat added with units of energy per unit mass. This result, the second law of thermodynamics for reversible processes, is the form that we see in any introductory thermodynamics course.

4-6 Further Consequences of Kelvin's Hypothesis

A characteristic of thermodynamics is the prediction of unexpected relations between quantities which might be considered independent, particularly since their empirical measurement is very different. These interrelationships all have the hypothesis of Kelvin as their origin, and it is through them that the hypothesis meets its most severe test. It is this interrelational aspect of thermodynamics, perhaps the aspect most exasperating to the student, which we examine now for the case of a binary mixture.

Gibbs' Relation

Equation (4-34a) gives us expression for the partial derivative of η with respect to ϵ. To obtain the expression for the partial of η with respect to α, we proceed by invoking the properties of Jacobians [an excellent summary of the properties is to be found in Kestin (1966, pp. 97–102)] to write

$$\left(\frac{\partial\eta}{\partial\alpha}\right)_{\epsilon,s} = \frac{\partial(\eta, \epsilon)}{\partial(\alpha, \epsilon)} = \frac{\partial(\eta, \epsilon)}{\partial(\eta, \alpha)}\frac{\partial(\eta, \alpha)}{\partial(\alpha, \epsilon)}$$

$$= -\left(\frac{\partial\epsilon}{\partial\alpha}\right)_{\eta,s}\left(\frac{\partial\eta}{\partial\epsilon}\right)_{\alpha,s}$$

and from (4-34a) and (4-34b),

$$\left(\frac{\partial\eta}{\partial\alpha}\right)_{\epsilon,s} = \frac{p}{T} \tag{4-41}$$

The common Jacobian notation has been used in which

$$J \equiv \frac{\partial(u, v)}{\partial(x, y)} \equiv \begin{vmatrix} \left(\frac{\partial u}{\partial x}\right)_y & \left(\frac{\partial v}{\partial x}\right)_y \\ \left(\frac{\partial u}{\partial y}\right)_x & \left(\frac{\partial v}{\partial y}\right)_x \end{vmatrix} \tag{4-42}$$

the symbols u, v, x, and y standing for arbitrary properties in this definition. Similarly, the partial derivative of η with respect to s is found,

$$\left(\frac{\partial\eta}{\partial s}\right)_{\alpha,\epsilon} = \frac{\partial(\eta, \epsilon)}{\partial(s, \epsilon)} = \frac{\partial(\eta, \epsilon)}{\partial(\eta, s)}\frac{\partial(\eta, s)}{\partial(s, \epsilon)}$$

$$= -\left(\frac{\partial\epsilon}{\partial s}\right)_{\alpha,\eta}\left(\frac{\partial\eta}{\partial\epsilon}\right)_{\alpha,s}$$

which from (4-34a) and (4-34c) gives

$$\left(\frac{\partial\eta}{\partial s}\right)_{\alpha,\epsilon} = -\frac{\tilde{\mu}}{T} \tag{4-43}$$

With these derivatives of η we can write that

$$d\eta = \frac{1}{T}\,d\epsilon + \frac{p}{T}\,d\alpha - \frac{\tilde{\mu}}{T}\,ds \tag{4-44}$$

the *Gibbs relation*, one of fundamental importance in thermodynamics. Clearly, this result also follows directly from combining (4-33) and (4-34a, b, c), but proceeding from (4-41) to (4-43) does represent an interesting consistency check.

Characteristic Functions or Thermodynamics Potentials

The Gibbs equation relates η to the independent variables ϵ, α, and s; η is said to be a characteristic function of these variables. Alternatively, with respect to the independent variables ϵ, α, and s, the entropy of a system is said to be a thermodynamic potential. If the Gibbs relation is rewritten as

$$d\epsilon = T\,d\eta - p\,d\alpha + \tilde{\mu}\,ds \tag{4-45}$$

then in terms of the independent variables η, α, and s, the thermodynamic potential or characteristic function of the system is the internal energy. A general technique for creating a potential is to begin with the differential of a potential, which by definition is exact, and add to it another exact differential. This type of transformation is known as a *Legendre differential transformation*.

Let us then transform (4-45) by adding $d(p\alpha)$ to it to obtain

$$d\tilde{h} \equiv d(\epsilon + p\alpha) = T\,d\eta + \alpha\,dp + \tilde{\mu}\,ds \tag{4-46}$$

by subtracting $d(T\eta)$ from it,

$$d\tilde{f} \equiv d(\epsilon - T\eta) = -\eta\,dT - p\,d\alpha + \tilde{\mu}\,ds \tag{4-47}$$

and by adding $d(p\alpha - T\eta)$ to it, we obtain

$$d\tilde{g} \equiv d(\epsilon + p\alpha - T\eta) = -\eta\,dT + \alpha\,dp + \tilde{\mu}\,ds \tag{4-48}$$

The function

$$\tilde{h} \equiv \epsilon + p\alpha \tag{4-49}$$

is called the *specific enthalpy* or *specific heat function* of the system; the function

$$\tilde{f} \equiv \epsilon - T\eta \tag{4-50}$$

is called the *specific free energy* or the *specific Helmholtz function* of the system; and the function

$$\tilde{g} \equiv \epsilon + p\alpha - T\eta \tag{4-51}$$

is called the *specific Gibbs potential* or *function* of the system. (Some authors call it the *specific thermodynamic potential,* a particularly confusing usage in view of the much more general meaning applied to the term "thermodynamic potential".) It follows from what has been said previously that the specific enthalpy $\tilde{h}$ is the thermodynamic potential with respect to the variables η, p, and s; the specific free energy $\tilde{f}$ is the thermodynamic potential with respect to T, α, and s; and the specific Gibbs potential $\tilde{g}$ is the thermodynamic potential with respect to T, p, and s. Clearly, $\tilde{h}$, $\tilde{f}$, and $\tilde{g}$ are all specific forms of extensive parameters and, therefore, from an application of Euler's theorem on homogeneous functions of degree one, we can write that for a binary mixture

$$\tilde{g}(T, p, s) = \tilde{\mu}_s s + \tilde{\mu}_w(1 - s) \tag{4-52}$$

(again see Kestin, 1966, pp. 326–329, for an excellent discussion of the application of Euler's theorem). Comparing (4-52) with (4-48), we obtain *Euler's identity*,

$$\epsilon + p\alpha - T\eta = \tilde{\mu}s + \tilde{\mu}_w \tag{4-53}$$

Differentiating this,

$$d\epsilon + p\,d\alpha + \alpha\,dp - T\,d\eta - \eta\,dT = \tilde{\mu}\,ds + s\,d\tilde{\mu} + d\tilde{\mu}_w$$

and substituting the Gibbs relation, (4-44), one obtains the applicable form of the well-known *Gibbs–Duhem relation*,

$$\eta\,dT - \alpha\,dp + s\,d\tilde{\mu} + d\tilde{\mu}_w = 0 \tag{4-54}$$

Heat Capacities

The heat capacity of a fluid is the amount of heat required to increase a unit mass of the fluid a unit degree of temperature. Two different heat capacities are defined—the *heat capacity at constant pressure*,

$$C_p \equiv T\left(\frac{\partial\eta}{\partial T}\right)_{p,s} \tag{4-55}$$

and the *heat capacity at constant volume*,

$$C_v \equiv T\left(\frac{\partial\eta}{\partial T}\right)_{\alpha,s} \tag{4-56}$$

As an example of unexpected and interesting relations that occur in thermodynamics, we can consider the expression for C_v in the system of variables (T, p, s). Using the known properties of Jacobians, one has that

$$\begin{aligned} C_v \equiv T\left(\frac{\partial\eta}{\partial T}\right)_{\alpha,s} &= T\,\frac{\partial(\eta,\alpha)}{\partial(T,\alpha)} \\ &= T\,\frac{\partial(\eta,\alpha)}{\partial(T,p)}\,\frac{1}{\partial(T,\alpha)/\partial(T,p)} \\ &= T\left(\frac{\partial\eta}{\partial T}\frac{\partial\alpha}{\partial p} - \frac{\partial\eta}{\partial p}\frac{\partial\alpha}{\partial T}\right)\frac{1}{\partial\alpha/\partial p} \end{aligned} \tag{4-57}$$

From (4-48) we obtain

$$\left(\frac{\partial\tilde{g}}{\partial T}\right)_{p,s} = -\eta \qquad \left(\frac{\partial\tilde{g}}{\partial p}\right)_{T,s} = \alpha \tag{4-58}$$

and taking the order of differentiation as immaterial,

$$-\left(\frac{\partial\eta}{\partial p}\right)_{T,s} = \left(\frac{\partial\alpha}{\partial T}\right)_{p,s} \tag{4-59}$$

in the (T, p, s) system of variables. Using this result and (4-55), equation (4-57) gives

$$C_v = C_p + T\,\frac{(\partial\alpha/\partial T)^2}{\partial\alpha/\partial p} \tag{4-60}$$

which permits the determination of C_v from C_p and the equation of state of the fluid. In this case it is possible to rigorously show (Kamenkovich, 1977) that which is intuitively obvious, namely

$$\left(\frac{\partial \alpha}{\partial p}\right)_{T,s} < 0$$

which gives

$$\frac{C_p}{C_v} \equiv \gamma > 1 \tag{4-61}$$

Adiabatic Lapse Rate of Temperature

Thomson (1857) defined an expression relating the temperature change with pressure in an adiabatic process,

$$\Gamma \equiv \left(\frac{\partial T}{\partial p}\right)_{\eta,s} \tag{4-62}$$

where Γ is called the *adiabatic lapse rate of temperature* or the *adiabatic temperature gradient*. In the (T, p, s) system, using Jacobians, one obtains

$$\Gamma = \frac{\partial(T, \eta)}{\partial(p, \eta)} = \frac{\partial(T, \eta)}{\partial(T, p)} \frac{1}{\partial(p, \eta)/\partial(T, p)}$$

which from (4-54) and (4-58) reduces to

$$\Gamma = \frac{T}{C_p}\left(\frac{\partial \alpha}{\partial T}\right)_{p,s} \tag{4-63}$$

Now the *potential temperature* Θ is defined as the temperature that a water sample would attain if raised adiabatically from a depth corresponding to the pressure p to the sea surface with the atmospheric pressure p_a. From this definition,

$$\Theta \equiv T(\eta, p_a, s) = T(\eta, p, s) - \int_{p_a}^{p} \Gamma(\eta, p, s)\, dp \tag{4-64}$$

The corresponding density for such a transition is called the *potential density*; that is, the potential density is the density that a parcel of fluid would have if it were determined from the *in situ* salinity and the potential temperature,

$$\begin{aligned}\rho_{\text{pot}} &\equiv \rho(\eta, p_a, s) = \rho(\Theta, p_a, s) \\ &= \rho(\eta, p, s) - \int_{p_a}^{p} \left(\frac{\partial \rho}{\partial p}\right)_{\eta,s} dp\end{aligned} \tag{4-65}$$

Compressibility

The compressibility or coefficient of compressibility of a fluid is the fractional decrease of volume with increasing pressure. Two different compressibilities are defined—an *isothermal compressibility*,

$$K_T \equiv -\frac{1}{\alpha}\left(\frac{\partial \alpha}{\partial p}\right)_{T,s} = \frac{1}{\rho}\left(\frac{\partial \rho}{\partial p}\right)_{T,s} \tag{4-66}$$

in the variables T, p, and s, and an *adiabatic compressibility*,

$$K_\eta \equiv -\frac{1}{\alpha}\left(\frac{\partial \alpha}{\partial p}\right)_{\eta, s} = \frac{1}{\rho}\left(\frac{\partial \rho}{\partial p}\right)_{\eta, s} \tag{4-67}$$

in the variables η, p, and s. K_η can be expressed in the (T, p, s) system. From application of Jacobians,

$$\left(\frac{\partial \rho}{\partial p}\right)_{\eta, s} = \frac{\partial(\rho, \eta)}{\partial(p, \eta)} = \frac{\partial(\rho, \eta)}{\partial(T, p)} \frac{1}{\partial(p, \eta)/\partial(T, p)}$$

$$= -\left(\frac{\partial \rho}{\partial T}\frac{\partial \eta}{\partial p} - \frac{\partial \rho}{\partial p}\frac{\partial \eta}{\partial T}\right)\frac{1}{\partial \eta/\partial T}$$

Substitution from (4-55), (4-59), and (4-63) yields

$$\left(\frac{\partial \rho}{\partial p}\right)_{\eta, s} = \Gamma\left(\frac{\partial \rho}{\partial T}\right)_{p, s} + \left(\frac{\partial \rho}{\partial p}\right)_{T, s} \tag{4-68}$$

and by (4-66) and (4-67),

$$K_\eta = \frac{\Gamma}{\rho}\left(\frac{\partial \rho}{\partial T}\right)_{p, s} + K_T \tag{4-69}$$

The first term on the right is merely the product of Γ and the negative of the *coefficient of thermal expansion* Δ, defined as

$$\Delta \equiv \frac{1}{\alpha}\left(\frac{\partial \alpha}{\partial T}\right)_{p, s}$$

which represents the relative change in volume of the system per unit change of temperature in a process during which $p =$ constant. K_η then becomes

$$K_\eta = -\Gamma\Delta + K_T \tag{4-70}$$

and using (4-62), (4-65), and (4-68),

$$\frac{K_\eta}{K_T} = \frac{C_v}{C_p} = \frac{1}{\gamma} \tag{4-71}$$

Since $C_v < C_p$, then $K_\eta < K_T$.

Speed of Sound

The adiabatic compressibility K_η is closely related to the speed of sound c in a fluid, where

$$c^2 \equiv \left(\frac{\partial p}{\partial \rho}\right)_{\eta, s} \tag{4-72}$$

From (4-67),

$$c^2 = \frac{1}{\rho K_\eta} \tag{4-73}$$

This result, along with (4-71), permits the determination of the ratio C_p/C_v from the known speed of sound and the equation of state of the fluid.

Väisälä or Stability Frequency

Suppose that in the sea the temperature, salinity, density, and all other thermodynamic parameters vary only in the vertical; that is, a column of water is vertically stratified. Assume that the fluid is in mechanical equilibrium. For convenience consider a small mass of fluid enclosed in a balloon-like membrane, the membrane being perfectly flaccid such that the pressures inside and outside are always the same and changes in fluid density can take place unimpeded. Assume also that these changes occur adiabatically. Take the density distribution of the undistrubed fluid as being given by $\rho_0(z)$. Now let the membrane and the small mass of fluid it contains be displaced from a depth z where its density is $\rho = \rho_0(z)$ to a depth $z + \xi$ (see Figure 4-1). The

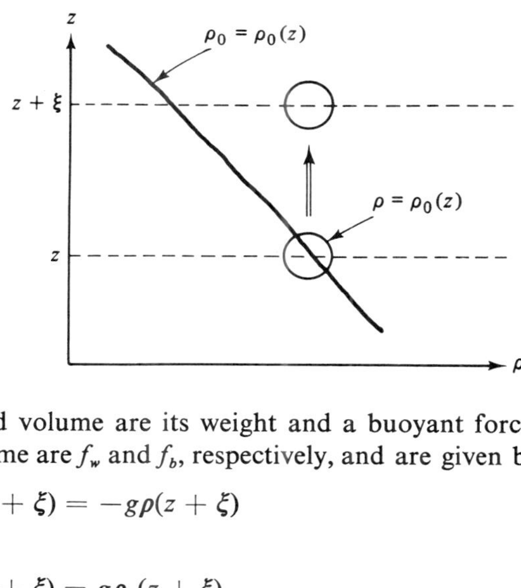

Fig. 4-1. Displacement of a small fluid mass adiabatically and without mixing from its equilibrium position.

forces acting on the displaced volume are its weight and a buoyant force, which expressed per unit volume are f_w and f_b, respectively, and are given by

$$f_w(z + \xi) = -g\rho(z + \xi)$$

and

$$f_b(z + \xi) = g\rho_0(z + \xi)$$

where g is gravity. The net force per unit volume f_v acting is then

$$\begin{aligned} f_v(z + \xi) &= f_b(z + \xi) - f_w(z + \xi) \\ &= g[\rho_0(z + \xi) - \rho(z + \xi)] \end{aligned} \tag{4-74}$$

Expanding in a Taylor series around depth z,

$$\rho(z + \xi) = \rho(z) + \left(\frac{d\rho}{dz}\right)_{\eta, s} \xi + \cdots \tag{4-75}$$

and

$$\rho_0(z + \xi) = \rho_0(z) + \frac{d\rho_0}{dz}\xi + \cdots \tag{4-76}$$

However, the first of these can also be written

$$\rho(z+\xi) = \rho_0(z) + \left(\frac{\partial \rho}{\partial p}\right)_{\eta, s} \left(\frac{dp}{dz}\right)\xi + \cdots \tag{4-77}$$

Now applying Newton's second law,

$$\mathbf{f}_v = \rho \mathbf{a}$$

we obtain, from (4-74),

$$\rho_0 \frac{d^2\xi}{dt^2} = g[\rho_0(z+\xi) - \rho(z+\xi)]$$

and, after substitution of (4-75) and (4-76),

$$\rho_0 \frac{d^2\xi}{dt^2} = g\frac{d\rho_0}{dz}\xi - g\left(\frac{\partial \rho}{\partial p}\right)_{\eta, s}\frac{dp}{dz}\xi \tag{4-78}$$

having dropped higher-order terms. Use of the hydrostatic equation

$$\frac{dp}{dz} = -\rho_0 g$$

results in

$$\rho_0 \frac{d^2\xi}{dt^2} = \left[g\frac{d\rho_0}{dz} + \rho_0 g^2 \left(\frac{\partial \rho}{\partial p}\right)_{\eta, s}\right]\xi$$

which from (4-71) can be written for adiabatic processes as

$$\frac{d^2\xi}{dt^2} + \left(-\frac{g}{\rho_0}\frac{d\rho_0}{dz} - \frac{g^2}{c^2}\right)\xi = 0 \tag{4-79}$$

The quantity in parentheses is defined as

$$\tilde{N}^2 \equiv -\frac{g}{\rho_0}\frac{d\rho_0}{dz} - \frac{g^2}{c^2} \tag{4-80}$$

where $\tilde{N}$ is called the *Väisälä frequency* or *stability frequency* and has units of reciprocal time.

Assuming that ρ_0 is expressed as a function of T, p, and s, then

$$\frac{d\rho_0}{dz} = \frac{\partial \rho_0}{\partial T}\frac{dT}{dz} + \frac{\partial \rho_0}{\partial p}\frac{dp}{dz} + \frac{\partial \rho_0}{\partial s}\frac{ds}{dz} \tag{4-81}$$

and $\tilde{N}^2$ is given by

$$\begin{aligned}\tilde{N}^2 &= -\frac{g}{\rho_0}\frac{d\rho_0}{dz} - g^2\left(\frac{\partial \rho_0}{\partial p}\right)_{\eta, s} \\ &= -\frac{g}{\rho_0}\left[\frac{d\rho_0}{dz} + \rho_0 g\Gamma\frac{\partial \rho_0}{\partial T} + \rho_0 g\frac{\partial \rho_0}{\partial p}\right]\end{aligned}$$

having used (4-68). Substituting for $d\rho_0/dz$ and employing the hydrostatic equation,

$$\tilde{N}^2 = -\frac{g}{\rho_0}\left[\left(\frac{\partial \rho_0}{\partial T}\right)_{p, s}\left(\frac{dT}{dz} + \rho_0 g\Gamma\right) + \left(\frac{\partial \rho_0}{\partial s}\right)_{T, p}\frac{ds}{dz}\right] \tag{4-82}$$

Another useful form of $\tilde{N}^2$ is obtained by starting with (4-80) and employing (4-81),

$$\tilde{N}^2 = -\frac{g}{\rho_0}\left(\frac{d\rho_0}{dz}\right)_p - \frac{g^2}{c^2} - \frac{g}{\rho_0}\left(\frac{\partial \rho_0}{\partial p}\right)_{T,s}\left(\frac{dp}{dz}\right)$$

which in view of the hydrostatic equation, the definition of K_T, (4-66), and (4-71) and (4-73) transforms to

$$\tilde{N}^2 = -\frac{g}{\rho_0}\left(\frac{d\rho_0}{dz}\right)_p + \frac{g^2}{c^2}\left(\frac{C_p}{C_v} - 1\right) \tag{4-83}$$

This result is convenient within homogeneous layers since for dT/dz and ds/dz small, $C_p \simeq C_v$ and the last term can be dropped. Further, since the density of seawater differs only slightly from the value of 1 g/cm³, it is useful to introduce the quantities

$$\sigma_{s,t,p} \equiv 10^3(\rho_{s,t,p} - 1 \text{ g/cm}^3) \tag{4-84}$$

and

$$\sigma_t \equiv 10^3(\rho_{s,t,0} - 1 \text{ g/cm}^3) \tag{4-85}$$

where $\rho_{s,t,p}$ is the density in gram per cubic centimeter of a seawater sample of salinity s, temperature t, at a gauge pressure, p; and $\rho_{s,t,0}$ is the density of a seawater sample in grams per cubic centimeter having a salinity s, a temperature t, and at atmospheric pressure. Taking

$$\left(\frac{d\rho_0}{dz}\right)_p \simeq 10^{-3}\frac{d\sigma_t}{dz}$$

an approximate form of (4-83) is

$$\tilde{N}^2 \simeq -10^{-3}\frac{g}{\rho_0}\frac{d\sigma_t}{dz} \tag{4-86}$$

Returning to the differential equation (4-79),

$$\frac{d^2\xi}{dt^2} + \tilde{N}^2\xi = 0 \tag{4-87}$$

the form of its solution depends upon the sign of $\tilde{N}^2$. If the stability frequency is real, $\tilde{N}^2 > 0$ and

$$\xi = Ae^{i\tilde{N}t} + Be^{-i\tilde{N}t}$$

or

$$\xi = A\cos\tilde{N}t$$

A and B constants. This states that the balloon-like membrane will oscillate with frequency $\tilde{N}$ as defined in (4-80). The condition for stability thus requires negative gradients, that is, $d\rho_0/dz < 0$, and that $|d\rho_0/dz| > \rho_0 g/c^2$. On the other hand, if $\tilde{N}^2$ is negative (i.e., $\tilde{N}$ is an imaginary number), (4-87) has its solution in the form $\xi = Ae^{\tilde{N}t}$ and the parcel of fluid will be unstable and continue to accelerate, thus departing widely from its initial position.

Maxwell's Reciprocal Relations

From (4-47) we find that

$$\tilde{\mu} = \left(\frac{\partial \tilde{g}}{\partial s}\right)_{p,T} \tag{4-88}$$

which gives upon differentiation with respect to T,

$$\left(\frac{\partial \tilde{\mu}}{\partial T}\right)_{p,s} = \left[\frac{\partial}{\partial T}\left(\frac{\partial \tilde{g}}{\partial s}\right)_{p,T}\right]_{p,s}$$

Then differentiating the first of (4-58) with respect to s,

$$\left[\frac{\partial}{\partial s}\left(\frac{\partial \tilde{g}}{\partial T}\right)_{p,s}\right]_{p,T} = -\left(\frac{\partial \eta}{\partial s}\right)_{p,T}$$

and taking the order of differentiation as immaterial, we have from the last two equations

$$\left(\frac{\partial \tilde{\mu}}{\partial T}\right)_{p,s} = -\left(\frac{\partial \eta}{\partial s}\right)_{p,T} \tag{4-89}$$

Differentiating (4-88) with respect to p,

$$\left(\frac{\partial \tilde{\mu}}{\partial p}\right)_{T,s} = \left[\frac{\partial}{\partial p}\left(\frac{\partial \tilde{g}}{\partial s}\right)_{p,T}\right]_{T,s}$$

and the second of (4-58) with respect to s,

$$\left[\frac{\partial}{\partial s}\left(\frac{\partial \tilde{g}}{\partial p}\right)_{T,s}\right]_{p,T} = \left(\frac{\partial \alpha}{\partial s}\right)_{p,T}$$

from which we see that

$$\left(\frac{\partial \tilde{\mu}}{\partial p}\right)_{T,s} = \left(\frac{\partial \alpha}{\partial s}\right)_{p,T} \tag{4-90}$$

having again taken the order of differentiation as being immaterial. Equations (4-59), (4-89), and (4-90) are the well-known *Maxwell reciprocal relations* for a binary fluid.

4-7 Determination of the Thermodynamic Properties of Seawater

It is possible to evaluate the thermodynamic properties (entropy, internal energy, relative chemical potential, etc.) of seawater from empirical data.

First, we note that empirical formulas of sufficient accuracy have been proposed for the equation of state $\alpha = \alpha(s, T, p)$ [Fofonoff (1962), Wilson and Bradley (1968), and others] and the specific heat C_p at atmospheric pressure [Cox and Smith (1959) and Fofonoff (1962)]. Since $\partial^3 \tilde{g}/\partial p\, \partial T^2 = \partial^3 \tilde{g}/\partial T^2\, \partial p$ (the subscripts have been dropped for simplicity since we will be using as independent variables s, T, and p), we obtain, from (4-55) and (4-58),

$$-\frac{1}{T}\frac{\partial C_p}{\partial p} = \frac{\partial^2 \alpha}{\partial T^2} \tag{4-91}$$

The specific heat C_p at any pressure may be computed after integration of this equation from atmospheric pressure p_0 to pressure p,

$$\frac{1}{T}[C_p(p_0) - C_p(p)] = \int_{p_0}^{p} \frac{\partial^2 \alpha}{\partial T^2}\, dp \tag{4-92}$$

Fofonoff and Froese (1958) have constructed a form of this result suitable for electronic computer applications. With these facts in mind we will assume that $\alpha = \alpha(s, T, p)$ and $C_p = C_p(s, T, p)$ are known.

To evaluate $\tilde{\mu} = \tilde{\mu}(s, T, p)$ from measurable quantities, its total differential,

$$d\tilde{\mu} = \frac{\partial \tilde{\mu}}{\partial s}\, ds + \frac{\partial \tilde{\mu}}{\partial T}\, dT + \frac{\partial \tilde{\mu}}{\partial p}\, dp \tag{4-93}$$

is written in the form

$$d\tilde{\mu} = \frac{\partial \tilde{\mu}}{\partial s}\, ds - \frac{\partial \eta}{ds}\, dT + \frac{\partial \alpha}{\partial s}\, dp \tag{4-94}$$

having substituted from the Maxwell reciprocal relations, (4-89) and (4-90). The coefficient $\partial\alpha/\partial s$ is readily determined from the empirical form of the equation of state. The coefficient $\partial\tilde{\mu}/\partial s$ can be found from measurements of the colligative properties of vapor-pressure lowering, freezing-point depression, boiling-point elevation, and/or osmotic pressure of a seawater solution. Without going into the details of the derivation, Fofonoff (1962) states that for seawater in equilibrium with water vapor at a given temperature, pressure, and salinity,

$$s\left(\frac{\partial \tilde{\mu}}{\partial s}\right) = \frac{R_v T}{1 - r}\frac{\partial r}{\partial s} \tag{4-95}$$

where $R_v \equiv R/M_v$, M_v being the gram-molecular weight of water vapor, R the universal gas constant ($R_v = 46.2 \times 10^5$ erg/g/°C), and r the fractional lowering of the vapor pressure of seawater,

$$r \equiv \frac{p_v^0 - p_v}{p_v^0}$$

where p_v is the vapor pressure of seawater and p_v^0 is the vapor pressure of pure water at the same temperature. The fractional lowering of vapor pressure has been determined to be approximately a linear function of salinity (Fofonoff, 1962). If, instead of vapor-pressure lowering, freezing-point depression is used, Foffonof states that for seawater in equilibrium with ice,

$$s\left(\frac{\partial \tilde{\mu}}{\partial s}\right) = -(\eta_W - \eta_{\text{ice}})\frac{\partial T}{\partial s} \tag{4-96}$$

where $T(\eta_W - \eta_{ice})$ is the heat of fusion of ice. For the boiling-point elevation,

$$s\left(\frac{\partial \tilde{\mu}}{\partial s}\right) = (\eta_v - \eta_W)\frac{\partial T}{\partial s} \tag{4-97}$$

where $T(\eta_v - \eta_W)$ is the heat of vaporization of seawater. All of this leaves as undetermined in (4-94), $\partial \eta/\partial s$, which quantity must be evaluated indirectly, but fortunately, need not be known for all values of s, T, and p. If the path of integration is taken from a reference state (s_0, T_0, p_0) to (s_0, T, p_0) to (s, T, p_0) to (s, T, p), as shown in Figure 4-2, then $\partial \eta/\partial s$ need be known only as a function of one variable (T), $\partial \tilde{\mu}/\partial s$ as a function of two variables (s and T), and $\partial \alpha/\partial s$ as a function of all three variables (s, T, and p). For such a path the integral relation for $\tilde{\mu}$ takes the form

$$\begin{aligned}\tilde{\mu}(s, T, p) = \tilde{\mu}_0 - \int_{T_0}^{T}\left(\frac{\partial \eta}{\partial s}\right)_{s_0, p_0} dT + \int_{s_0}^{s}\left(\frac{\partial \tilde{\mu}}{\partial s}\right)_{p_0} ds \\ + \int_{p_0}^{p}\left(\frac{\partial \alpha}{\partial s}\right) dp\end{aligned} \tag{4-98}$$

where $\tilde{\mu}_0$ is an arbitrary constant appropriate to the reference state (s_0, T_0, p_0). Note that $\partial \eta/\partial s$ must be known only as a function of T for the reference salinity and pressure, s_0 and p_0; that $\partial \tilde{\mu}/\partial s$ must be known as a function of s and T for the reference pressure; and that $\partial \alpha/\partial s$ must be known as a function of all three variables. This still leaves us with the problem of evaluating $\partial \eta/\partial s$. From (4-55) we have that

$$\frac{\partial \eta}{\partial T} = \frac{C_p}{T}$$

Differentiating with respect to s and exchanging the order yields

$$\frac{\partial}{\partial T}\left(\frac{\partial \eta}{\partial s}\right) = \frac{1}{T}\left(\frac{\partial C_p}{\partial s}\right) \tag{4-99}$$

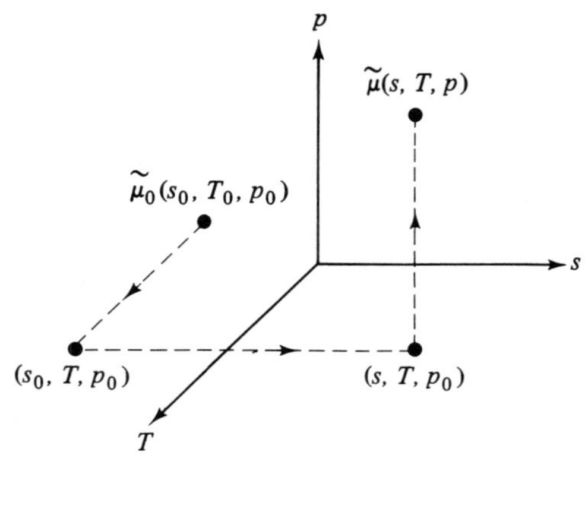

Fig. 4-2. Path of integration of (4-93) in the s, T, p plane.

Integrating with respect to T with fixed salinity s_0 and fixed pressure p_0 gives

$$\left(\frac{\partial \eta}{\partial s}\right)_{p_0, s_0} = b + \int_{T_0}^{T} \frac{1}{T}\left(\frac{\partial C_p}{\partial s}\right)_{p_0, s_0} dT \tag{4-100}$$

where b is an arbitrary constant. The relation for $\tilde{\mu}$, (4-98), then takes the form

$$\begin{aligned}\tilde{\mu}(s, T, p) = \tilde{\mu}_0 - bT - \int_{T_0}^{T}\int_{T_0}^{T} \frac{1}{T'}\left(\frac{\partial C_p}{\partial s}\right)_{p_0, s_0} dT'\, dT \\ + \int_{s_0}^{s}\left(\frac{\partial \tilde{\mu}}{\partial s}\right)_{p_0} ds + \int_{p_0}^{p}\left(\frac{\partial \alpha}{\partial s}\right) dp\end{aligned} \tag{4-101}$$

From all of this we see that $\tilde{\mu}$ may be computed from a limited amount of empirical data apart from the function $\tilde{\mu}_0 - bT$, since both $\tilde{\mu}_0$ and b are arbitrary for the reference state.

Proceeding in a similar manner to determine the equation for the specific entropy, $\eta = \eta(s, T, p)$, its total differential,

$$d\eta = \frac{\partial \eta}{\partial s} ds + \frac{\partial \eta}{\partial T} dT + \frac{\partial \eta}{\partial p} dp \tag{4-102}$$

can be written

$$d\eta = -\frac{\partial \tilde{\mu}}{\partial T} ds + \frac{C_p}{T} dT - \frac{\partial \alpha}{\partial T} dp \tag{4-103}$$

after substitution from (4-55), (4-59), and (4-89). Of the three coefficients, only the thermal rate of change of the relative chemical potential is not completely specified, and integration of (4-103),

$$\begin{aligned}\eta(s, T, p) = \eta_0 - \int_{s_0}^{s}\left(\frac{\partial \tilde{\mu}}{\partial T}\right) ds + \int_{T_0}^{T}\left(\frac{C_p}{T}\right) dT \\ - \int_{p_0}^{p}\left(\frac{\partial \alpha}{\partial T}\right) dp\end{aligned} \tag{4-104}$$

where η_0 is an arbitrary constant, will give us the entropy apart from a linear function of s.

An expression for the specific internal energy ϵ is also readily found. In the system of variables s, T, and p,

$$d\epsilon = \frac{\partial \epsilon}{\partial s} ds + \frac{\partial \epsilon}{\partial T} dT + \frac{\partial \epsilon}{\partial p} dp \tag{4-105}$$

Differentiating (4-51) with respect to T, we obtain

$$\frac{d\tilde{g}}{dT} = \frac{\partial \epsilon}{\partial T} + p\frac{\partial \alpha}{\partial T} - \eta - T\frac{\partial \eta}{\partial T} \tag{4-106}$$

which, from (4-55) and (4-58), simplifies to

$$\frac{\partial \epsilon}{\partial T} = -p\frac{d\alpha}{\partial T} + C_p \tag{4-107}$$

Analogously, differentiating (4-51) with respect to p,

$$\frac{\partial \tilde{g}}{\partial p} = \frac{\partial \epsilon}{\partial p} + \alpha + p\,\frac{\partial \alpha}{\partial p} - T\,\frac{\partial \eta}{\partial p} \tag{4-108}$$

and with respect to s,

$$\frac{\partial \tilde{g}}{\partial s} = \frac{\partial \epsilon}{\partial s} + p\,\frac{\partial \alpha}{\partial s} - T\,\frac{\partial \eta}{\partial s} \tag{4-109}$$

The first of these after simplifying via (4-58) and (4-59) reduces to

$$\frac{\partial \epsilon}{\partial p} = -p\,\frac{\partial \alpha}{\partial p} + T\,\frac{\partial \alpha}{\partial T} \tag{4-110}$$

and utilizing (4-48) and (4-89) the second becomes

$$\frac{\partial \epsilon}{\partial s} = \tilde{\mu} - p\,\frac{\partial \alpha}{\partial s} - T\,\frac{\partial \tilde{\mu}}{\partial T} \tag{4-111}$$

Substituting (4-107), (4-110), and (4-111) in (4-105), we obtain

$$\begin{aligned} d\epsilon = \left(\tilde{\mu} - p\,\frac{\partial \alpha}{\partial s} - T\,\frac{\partial \tilde{\mu}}{\partial T}\right) ds + \left(C_p - p\,\frac{\partial \alpha}{\partial T}\right) dT \\ + \left(T\,\frac{\partial \alpha}{\partial T} - p\,\frac{\partial \alpha}{\partial P}\right) dp \end{aligned} \tag{4-112}$$

integration of which will give $\epsilon(s, T, p)$ to within a linear function of s. Thus, from the same empirical data that permit the determination of $\tilde{\mu}(s, T, p)$ and $\eta(s, T, p)$, one may find the specific internal energy.

Finally, in a similar fashion it is readily shown that the specific Gibbs potential $\tilde{g}$, as defined by (4-51), can be found as a function of the independent variables s, T, and p apart from a function of the form $b_1 + b_2 T + b_3 s + b_4 s T$, where b_1, b_2, b_3, and b_4 are arbitrary constants.

FIVE

Large-Scale Ocean Waves in the Absence of Tidal Forces

5-1 Introduction

Ocean waves in general occur in a wide range of frequencies (or wave lengths); they may be represented as to type according to various criteria. Figure 5-1 illustrates three of these schemes—classification according to frequency, according to the primary disturbing force, and according to the primary restoring force. The lower curve schematically indicates the approximate relative distribution of energy in the waves of differing period. For example, the greatest amount of energy occurs in wind waves of periods in the range of from 1 to 30 sec, with a peak between 8 to 10 sec. At the right end of the spectrum are to be found *capillary waves*; waves of very short length and period whose motion is controlled to all intents and purposes by surface tension. Wind-generated ordinary *gravity waves* are perhaps the most conspicuous feature of the oceans. As the name implies, the force of gravity is the primary disturbing force. In turn, a distinction is made between sea and swell—*sea* referring to ordinary gravity waves within a generating area (i.e., still being acted on by the wind); *swell*, on the other hand, consists of waves outside the generating area that have acquired some smoothness and uniformity.

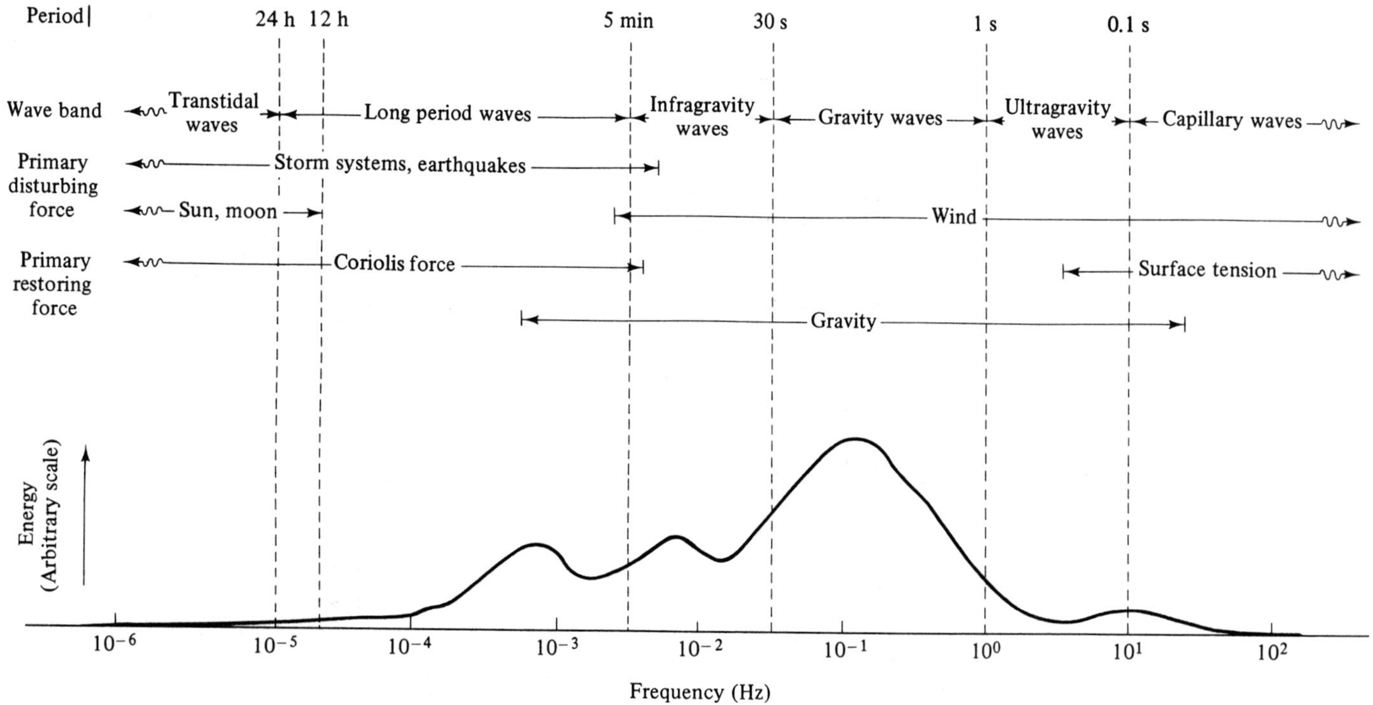

Fig. 5-1. Classification of surface waves and schematic representation of their power spectrum. Reprinted with permission from Kinsman, *Wind Waves*, Englewood Cliffs, NJ: Prentice-Hall, Inc. 1965, p. 23.

Infragravity waves can be thought of as marking the beginning of the "invisible" part of the ocean wave spectrum in that fluctuations having a period of more than about 30 sec are not generally perceived by an observer to be a wave type of phenomenon. Infragravity waves are comprised of *groundswell* generated by wind and atmospheric pressure during periods of storm, *surf beat*, of a period 1 to 5 min, due to the superposition of swell originating from more than one source, and *seiches*, resulting from the reflection of ground swell in an enclosed or semienclosed body of water with the development thereby of a resonant or near-resonant standing-wave system. In the *long-period wave band*, barometric fluctuations tend to create long-period sea-level disturbances. *Storm tides*, surges generated by moving low-pressure regions of considerable intensity such as hurricances or typhoons, give rise to long-period waves. Less frequently, submarine earthquakes are the origin of the *seismic sea wave* or *tsunami*. In the higher-frequency portion of this band, both gravity and the Coriolis force control the resulting motion. Astronomical *tides* are forced waves (in a *forced wave* the disturbing force is applied continuously in distinction to a *free wave* which is generated by an instantaneous application of the disturbing force) whose origin is the gravitational attraction of the sun and moon. The tides are represented as strong, relatively sharp lines in the wave spectrum with principal peaks of about 12 and 24 hr. Such things as slow-moving barometic depressions, long-term secular changes in sea level, or the rising and falling of land masses result in waves in the *transtidal* zone, waves whose motion is almost exclusively governed by the Coriolis force.

Classical papers on tidal theory [Laplace (1799) and Hough (1897, 1898), to mention only a couple] have discussed the free motion of a thin layer of fluid, when the fluid is assumed to cover the entire earth, under the influence of gravitational and Coriolis forces. Hough shows that the resulting motion can be essentially divided into two classes, these classes being distinguishable depending on the behavior of the motion as the rate of rotation goes to zero. In the *first class* are gravity waves and certain long-period waves whose motions may be modified to a certain extent by the rotation of the earth. However, first-class waves would still be waves even though the earth were to stop rotating. In the *second class* are periodic motions that degenerate into steady currents as the rotational rate goes to zero. For waves of the second class it is, in fact, the rotation of the earth that makes the motion periodic in time, the period usually being greater than 24 hr.

It is our intent in this section to present an introduction to the mathematics of large-scale ocean waves in the absence of tidal forces. We will develop the long-wave hydrodynamic equations and apply them to examine in detail the properties of various first-class waves—the plane Sverdrup wave, the plane Poincaré wave, and plane Kelvin and Proudman waves. In this we will follow the very lucid development of Platzman (1971). Finally, an elementary derivation and analysis of Rossby waves, a wave of the second class, will be presented.

5-2 The Long-Wave Equations

We begin with the x and y components of the equation of motion as given by (2-197) and (2-198). Assuming frictionless flow, a linear variation in the Coriolis parameter f, dropping the inertial terms, and taking $2w\Omega \cos \theta$ as negligible compared to the other quantities, we have that

$$\frac{\partial u}{\partial t} - fv = -\frac{1}{\rho}\frac{\partial p}{\partial x} \tag{5-1}$$

$$\frac{\partial v}{\partial t} + fu = -\frac{1}{\rho}\frac{\partial p}{\partial y} \tag{5-2}$$

and, in the vertical almost hydrostatic conditions such that

$$\frac{\partial p}{\partial z} \doteq -\rho g \tag{2-217}$$

With reference to the definition sketch of Figure 5-2, with the fluid homogeneous, the pressure at any depth z can be found by a vertical integration of (2-217),

$$\int_z^\eta \frac{\partial p}{\partial z}\, dz = -\rho g \int_z^\eta dz$$

$$\int_{p_z}^{p_\eta} dp = -\rho g(\eta - z)$$

or

$$p_z = p_\eta + \rho g(\eta - z) \tag{5-3}$$

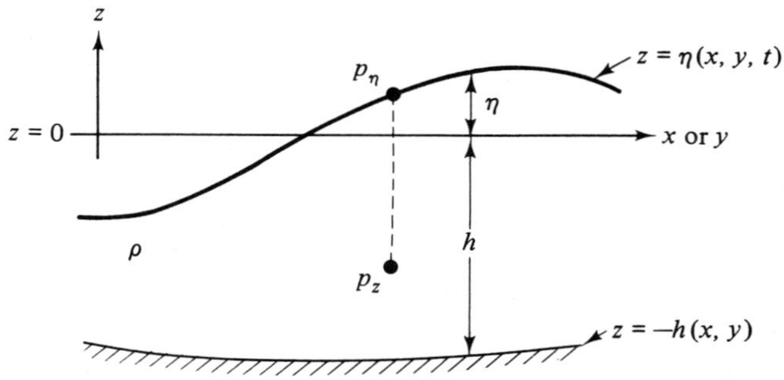

Fig. 5-2. Definition sketch for the long-wave equations.

Differentiating with respect to x and y,

$$\frac{1}{\rho}\frac{\partial p}{\partial x} = g\frac{\partial \eta}{\partial x} \qquad \frac{1}{\rho}\frac{\partial p}{\partial y} = g\frac{\partial \eta}{\partial y} \tag{5-4}$$

and substituting into (5-1) and (5-2) gives one form of the well-known *long-wave equations*:

$$\frac{\partial u}{\partial t} - fv = -g\frac{\partial \eta}{\partial x} \tag{5-5}$$

and

$$\frac{\partial v}{\partial t} + fu = -g\frac{\partial \eta}{\partial y} \tag{5-6}$$

The continuity equation for an incompressible fluid,

$$\frac{\partial u}{\partial x} + \frac{\partial v}{\partial y} + \frac{\partial w}{\partial z} = 0 \tag{2-135}$$

when vertically integrated yields

$$\int_{-h(x,y)}^{\eta(x,y,t)} \left(\frac{\partial u}{\partial x} + \frac{\partial v}{\partial y} + \frac{\partial w}{\partial z}\right) = 0 \tag{5-7}$$

Applying the Leibnitz rule of (2-136) to the first and second terms of this result,

$$\int_{-h(x,y)}^{\eta(x,y,t)} \frac{\partial u}{\partial x}\,dz = \frac{\partial}{\partial x}\int_{-h}^{\eta} u\,dz - u\bigg|_{z=\eta}\frac{\partial \eta}{\partial x} - u\bigg|_{z=-h}\frac{\partial h}{\partial x}$$

$$\int_{-h(x,y)}^{\eta(x,y,t)} \frac{\partial v}{\partial y}\,dz = \frac{\partial}{\partial y}\int_{-h}^{\eta} v\,dz - v\bigg|_{z=\eta}\frac{\partial \eta}{\partial y} - v\bigg|_{z=-h}\frac{\partial h}{\partial y}$$

gives the integrated continuity equation in the form

$$\begin{aligned} \frac{\partial}{\partial x}\int_{-h}^{\eta} u\,dz - u\bigg|_{z=\eta}\frac{\partial \eta}{\partial x} - u\bigg|_{z=-h}\frac{\partial h}{\partial x} \\ + \frac{\partial}{\partial y}\int_{-h}^{\eta} v\,dz - v\bigg|_{z=\eta}\frac{\partial \eta}{\partial y} - v\bigg|_{z=-h}\frac{\partial h}{\partial y} \\ + w\bigg|_{z=\eta} - w\bigg|_{z=-h} = 0 \end{aligned} \tag{5-8}$$

Now in Section 2-16 it was shown that the kinematic boundary condition for a free surface could be written

$$-\frac{\partial \eta}{\partial t} - u\bigg|_{z=\eta}\frac{\partial \eta}{\partial x} - v\bigg|_{z=\eta}\frac{\partial \eta}{\partial y} + w\bigg|_{z=\eta} = 0 \tag{2-98}$$

In a similar way for the fixed bottom, $z = -h(x, y)$,

$$z + h(x, y) = 0$$

and differentiating,

$$\frac{D}{Dt}[z + h(x, y)] = 0$$

or

$$w\Big|_{z=-h} + u\Big|_{z=-h}\frac{\partial h}{\partial x} + v\Big|_{z=-h}\frac{\partial h}{\partial y} = 0 \tag{5-9}$$

Equation (5-8), in view of (2-98) and (5-9), reduces to

$$\frac{\partial \eta}{\partial t} + \frac{\partial}{\partial x}\int_{-h}^{\eta} u\,dz + \frac{\partial}{\partial y}\int_{-h}^{\eta} v\,dz = 0 \tag{5-10}$$

which expresses the conservation of mass for a fixed vertical column extending from the sea floor to the air–sea interface.

In long-wave theory it is generally taken that the horizontal velocities u and v are independent of the depth z. This follows from (5-4) and (5-5) and (5-6) in that the x and y components of the acceleration of the water particles are independent of z, and hence u and v are also independent of z if they were at any time, say $t = 0$. This is true, for example, in the special case in which the water was at rest at $t = 0$; we shall assume that this is true in all cases. Equation (5-10) then integrates to

$$\frac{\partial \eta}{\partial t} + \frac{\partial}{\partial x}[u(h + \eta)] + \frac{\partial}{\partial y}[v(h + \eta)] = 0$$

Moreover, except in very shallow regions, $|\eta| \ll h$, and for most purposes this last result can be approximated by

$$\frac{\partial \eta}{\partial t} + \frac{\partial uh}{\partial x} + \frac{\partial vh}{\partial y} = 0 \tag{5-11}$$

a continuity equation.

An equation that permits a determination of the vertical velocity w from other dependent variables is obtained from an integration of (2-135) from the arbitrary depth z up to the surface $z = \eta$. Doing so,

$$-\int_{z}^{\eta}\frac{\partial w}{\partial z}\,dz = \int_{z}^{\eta}\frac{\partial u}{\partial x}\,dz + \int_{z}^{\eta}\frac{\partial v}{\partial y}\,dz$$

$$-\int_{w}^{w_\eta} dw = (\eta - z)\left(\frac{\partial u}{\partial x} + \frac{\partial v}{\partial y}\right)$$

$$w = w_\eta + (\eta - z)\left(\frac{\partial u}{\partial x} + \frac{\partial v}{\partial y}\right)$$

$$w = \frac{\partial \eta}{dt} + (\eta - z)\left(\frac{\partial u}{\partial x} + \frac{\partial v}{\partial y}\right)$$

or ignoring second-order terms,

$$w \doteq \frac{\partial \eta}{\partial t} - z\left(\frac{\partial u}{\partial x} + \frac{\partial v}{\partial y}\right) \tag{5-12}$$

5-3 Waves of the First Class

To summarize, the equations whose periodic solution we seek are the long-wave equations, (5-5) and (5-6), and a continuity condition as expressed by (5-11). These equations are homogeneous and linear in the dependent variables u, v, and η. The horizontal axes of our coordinate system can be considered to be of arbitrary azimuth (y not necessarily to the north and x not necessarily to the east) since the only planetary rotation is that around the local vertical, as given by the Coriolis parameter f. Both f and h are still considered to be functions of x and y. The vertical velocity w can be determined from the diagnostic equation, (5-12).

Let us assume harmonic solutions of the form

$$\begin{aligned} \eta &\equiv \mathcal{R}[\mathcal{Z}(x, y)e^{-i\sigma t}] \\ u &\equiv \mathcal{R}[\mathcal{U}(x, y)e^{-i\sigma t}] \\ v &\equiv \mathcal{R}[\mathcal{V}(x, y)e^{-i\sigma t}] \\ w &\equiv \mathcal{R}[\mathcal{W}(x, y, z)e^{-i\sigma t}] \end{aligned} \tag{5-13}$$

where the script $\mathcal{R}$ indicates the real portion of the complex function. Substitution of (5-13) into (5-5) and (5-6) gives

$$-i\sigma\mathcal{U} - f\mathcal{V} = -g\frac{\partial \mathcal{Z}}{\partial x}$$

and

$$-i\sigma\mathcal{V} + f\mathcal{U} = -g\frac{\partial \mathcal{Z}}{\partial y}$$

Solving algebraically for $\mathcal{U}$ and $\mathcal{V}$ yields what Platzman calls the *polarization equations*,

$$\begin{aligned} \mathcal{U} &= \frac{g}{\sigma^2 - f^2}\left(-i\sigma\frac{\partial \mathcal{Z}}{\partial x} + f\frac{\partial \mathcal{Z}}{\partial y}\right) \\ \mathcal{V} &= \frac{g}{\sigma^2 - f^2}\left(-i\sigma\frac{\partial \mathcal{Z}}{\partial y} - f\frac{\partial \mathcal{Z}}{\partial x}\right) \end{aligned} \tag{5-14}$$

and from substitution of (5-11) and (5-13) in (5-12) the corresponding equation for $\mathcal{W}$ is obtained as

$$\mathcal{W} = -i\sigma\mathcal{Z}\left(\frac{h + z}{h}\right) + \frac{z}{h}\left(\mathcal{U}\frac{\partial h}{\partial x} + \mathcal{V}\frac{\partial h}{\partial y}\right) \tag{5-15}$$

Thus, a knowledge of the function $\mathcal{Z}$ is all that is required to find the functions $\mathcal{U}$, $\mathcal{V}$, and $\mathcal{W}$, and in turn the fluid particle velocities u, v, and w, and to establish the Lagrangian description of the motion of fluid.

To find $\mathcal{Z}$ we eliminate u and v from the continuity equation, (5-11), by means of (5-13) and (5-14) while making the assumption that the depth h

and the Coriolis parameter f are each uniform. Proceeding to do this,

$$\nabla^2 \mathcal{Z} + \left(\frac{\sigma^2 - f^2}{gh}\right)\mathcal{Z} = 0 \tag{5-16}$$

where

$$\nabla^2 \equiv \frac{\partial^2}{\partial x^2} + \frac{\partial^2}{\partial y^2}$$

The assumption of h and f uniform, although pragmatic, has far-reaching consequences. In effect, we are "filtering out," and therefore removing from consideration certain kinds of wave motion. All waves that depend for their existence on a variation of the planetary potential vorticity, that is, on the gradient of f/h, are in this category. These waves are of the second class, have a very low frequency, and include the *planetary* or *Rossby waves*, which will be discussed later. Topographic waves will be excluded; this includes all wave motion that depends upon a gradient of the planetary potential vorticity through the spatial variation of h, an example of which is the edge wave associated with the continental shelf. Equation (5-16) then describes long gravity waves that are modified by the rotation of the earth but would still be waves without this rotation, that is, waves of class I.

No explicit simplification occurs in the polarization equations, (5-14), for h and f uniform. However, from (5-15), the corresponding equation for $\mathcal{W}$ reduces to

$$\mathcal{W} = -i\sigma\left(\frac{h + z}{h}\right)\mathcal{Z} \tag{5-17}$$

from which it is apparent that the vertical velocity w varies linearly with z, from zero at the bottom ($z = -h$) to, from (5-12), $\partial\eta/\partial t$ at the surface ($z \doteq 0$).

Finally, for the purpose of reference in the future, the average potential and kinetic energy densities and average energy fluxes for the wave are formulated. From Figure 5-3, with respect to the reference level of $z = -h$, the potential energy of an elemental column of fluid of height $h + \eta$, length

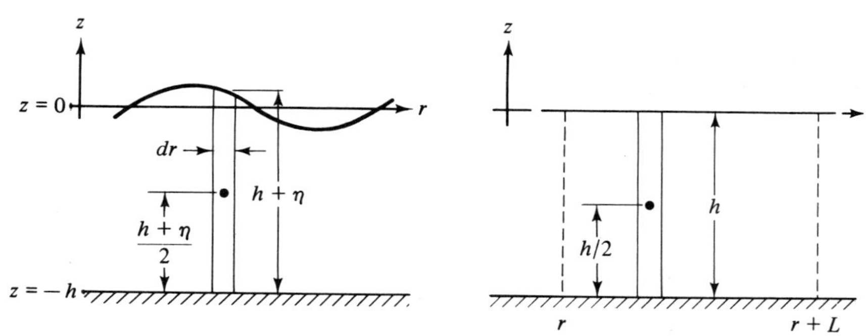

Fig. 5-3. Definition sketch for potential energy determination.

dr, and of unit width along the crest is

$$dE_{P_1} = \left[\rho g\left(\frac{h+\eta}{2}\right)\right][(h+\zeta)(1)\,dr]$$

The average potential energy per unit horizontal area, sometimes called the *average potential energy density*, of all the fluid above $z = -h$ is

$$\overline{E_{P_1}} = \frac{\rho g}{2L}\int_r^{r+L} (h+\eta)^2\,dr$$

where L is the wavelength. The average potential energy density in the absence of a wave would be

$$\overline{E_{P_2}} = \frac{\rho g}{2L}\int_r^{r+L} h^2\,dr = \frac{\rho g h^2}{2}$$

The average potential energy density attributable to the presence of the wave on the free surface is then

$$\begin{aligned}\overline{E_P} &\equiv \overline{E_{P_1}} - \overline{E_{P_2}}\\ &= \frac{\rho g}{2L}\int_r^{r+L}(2h\eta + \eta^2)\,dr\\ &= \frac{\rho g}{2}(h\bar{\eta} + \overline{\eta^2})\end{aligned}$$

or

$$\overline{E_P} = \tfrac{1}{2}\rho g\overline{\eta^2} \tag{5-18}$$

since η averaged over one wave length, $\bar{\eta}$, is zero.

From Figure 5-4, the kinetic energy of an elemental volume of fluid dr long, dz high, of unit width along the crest, and with velocity components u and v in the horizontal (w taken as negligible), is

$$dE_K = \tfrac{1}{2}\rho(u^2 + v^2)(1)\,dz\,dr$$

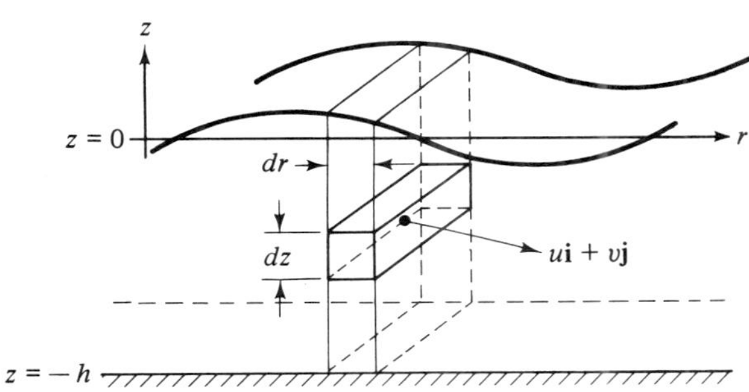

Fig. 5-4. Definition sketch for kinetic energy determination.

The average kinetic energy per unit horizontal surface area, or the *average kinetic energy density*, is given by

$$\overline{E_K} = \frac{\rho}{2L}\int_r^{r+L}\int_{-h}^{\eta}(u^2 + v^2)\,dz\,dr$$

Since in general we have taken that $|\eta| \ll h$ [see (5-11), for example], we approximate the upper limit of the second integral by zero and thus obtain

$$\overline{E_K} = \frac{\rho}{2L}\int_r^{r+L}\int_{-h}^{0}(u^2 + v^2)\,dz\,dr$$

$$= \frac{\rho h}{2L}\int_r^{r+L}(u^2 + v^2)\,dr$$

or

$$\overline{E_K} = \tfrac{1}{2}\rho h\overline{(u^2 + v^2)} \tag{5-19}$$

For a wave to propagate into undisturbed water, it must do work on the previously undisturbed fluid. We are interested in finding an expression for the average rate at which work is done by a wave over a wave period, or the wave *power*. At any depth, the variation of the pressure from the mean sea-level hydrostatic pressure is, from (5-3), $\rho g\eta$. The time rate at which work is done on a vertical surface element perpendicular to the x axis, of height dz, and of unit width is

$$dP_x = (\rho g\eta u)(1)(dz)$$

Integrating over the depth of the fluid and averaging over a wave period, we obtain as the average power per unit horizontal distance, or the *average energy flux*, in the x direction,

$$\overline{P_x} = \frac{\rho g}{T}\int_t^{t+T}\int_{-h}^{0}\eta u\,dz\,dt$$

$$= \frac{gh}{T}\int_t^{t+T}\eta u\,dt$$

or

$$\overline{P_x} = \rho gh\overline{\eta u} \tag{5-20}$$

Similarly, the average energy flux in the y direction will be given by

$$\overline{P_y} = \rho gh\overline{\eta v} \tag{5-21}$$

5-4 Plane Sverdrup Waves

Waves are said to be *plane* (or *long-crested*) if the disturbance is independent of the coordinate normal to the direction of wave propagation. Wave crests can then be represented by straight lines in a horizontal plane.

Equation (5-16) is a classic differential equation and serves as a starting point for mechanical radiation problems. It is the *reduced wave equation*, reduced since the time dependency of the wave equation has been removed due to the adoption of (5-13). We can obtain a plane wave solution of it through the separation of variables process. Thus, we try

$$\mathcal{Z} \equiv X(x)\,Y(y) \tag{5-22}$$

where initially X and Y are arbitrary functions of only x and only y, respectively. To find the conditions they must satisfy in order that $\mathcal{Z}$ satisfies (5-16), we substitute into (5-16) to obtain

$$\frac{X''}{X} + \frac{\sigma^2 - f^2}{gh} = -\frac{Y''}{Y}$$

where the primes indicate the appropriate differentiation. Using the same technique as before, we see that both sides must be equal to the same constant, which we will call

$$\frac{\sigma^2 - f^2}{gh} - k^2$$

This means that

$$X'' + k^2 X = 0 \tag{5-23}$$

Similarly,

$$Y'' + \left(\frac{\sigma^2 - f^2}{gh} + k^2\right) Y = 0$$

For convenience we make the definition

$$\ell^2 \equiv \frac{\sigma^2 - f^2}{gh} - k^2 \tag{5-24}$$

and rewrite the last differential equation as

$$Y'' + \ell^2 Y = 0 \tag{5-25}$$

The solutions of the second-order linear equations for X and Y are, respectively,

$$\begin{aligned} X &= A_1 e^{-ikx} + B_1 e^{ikx} \\ Y &= A_2 e^{-i\ell y} + B_2 e^{i\ell y} \end{aligned} \tag{5-26}$$

where the A's and B's are arbitrary constants. Clearly, the number of possible solutions of the reduced wave equation, (5-16), in the form of (5-22) is very large. The plane wave resulting from taking $A_1 = A_2 = 0$ is one of these and (5-22) will take the form

$$\mathcal{Z} = h e^{i(kx + \ell y)} \tag{5-27}$$

where, duplicating Platzman, the amplitude $B_1 B_2$ of the wave has been arbitrarily scaled so that it is equal to h, the depth of the fluid. Solving (5-24) for the angular frequency σ, we get the dispersion relation

$$\sigma = \pm[f^2 + (k^2 + \ell^2)gh]^{1/2} \tag{5-28}$$

and the polarization equations of (5-14) and (5-15) give, in the light of (5-24) and (5-27),

$$\mathcal{U} = Ae^{i(kx+\ell y)} \tag{5-29a}$$

$$\mathcal{V} = Be^{i(kx+\ell y)} \tag{5-29b}$$

$$\mathcal{W} = Ce^{i(kx+\ell y)} \tag{5-29c}$$

where

$$A \equiv \frac{\sigma k + if\ell}{k^2 + \ell^2} \tag{5-30a}$$

$$B \equiv \frac{\sigma \ell - ifk}{k^2 + \ell^2} \tag{5-30b}$$

$$C \equiv -i\sigma(z + h) \tag{5-30c}$$

From (5-13), the complete plane wave solution is then

$$\eta = \mathcal{R}[he^{i(kx+\ell y-\sigma t)}] \tag{5-31a}$$

$$u = \mathcal{R}[Ae^{i(kx+\ell y-\sigma t)}] \tag{5-31b}$$

$$v = \mathcal{R}[Be^{i(kx+\ell y-\sigma t)}] \tag{5-31c}$$

$$w = \mathcal{R}[Ce^{i(kx+\ell y-\sigma t)}] \tag{5-31d}$$

The wave described by (5-31), for the case of k and ℓ real, Platzman calls a *Sverdrup wave*. From (5-28) it is seen that the minimum frequency possible for a Sverdrup wave is simply f, which increases from zero at the equator to two cycles per sidereal day at the pole, and that as a consequence Sverdrup waves of a given frequency $|\sigma|$ cannot exist poleward of the latitude where $|\sigma| = f$.

The kinetic energy density of the Sverdrup wave is found from (5-19). The real parts of (5-31b) and (5-31c) gives us

$$u = \frac{\sigma k \cos(kx + \ell y - \sigma t) - f\ell \sin(kx + \ell y - \sigma t)}{k^2 + \ell^2}$$

$$v = \frac{\sigma \ell \cos(kx + \ell y - \sigma t) + fk \sin(kx + \ell y - \sigma t)}{k^2 + \ell^2}$$

Then

$$u^2 + v^2 = \frac{\sigma^2 \cos^2(kx + \ell y - \sigma t) + f^2 \sin^2(kx + \ell y - \sigma t)}{k^2 + \ell^2}$$

and

$$\overline{u^2 + v^2} = \frac{1}{2}\left(\frac{\sigma^2 + f^2}{k^2 + \ell^2}\right) = \frac{1}{2}gh\left(\frac{\sigma^2 + f^2}{\sigma^2 - f^2}\right)$$

Substituting into (5-19),

$$\overline{E_K} = \tfrac{1}{4}\rho g h\left(\frac{\sigma^2 + f^2}{\sigma^2 - f^2}\right) \tag{5-32}$$

Similarly, the average potential energy density from (5-18) is

$$\overline{E_P} = \tfrac{1}{4}\rho g h^2 \tag{5-33}$$

since

$$\eta = h \cos(kx + \ell y - \sigma t)$$

and $\overline{\eta^2} = \frac{1}{2}h^2$. We see that for $f = 0$ there is an equipartition of energy between kinetic and potential energy,

$$\overline{E_K} = \overline{E_P} = \tfrac{1}{4}\rho g h^2$$

this result being consistent with that arising from the theory of simple harmonic waves. However, in the presence of the Coriolis force, the kinetic energy density is greater than the potential energy density in the ratio $(\sigma^2 + f^2)/(\sigma^2 - f^2)$.

The details of the Sverdrup wave can be more conveniently examined if the coordinate axes are aligned such that $\ell = 0$ and $k > 0$. Taking the real parts of (5-31), we obtain

$$\eta = h \cos(kx - \sigma t) \tag{5-34}$$

$$u = \frac{\sigma}{k}\cos(kx - \sigma t) \tag{5-35a}$$

$$v = \frac{f}{k}\sin(kx - \sigma t) \tag{5-35b}$$

$$w = \sigma(z + h)\sin(kx - \sigma t) \tag{5-35c}$$

and from (5-28),

$$\sigma = \pm(f^2 + ghk^2)^{1/2} \tag{5-36}$$

Equation (5-35b) tells us that the inclusion of the Coriolis force produces a wave having a component of its motion horizontally transverse to the direction of wave propagation. Also, from (5-36) we see that its inclusion increases the wave frequency σ for any given wave number k and makes the wave dispersive since the wave speed or celerity, σ/k, varies with k. Depending upon the sign adopted in (5-36), we have, for $\sigma > 0$, a wave that travels in the direction of increasing values of the phase angle $(kx - \sigma t)$ or, for $\sigma < 0$, a wave that travels in the opposite direction; in either case the speed is identical. Figure 5-5 illustrates the behavior of the relation, as given by (5-36), between $|\sigma|$ and k for an assumed value of $f = 1$ cycle per sidereal day corresponding to a latitude of 29.91°. Figure 5-6 shows the phase speed, $|\sigma|/k$, and the group speed, $|d\sigma/dk|$, as functions of $|\sigma|$ for $f = 1$ cycle per sidereal day. Solving for k from (5-36) and substituting, the phase speed is

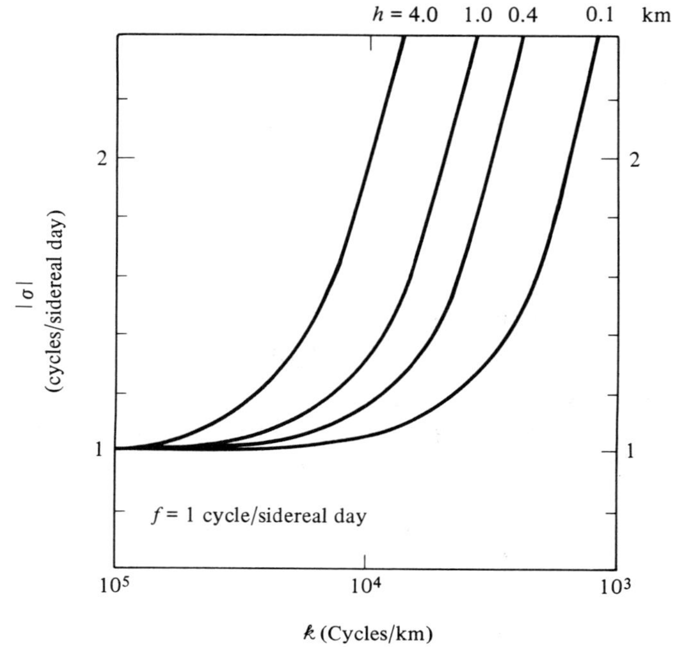

Fig. 5-5. Plot of (5-36) illustrating the variation of frequency with wave number for pure Sverdrup waves. Adapted from "Ocean Tides and Related Waves," *Mathematical Problems in the Geophysical Sciences*, Lectures in Applied Mathematics, Vol. 14, page 274, by permission of the Amer. Math. Soc. © 1971 by the Amer. Math. Soc.

given as

$$c \equiv \frac{|\sigma|}{k} = \frac{|\sigma|}{(\sigma^2 - f^2)^{1/2}}(gh)^{1/2} \tag{5-37}$$

and differentiating to give the group speed, defined as the speed with which the wave energy is propagated, as

$$c_g \equiv \left|\frac{d\sigma}{dk}\right| = \frac{(\sigma^2 - f^2)^{1/2}}{|\sigma|}(gh)^{1/2} \tag{5-38}$$

The Lagrangian description of the fluid motion is found by defining a new coordinate system, call it the (ξ, ν, ζ) system, whose axes are parallel to the (x, y, z) system, but whose origin is the undisturbed or equilibrium position of a fluid particle (Figure 5-7). Let the coordinates of the undisturbed particle be $x = x_1, y = y_1, z = z_1$, or $\xi = 0, \nu = 0, \zeta = 0$. In the presence of a wave the particle will occupy the instantaneous position $x = x_1 + \xi_1$, $y = y_1 + \nu_1$, $z = z_1 + \zeta_1$ in the (x, y, z) system or $\xi = \xi_1$, $\nu = \nu_1$, $\zeta = \zeta_1$ in

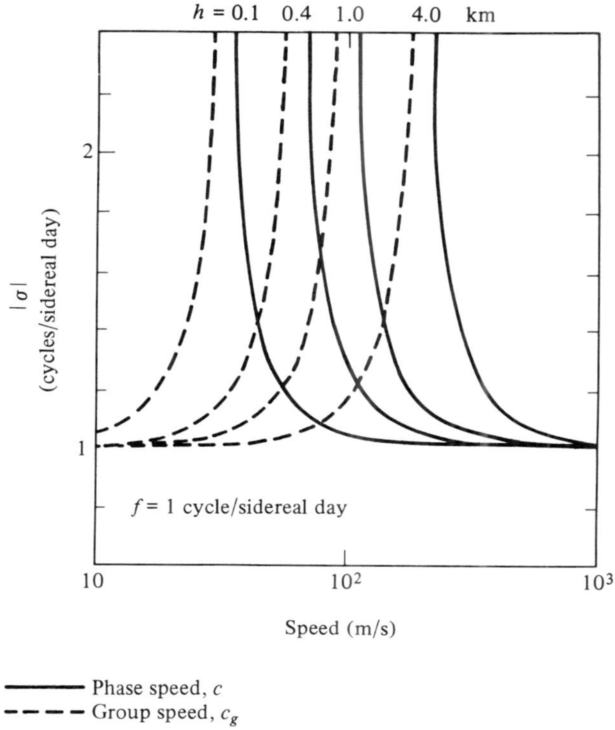

Fig. 5-6. Plot of (5-37) and (5-38) illustrating the variation of phase speed and group speed with frequency for plane Sverdrup waves. Adapted from "Ocean Tides and Related Waves," *Mathematical Problems in the Geophysical Sciences*, Lectures in Applied Mathematics, Vol. 14, p. 275, by permission of the Amer. Math. Soc. © 1971 by the Amer. Math. Soc.

the (ξ, ν, ζ) system. We have that

$$u = \frac{\partial x_1}{dt} = \frac{d\xi_1}{dt} \tag{5-39a}$$

$$v = \frac{dy_1}{dt} = \frac{d\nu_1}{dt} \tag{5-39b}$$

$$w = \frac{dz_1}{dt} = \frac{d\zeta_1}{dt} \tag{5-39c}$$

and substituting from equations (5-35) while making the approximation that $x_1 \simeq x_1 + \xi_1$ and $z_1 \simeq z_1 + \zeta_1$, we obtain

$$\frac{d\xi_1}{dt} = \frac{\sigma}{k} \cos(kx_1 - \sigma t) \tag{5-40a}$$

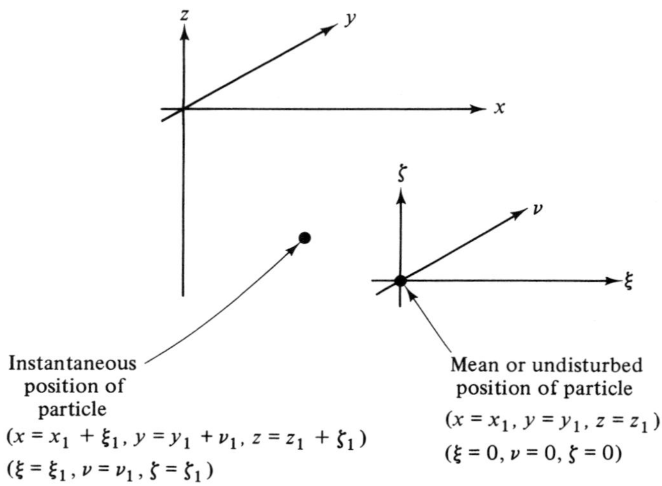

Fig. 5-7. Coordinate system for the determination of the particle path for a plane Sverdrup wave.

$$\frac{d\nu_1}{dt} = \frac{f}{k} \sin (kx_1 - \sigma t) \qquad (5\text{-}40b)$$

$$\frac{d\zeta_1}{dt} = \sigma(z_1 + h) \sin (kx_1 - \sigma t) \qquad (5\text{-}40c)$$

Integrating and dropping the subscripts yields

$$\xi = -\frac{1}{k} \sin (kx - \sigma t) \qquad (5\text{-}41a)$$

$$\nu = \frac{f}{\sigma k} \cos (kx - \sigma t) \qquad (5\text{-}41b)$$

$$\zeta = (z + h) \cos (kx - \sigma t) \qquad (5\text{-}41c)$$

the set of equations describing the trajectory of a particle in terms of its equilibrium position and of time. Squaring (5-41a) and (5-41b) and summing, we find that

$$\frac{\xi^2}{(1/k)^2} + \frac{\nu^2}{(f/\sigma k)^2} = 1 \qquad (5\text{-}42)$$

which is the equation of an ellipse with semiaxes k^{-1} and $|f/\sigma|k^{-1}$, representing the projection of the particle orbit in the x, y plane. Doing the same with (5-41a) and (5-41c), the projection of the orbit in the x, z plane is the ellipse

$$\frac{\xi^2}{(1/k)^2} + \frac{\zeta^2}{(z + h)^2} = 1 \qquad (5\text{-}43)$$

with semiaxes k^{-1} and $(z + h)^{-1}$. In the y, z plane, (5-41b) and (5-41c) represent simple harmonic motion having an amplitude $[(f/\sigma k)^2 + (z + h)^2]^{1/2}$. Putting this all together, what we have in the Sverdrup wave is a wave polarized in a plane perpendicular to the wave fronts with the fluid particles moving in closed elliptic orbits, the plane of which is inclined upward to the left of the propagation direction (Figure 5-8). In the Northern Hemisphere, $f > 0$ and the particle traverses the orbit in a direction that gives clockwise rotation when projected on the x, y plane. From (5-42), the horizontal *aspect ratio* (defined as the ratio of the v semiaxis to the ξ semiaxis) is $|f/\sigma|$, which from writing (5-36) as

$$\left|\frac{\sigma}{f}\right| = \left(1 + \frac{k^2}{f^2}gh\right)^{1/2} > 1$$

is seen to be less than 1 and approaches zero in the high-frequency limit. In this limit the orbital plane is vertical, indicating that the Coriolis force has no effect on the wave motion. In the low-frequency limit $|f/\sigma|$ approaches 1 and the strong influence of the Coriolis force, from (5-42), results in an almost circular and almost horizontal orbital path. The vertical aspect ratio (defined as the ratio of the ζ semiaxis to the ξ semiaxis) is $k(z + h)$ and has its greatest value of kh at the free surface, kh for the long-wave approximation being much less than 1.

Figure 5-9 illustrates the surface profile of the Sverdrup wave, the crests and troughs of which are horizontal. The streamline pattern in the x, y and x, z planes obtained from integrating (2-12) is shown in Figure 5-10; note their convergence forward of a crest and divergence forward of a trough, resulting in the crests and troughs propagating in the $+x$ direction.

Finally, the energy flux of these waves is given, from the substitution of (5-34) and (5-35a) in (5-20) for the x direction, as

$$\begin{aligned}\overline{P_x} &= \rho g h \overline{\eta u} \\ &= \frac{\rho g h}{T}\int_t^{t+T} \frac{h\sigma}{k}\cos^2(kx - \sigma t)\,dt \qquad \text{(5-44)} \\ &= \tfrac{1}{2}\rho g h^2\left(\frac{\sigma}{k}\right)\end{aligned}$$

and from the substitution of (5-34) and (5-35b) for the y direction as

$$\begin{aligned}\overline{P_y} &= \rho g h \overline{\eta v} \\ &= \frac{\rho g h}{T}\int_t^{t+T} \frac{hf}{k}\sin(kx - \sigma t)\cos(kx - \sigma t)\,dt = 0 \qquad \text{(5-45)}\end{aligned}$$

showing that, despite the existence of a transverse velocity, no energy is transmitted parallel to the wave crests. To check the consistency of these

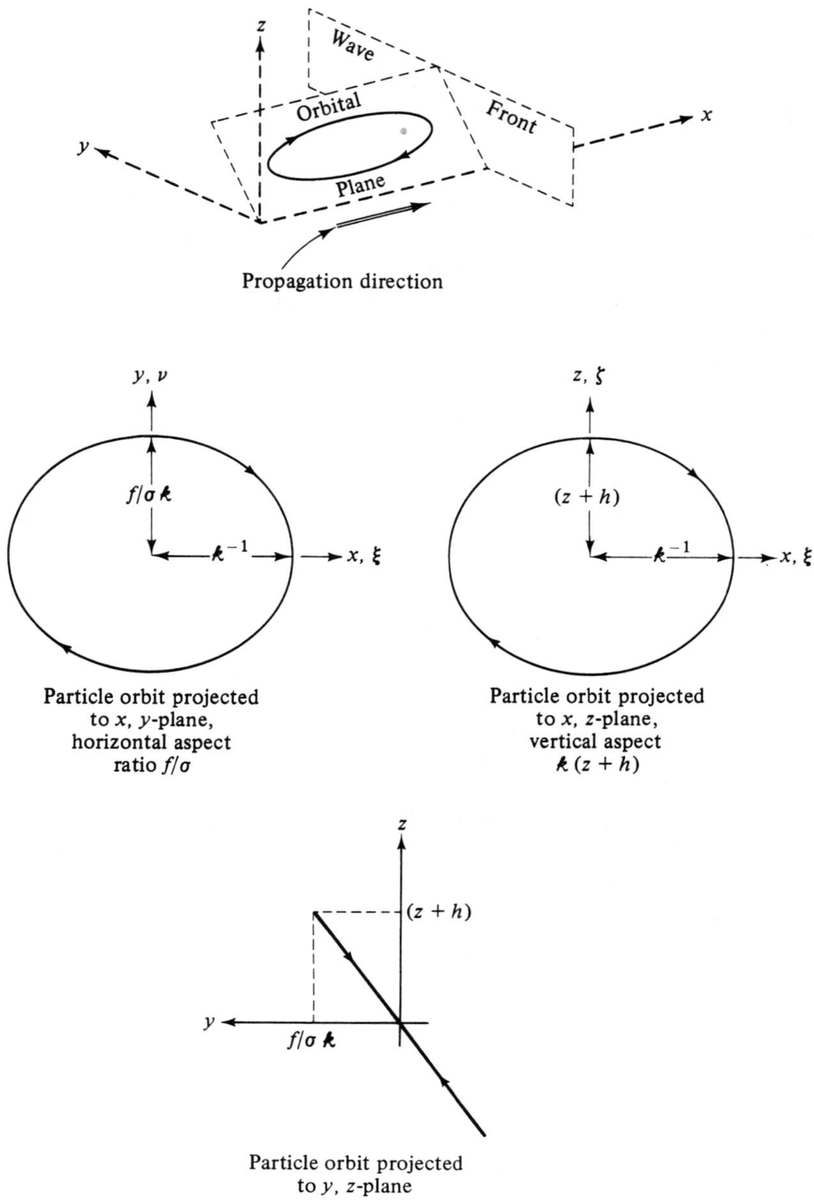

Fig. 5-8. Orbital motion of a plane Sverdrup wave in the Northern Hemisphere (not to scale). Adapted from "Ocean Tides and Related Waves," *Mathematical Problems in the Geophysical Sciences*, Lectures in Applied Mathematics, Vol. 14, p. 275, by permission of the Amer. Math. Soc. © 1971 by the Amer. Math. Soc.

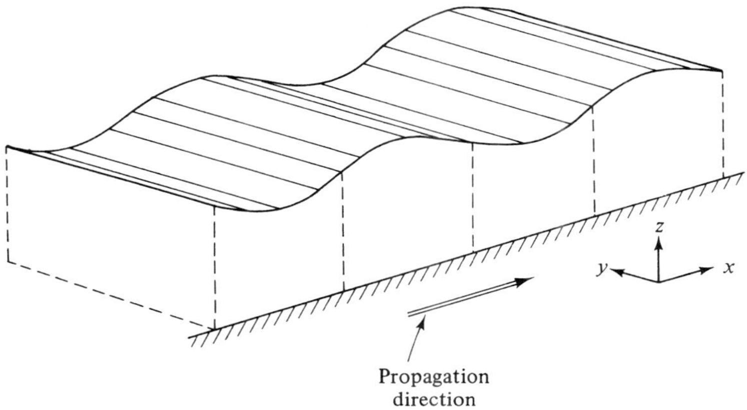

Fig. 5-9. Plane Sverdrup wave form. Adapted from "Ocean Waves and Related Tides," *Mathematical Problems in the Geophysical Sciences*, Lectures in Applied Mathematics, Vol. 14, p. 276, by permission of the Amer. Math. Soc. © 1971 by the Amer. Math. Soc.

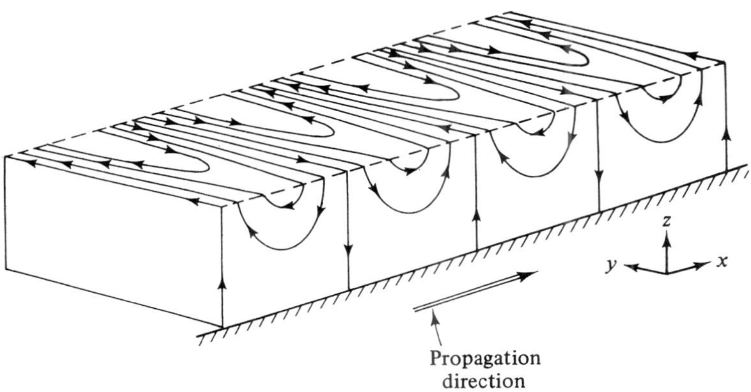

Fig. 5-10. Streamline pattern for plane Sverdrup waves in the x, y and x, z planes. Adapted from "Ocean Tides and Related Waves," *Mathematical Problems in the Geophysical Sciences*, Lectures in Applied Mathematics, Vol. 14, p. 276, by permission of the Amer. Math. Soc. © 1971 by the Amer. Math. Soc.

results, we note that, from the definition, the group speed should be given by the ratio of the energy flux to the mechanical energy density, $\bar{E}$, where

$$\bar{E} \equiv \overline{E_K} + \overline{E_P}$$

Therefore, for this particular case, applying (5-32), (5-33), and (5-44),

$$c_g = \frac{\overline{P_x}}{\overline{E_K} + \overline{E_P}}$$

$$= \frac{\frac{1}{2}\rho g h^2\left(\frac{\sigma}{k}\right)}{\frac{1}{4}\rho g h^2\left(\frac{\sigma^2 + f^2}{\sigma^2 - f^2}\right) + \frac{1}{4}\rho g h^2} \tag{5-46}$$

$$= gh\left(\frac{k}{\sigma}\right)$$

having used (5-36), this result is identical to that obtained in (5-38), in which c_g was defined mathematically as equal to $|d\sigma/dk|$.

5-5 Plane Poincaré Waves

Another wave of the first class, one that is dynamically identical to the plane Sverdrup wave, results from the reflection of a Sverdrup wave from a straight coast. Following Platzman, we take the coast as being along the x axis with the y axis directed offshore and assume that the wave is moving in the $+x$ direction with a trace velocity (defined as the component of the phase velocity in a particular direction) in that direction of σ/k with both σ and k greater than zero. From continuity considerations the incident and reflected wave must have the same frequency and trace velocity. If the incident wave has a vector wave number $\mathbf{K}_i$,

$$\mathbf{K}_i \equiv k\mathbf{i} - \ell\mathbf{j} \tag{5-47}$$

the reflected wave will have a vector wave number $\mathbf{K}_r$,

$$\mathbf{K}_r \equiv k\mathbf{i} + \ell\mathbf{j} \tag{5-48}$$

with the requirement that $\ell > 0$ (Figure 5-11). The wave length of the incident and reflected wave is given by

$$L = \frac{2\pi}{|\mathbf{K}_i|} = \frac{2\pi}{|\mathbf{K}_r|} = \frac{2\pi}{(k^2 + \ell^2)^{1/2}} \tag{5-49}$$

From (5-30) and (5-31) the incident wave is characterized by the equations

$$\eta_i = \Re[\tfrac{1}{2}he^{i(kx - \ell y - \sigma t)}] \tag{5-50a}$$

$$u_i = \Re[\tfrac{1}{2}A^*e^{i(kx - \ell y - \sigma t)}] \tag{5-50b}$$

$$v_i = \Re[-\tfrac{1}{2}B^*e^{i(kx - \ell y - \sigma t)}] \tag{5-50c}$$

$$w_i = \Re[\tfrac{1}{2}Ce^{i(kx - \ell y - \sigma t)}] \tag{5-50d}$$

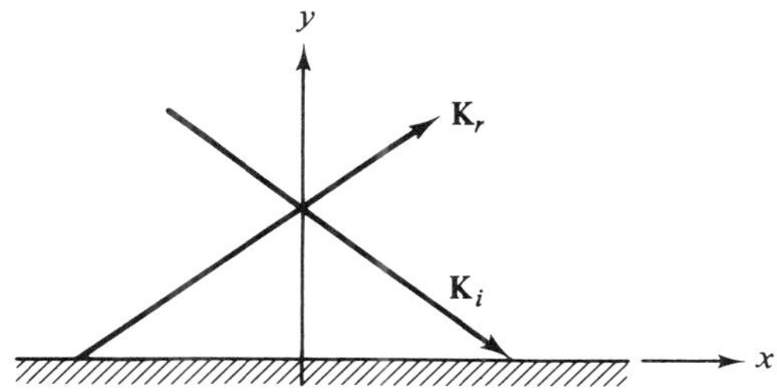

Fig. 5-11. Wave number vectors of incident and reflected waves.

where the asterisk designates the conjugate of the polarization coefficients of (5-30) and the factor $\frac{1}{2}$ is introduced for later convenience. In like fashion, the reflected wave is characterized by

$$\eta_r = \Re[\tilde{K}\tfrac{1}{2}he^{i(kx+\ell y-\sigma t)}] \tag{5-51a}$$

$$u_r = \Re[\tilde{K}\tfrac{1}{2}Ae^{i(kx+\ell y-\sigma t)}] \tag{5-51b}$$

$$v_r = \Re[\tilde{K}\tfrac{1}{2}Be^{i(kx+\ell y-\sigma t)}] \tag{5-51c}$$

$$w_r = \Re[\tilde{K}\tfrac{1}{2}Ce^{i(kx+\ell y-\sigma t)}] \tag{5-51d}$$

where $\tilde{K}$ is the reflection coefficient whose magnitude is the ratio of the amplitude of the reflected wave to the amplitude of the incident wave.

For perfect reflection $|\tilde{K}| = 1$. The implied rigid and impervious coastal conditions requiring no mass or energy flux in the y direction at $y = 0$ are satisfied when $v|_{y=0} = 0$, where v is the y-component velocity of the composite wave,

$$v \equiv v_i + v_r$$

From (5-50c) and (5-51c), then,

$$v|_{y=0} = (v_i + v_r)|_{y=0} = -B^* + \tilde{K}B = 0$$

or

$$\tilde{K} = \frac{B^*}{B} \tag{5-52}$$

But B and B^* can be written as

$$B = \text{mod } Be^{-i \arg B}$$

and

$$B^* = \text{mod } Be^{i \arg B}$$

where the absolute value or modulus of B, written mod B, is

$$\text{mod } B = \frac{[(\sigma\ell)^2 + (fk)^2]^{1/2}}{k^2 + \ell^2}$$

and the amplitude or argument of B, written arg B, is

$$\arg B = \tan^{-1}\left(-\frac{fk}{\sigma\ell}\right)$$

Thus, (5-52) becomes

$$\tilde{K} = e^{-2i \arg B} \tag{5-53}$$

which when substituted into the equations (5-51) shows that, because of the earth's rotation, the reflected wave is shifted in phase by an amount $-2 \arg B = -2 \arg (\sigma\ell - if k)$.

The profile of the composite wave is found by adding (5-50a) and (5-51a),

$$\begin{aligned}\eta &\equiv \eta_i + \eta_r \\ &= \Re[\tfrac{1}{2}h(e^{i(kx-\ell y-\sigma t)} + e^{i(kx+\ell y-\sigma t-2\arg B)})] \\ &= \Re[\tfrac{1}{2}h(e^{i(kx-\arg B-\ell y+\arg B-\sigma t)} + e^{i(kx-\arg B+\ell y-\arg B-\sigma t)})]\end{aligned}$$

which can be simplified through a shifting of the origin of the x coordinate such that

$$kx' \equiv kx - \arg B$$

to give

$$\eta = \Re[\tfrac{1}{2}h(e^{i(kx'-\ell y+\arg B-\sigma t)} + e^{i(kx'+\ell y-\arg B-\sigma t)})]$$

Taking the real part and dropping the prime on the x as unnecessary,

$$\begin{aligned}\eta = \tfrac{1}{2}h &\cos[(kx - \sigma t) - (\ell y - \arg B)] \\ &+ \tfrac{1}{2}h \cos[(kx - \sigma t) + (\ell y - \arg B)]\end{aligned}$$

and finally, after expanding the cosines,

$$\eta = h \cos(\ell y - \arg B) \cos(kx - \sigma t) \tag{5-54a}$$

Proceeding in a similar manner with the remaining equations (5-50) and (5-51), the fluid velocity components for the composite wave are found to be

$$u = |A| \cos(\ell y - \arg B + \arg A) \cos(kx - \sigma t) \tag{5-54b}$$

$$v = -|B| \sin \ell y \sin(kx - \sigma t) \tag{5-54c}$$

$$w = \sigma(z + h) \cos(\ell y - \arg B \sin(kx - \sigma t) \tag{5-54d}$$

Equations (5-54), resulting from the superposition of a incident plane Sverdrup wave with its perfectly reflected counterpart, is called a *Poincaré wave*, the profile of which is shown in Figure 5-12. From (5-54a) it is apparent that the wave crests and troughs are straight lines perpendicular to the coast and that nodal lines occur parallel to the coast.

Platzman refers to Poincaré waves as being *right-bounded* or *left-bounded*, depending upon whether the coast is to the right or left, respectively, of the direction of propagation. In the context of the coordinate axes of Figure 5-10, $\sigma > 0$ in equations (5-54) produces a right-bounded wave, while $\sigma < 0$ yields a left-bounded wave, since then the direction of propagation is reversed from that shown in Figure 5-12. Right-bounded and left-bounded Poincaré waves derived from the same wave numbers, k and ℓ, show a

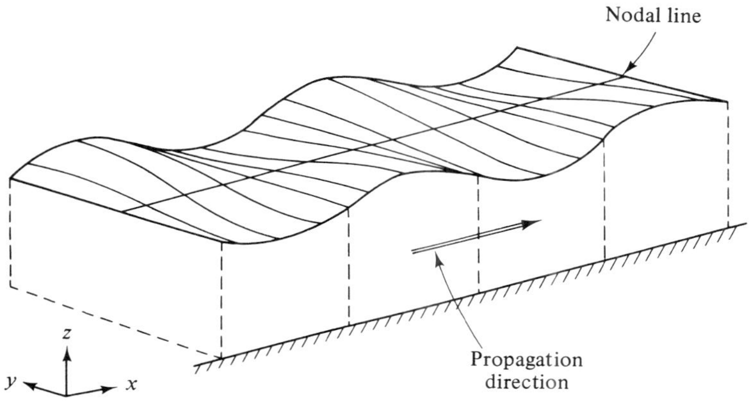

Fig. 5-12. Profile of a Poincaré wave. Adapted from "Ocean Tides and Related Waves," *Mathematical Problems in the Geophysical Sciences*, Lectures in Applied Mathematics, Vol. 14, p. 279, by permission of the Amer. Math. Soc. © 1971 by the Amer. Math. Soc.

marked asymmetry. For either type, nodal lines paralleling the coast are spaced a distance of π/ℓ apart; however, their respective distances from the coast are different for the two types of waves. For the right-bounded wave the quantity arg $B = \arg(\sigma\ell - ifk)$ is in the fourth quadrant of the argand diagram, while for a left-bounded wave it is in the third quadrant (see Figure 5-13). Along a nodal line we require that $\eta = 0$ regardless of the value of $\cos(kx - \sigma t)$ in (5-54a), and therefore

$$\cos(\ell y - \arg B) = 0$$

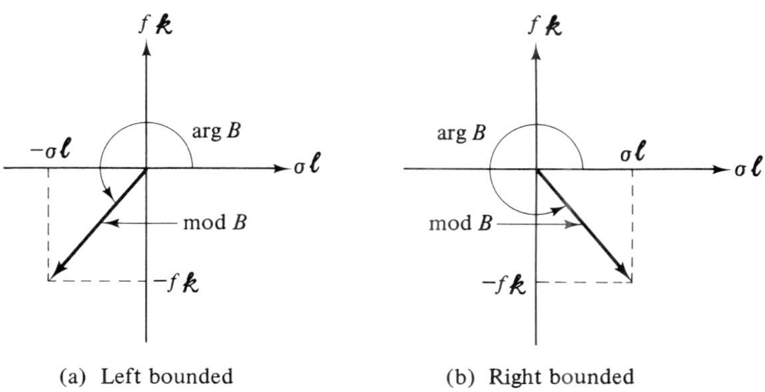

Fig. 5-13. Argand diagram for left- and right-bounded Poincaré waves with the same wave numbers.

along a nodal line; or at the first offshore nodal line,

$$\ell y - \arg B = \frac{\pi}{2}$$

$$\ell y - \tan^{-1} -\frac{fk}{\sigma \ell} = \frac{\pi}{2}$$

from which

$$y = \frac{1}{\ell}\left(\frac{\pi}{2} + \tan^{-1} -\frac{fk}{\sigma \ell}\right)$$

$$= \frac{1}{\ell}\left(\frac{\pi}{2} - \tan^{-1} \frac{fk}{\sigma \ell}\right)$$

Thus, for $\sigma > 0$, the situation for a right-bounded wave, $y < \pi/2\ell$, representing a distance less than one-fourth the transverse wave length, while for $\sigma < 0$, a left-bounded wave, $y > \pi/2\ell$. This transverse asymmetry is illustrated in Figure 5-14. An interesting consequence of this asymmetry, as pointed out by Platzman, is that a standing-wave pattern will not result from the superposition of forward and backward Poincaré waves, and thus the problem of the reflection of a Poincaré wave from a barrier perpendicular to the coast cannot be solved. This fact greatly complicates the mathematical analysis of waves in a rotating basin when sharp corners are present.

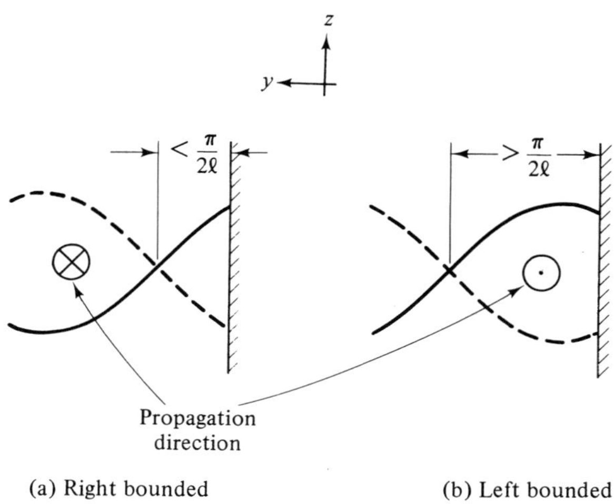

Fig. 5-14. Transverse profiles of Poincaré waves in two sections one-half wave length apart. Adapted from "Ocean Tides and Related Waves," *Mathematical Problems in the Geophysical Sciences*, Lectures in Applied Mathematics, Vol. 14, p. 280, by permission of the Amer. Math. Soc. © 1971 by the Amer. Math. Soc.

The dispersion relation for a Poincaré wave, although identical to that for the constituent plane Sverdrup waves, as given by (5-28), can be expressed in terms of the longshore wave number k and the direction α of the constituent Sverdrup waves. With α defined as in Figure 5-15, we see that

$$\sec^2 \alpha = \frac{1}{\cos^2 \alpha} = \frac{k^2 + \ell^2}{k^2}$$

which upon substitution in (5-28) gives

$$\sigma = \pm[f^2 + ghk^2 \sec^2 \alpha]^{1/2} \tag{5-55}$$

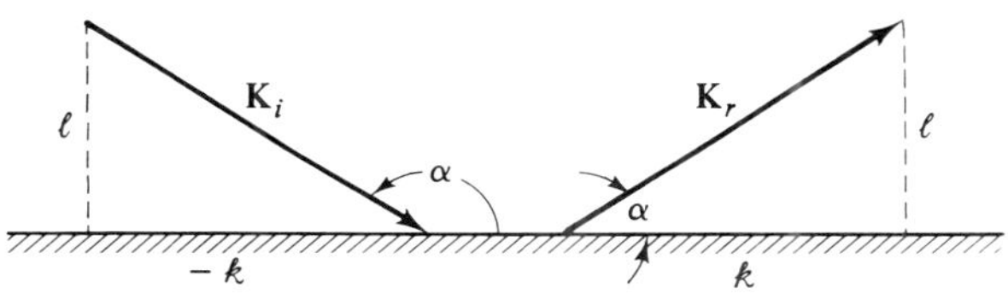

Fig. 5-15. Definition sketch for direction angle α for the relection of a plane Sverdrup wave.

A plot of this equation, for various values of α, is shown in Figure 5-16. The spectrum is seen to be continuous (shaded region) above the cutoff at $(f^2 + ghk^2)^{1/2}$, along which $\ell = 0$, $\alpha = 0$, corresponding to grazing incidence. A standing Sverdrup wave results when $\alpha = 90°$ (normal incidence), $k = 0$.

So-called *guided mode solutions* for the progression of Sverdrup waves through a channel with straight, parallel sides are possible in the form of Poincaré waves. From (5-54c) we see that the boundary condition $v = 0$ at $y = Y$, where Y is the width of the channel, is satisfied for any Poincaré wave having an offshore wave number ℓ such that

$$\ell = \frac{n\pi}{Y} \qquad (n = 1, 2, 3, \ldots)$$

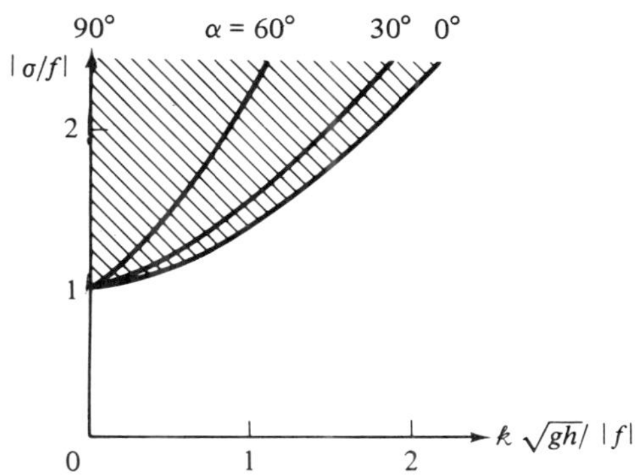

Fig. 5-16. Relation for Poincaré waves between frequency σ, longshore wave number k, and incidence angle α, (5-55). Adapted from "Ocean Tides and Related Waves," *Mathematical Problems in the Geophysical Sciences*, Lectures in Applied Mathematics, Vol. 14, p. 280, by permission of the Amer. Math. Soc. © 1971 by the Amer. Math. Soc.

The frequency spectrum for these guided modes is continuous in k but otherwise discrete, the cutoff frequency being that of the lowest mode, $n = 1$, which from (5-28) is

$$\left|\frac{\sigma}{f}\right| = \left(1 + \frac{ghk^2}{f^2} + \frac{\pi^2 gh}{f^2 Y^2}\right)^{1/2} \qquad (5\text{-}56)$$

Taking for convenience that

$$\frac{\pi\sqrt{gh}}{|f|Y} = 1$$

we obtain

$$\left|\frac{\sigma}{f}\right| = \left(2 + \frac{ghk^2}{f^2}\right)$$

Platzman plotted this result, shown in Figure 5-17, as well as that for the mode of $n = 2$.

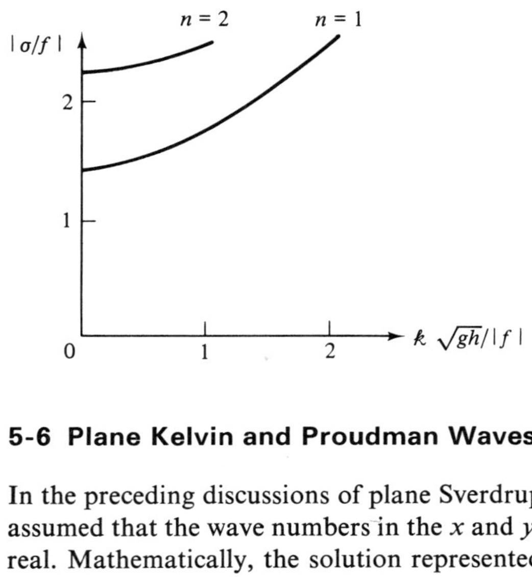

Fig. 5-17. Semidiscrete spectrum of Poincaré waves guided along a channel with straight, parallel sides; (5-56) with $\pi\sqrt{gh}/|f|Y = 1$. Adapted from "Ocean Tides and Related Waves," *Mathematical Problems in the Geophysical Sciences*, Lectures in Applied Mathematics, Vol. 14, p. 281, by permission of the Amer. Math. Soc. © 1971 by the Amer. Math. Soc.

5-6 Plane Kelvin and Proudman Waves

In the preceding discussions of plane Sverdrup and Poincaré waves, we have assumed that the wave numbers in the x and y directions, k and ℓ, were both real. Mathematically, the solution represented by equations (5-27) to (5-31) are valid when k and ℓ are complex. Suppose that we take k as real and positive and ℓ as pure imaginary, say $\ell = -i\ell'$. This is a satisfactory choice since $(k^2 + \ell^2)$ will still be real and, consequently, from (5-28), σ will be real. A real σ is required if we are to continue to treat purely harmonic waves, since a complex σ introduces the concept of a damped oscillation, while a purely imaginary σ represents the exponential case in which the disturbance grows or decays exponentially but does not oscillate. Substituting then in (5-28),

$$\sigma = \pm[f^2 + (k^2 - \ell'^2)gh]^{1/2} \qquad (5\text{-}57)$$

in (5-30a, b),

$$A \equiv \frac{\sigma k + f\ell'}{k^2 - \ell'^2} \tag{5-58a}$$

$$B \equiv \frac{-i(\sigma\ell' + fk)}{k^2 - \ell'^2} \equiv -iB' \tag{5-58b}$$

and in (5-31),

$$\eta = \Re[he^{i(kx-\sigma t)}e^{\ell' y}] \tag{5-59a}$$

$$u = \Re[Ae^{i(kx-\sigma t)}e^{\ell' y}] \tag{5-59b}$$

$$v = \Re[Be^{i(kx-\sigma t)}e^{\ell' y}] \tag{5-59c}$$

$$w = \Re[Ce^{i(kx-\sigma t)}e^{\ell' y}] \tag{5-59d}$$

Taking the real part of (5-59a),

$$\eta = he^{\ell' y} \cos(kx - \sigma t) \tag{5-60a}$$

of (5-59b),

$$u = Ae^{\ell' y} \cos(kx - \sigma t) \tag{5-60b}$$

of (5-59c),

$$v = B'e^{\ell' y} \sin(kx - \sigma t) \tag{5-60c}$$

where B' is defined in (5-58b). Finally, the real part of (5-59d) is

$$w = C'e^{\ell' y} \sin(kx - \sigma t) \tag{5-60d}$$

where $C' \equiv \sigma(z + h)$. These equations are said to describe a *nonhomogeneous plane wave* since, from (5-59a), the amplitude of the wave varies exponentially in the transverse (in this case, y) direction.

Figure 5-18, a schematic representation of the dispersion relation of (5-57) when written as

$$\left|\frac{\sigma}{f}\right| = \left[1 + (k^2 - \ell'^2)\frac{gh}{f^2}\right]^{1/2} \tag{5-61}$$

shows the dependence of $|\sigma|$ on k for various values of ℓ'^2. The isoline $\ell' = 0$ represents a plane Sverdrup wave propagating in the $+x$ direction [compare (5-34) and (5-36) with (5-57) and (5-60a)]; the shaded portion where $\ell'^2 < 0$ (this is possible if ℓ' is in itself imaginary, which is the same as saying that ℓ is real) corresponds to plane Sverdrup waves having both an x and a y component in their direction of propagation as in (5-28) and (5-31a). We see, then, that for a given value of k, nonhomogeneous plane waves can exist at frequencies less than the plane Sverdrup wave frequency as given by (5-36).

It is apparent that equations (5-57) through (5-60) represent the four possible nonhomogeneous wave cases of waves propagating in the $+x$ direction ($\sigma > 0$) with the amplitude decreasing ($\ell' < 0$) to the left or increasing ($\ell' > 0$) to the left of the direction of propagation, and alternatively, of waves

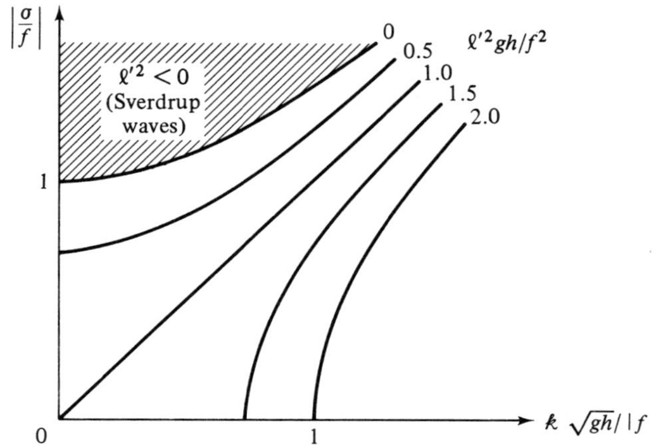

Fig. 5-18. Schematic plot of the dispersion relation, (5-61), for inhomogeneous plane waves. Adapted from "Ocean Tides and Related Waves," *Mathematical Problems in the Geophysical Sciences*, Lectures in Applied Mathematics, Vol. 14, p. 283, by permission of the Amer. Math. Soc. © by the Amer. Math. Soc.

propagating in the $-x$ direction ($\sigma < 0$) with the amplitude decreasing or increasing to the right of the direction of propagation. Nothing new is encountered in examining both forward and backward moving waves, so we will consider only the former. Following Platzman, we designate those forward-moving waves with amplitude decreasing to the left as *left-decreasing waves* and those with amplitude increasing to the left as *right decreasing waves*. Profiles of both types are illustrated in Figure 5-19. It is instructive to make a dynamic comparison of these two wave types through an examination of the balance of forces in the transverse direction as given by (5-6),

$$\frac{\partial v}{\partial t} = -g\frac{\partial \eta}{\partial y} - fu \tag{5-6}$$

For a wave in perfect geostrophic balance in the transverse direction, the transverse pressure gradient force due to the transverse surface slope will be balanced by the transverse Coriolis force, that is, $(-g\,\partial\eta/\partial y)/fu = 1$, and $\partial v/\partial t = 0$. In the plane Sverdrup wave, $\partial\eta/\partial y = 0$, this ratio being zero regardless of the wave frequency σ, since η is not a function of y. On the other hand, for the inhomogeneous plane wave we obtain, from (5-60a) and (5-60b),

$$\frac{-g\,\partial\eta/\partial y}{fu} = -\left(\frac{gh}{f}\right)\left(\frac{\ell'}{A}\right) \tag{5-62}$$

whose value does, in fact, depend upon the frequency σ. Figure 5-20, taken from Platzman, illustrates this frequency dependence for a fixed value of

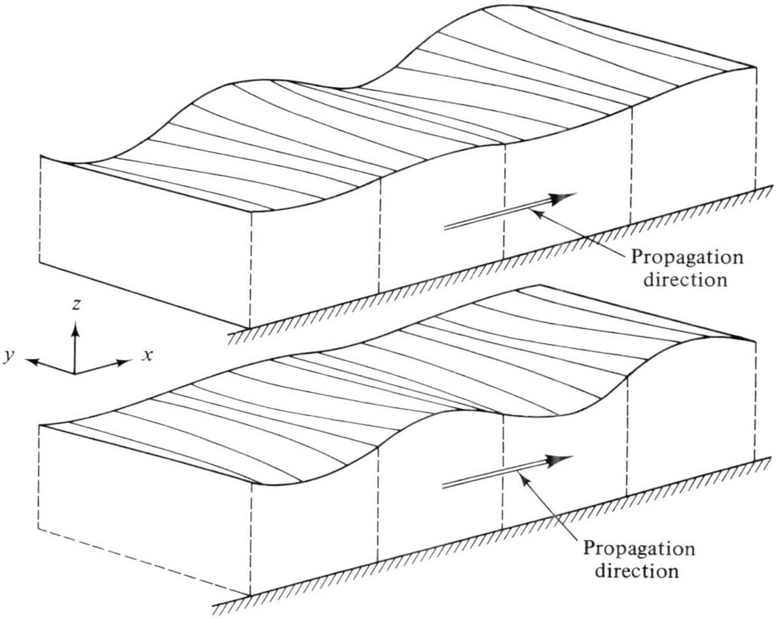

Fig. 5-19. Profiles of right-decreasing (top) and left-decreasing (bottom) inhomogeneous plane waves. Adapted from "Ocean Tides and Related Waves," Mathematical Problems in the Geophysical Sciences, Lectures in Applied Mathematics, Vol. 14, p. 283, by permission of the Amer. Math. Soc. © 1971 by the Amer. Math. Soc.

$k\sqrt{gh} \equiv (a)(f)$, where a is a numerical constant, and where for simplicity we have assumed that $f > 0$ (Northern Hemisphere). The upper branch of the curve corresponds to $\ell' < 0$ ($f\ell' < 0$ in the Northern Hemisphere) and the lower branch to $\ell' > 0$ ($f\ell' > 0$ in the Northern Hemisphere), which Platzman refers to as *Kelvin waves* and *Proudman waves*, respectively. We will examine these two branches separately.

The Kelvin Wave. With $\ell' < 0$ and $f > 0$, the Kelvin wave is left-decreasing (in the Southern Hemisphere $f < 0$ and $\ell' > 0$ results in a right-decreasing wave). At point 1, $(gh\ell')/fA = 0$, giving that $\ell' = 0$ and, from (5-57), that the maximum Kelvin wave frequency is $\sigma = (f^2 + k^2gh)^{1/2}$. But with $\ell' = 0$, equations (5-57) and (5-60) are identical to those of (5-34) and (5-36); we conclude that point 1 represents a plane Sverdrup wave progressing in the $+x$ direction. Moving on the curve in a clockwise direction from point 1, ℓ' becomes increasingly negative, the index $-gh\ell'/fA$ becomes increasing positive, and σ decreases. The transverse slope is said to be *quasigeostrophic* at all

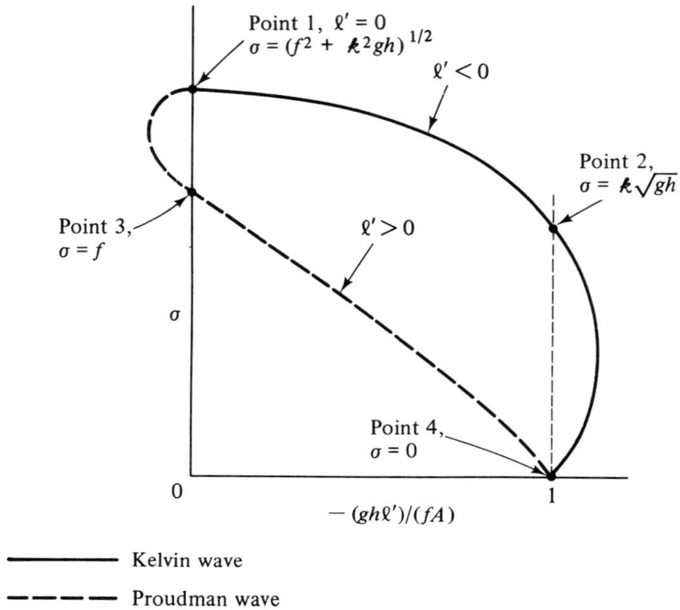

Fig. 5-20. Index of geostrophic balance in the transverse slope of the wave profile of (5-62) versus frequency, for the Northen Hemisphere, $k\sqrt{gh} = af$. Adapted from "Ocean Tides and Related Waves," *Mathematical Problems in the Geophysical Sciences*, Lectures in Applied Mathematics, Vol. 14, p. 284, by permission of the Amer. Math. Soc. © 1971 by the Amer. Math. Soc.

frequencies for which the index is positive but not equal to 1. As σ decreases, a value is eventually reached such that $\sigma = k\sqrt{gh}$, point 2 on the diagram. At point 2, equation (5-57) tells us that $\ell' = -f/\sqrt{gh}$ and the index is

$$\frac{-gh\ell'}{fA} = \frac{(-gh)(-f/\sqrt{gh})}{(f)\left(\dfrac{k^2\sqrt{gh} - f^2/\sqrt{gh}}{k^2 - f^2/gh}\right)} \tag{5-63}$$

indicating that the balance is *exactly geostrophic*. At this point, from (5-58), $A = \sqrt{gh}$, $B' = 0$, and the wave description from (5-60) is given by

$$\eta = he^{-fy/\sqrt{gh}} \cos (kx - \sigma t) \tag{5-64a}$$

$$u = \sqrt{gh}\, e^{-fy/\sqrt{gh}} \cos (kx - \sigma t) \tag{5-64b}$$

$$v = 0 \tag{5-64c}$$

$$w = \sigma(z + h)e^{-fy/\sqrt{gh}} \sin (kx - \sigma t) \tag{5-64d}$$

where $k = \sigma/\sqrt{gh}$. This particular or special type of Kelvin wave, termed the *prototype Kelvin wave* by Platzman, was first discussed by Thomson

(1879) and for this reason has been referred to in most references as the Kelvin wave. We see here, however, that it is a special case of a more general solution. Its particular utilization has been that of a boundary wave in long canals of constant depth. Our solution shows that this prototype is vertically polarized ($v = 0$) and that the earth's rotation exerts an influence on the displacement η and the fluid velocity in the x direction (u), but not on the nondispersive propagation speed $c = \sqrt{gh}$.

The Proudman Wave. Between points 1 and 3 on the Proudman branch of the curve of Figure 5-20, the index of (5-62) is negative, ℓ' is positive (increasing from point 1 to point 2), and the transverse slope is said to be *ageostrophic*. In the limit as σ approaches f, k^2 approaches ℓ'^2 [from (5-57)], A and B' [equations (5-58)] are infinite, B'/A approaches 1, and the index of $-(gh/f)(\ell'/A)$ approaches zero. Dividing equations (5-60) by A and taking the limit as $\sigma \longrightarrow f$, we obtain at this singular frequency,

$$\eta = 0 \tag{5-65a}$$

$$u = e^{\ell' y} \cos(kx - ft) \tag{5-65b}$$

$$v = e^{\ell' y} \sin(kx - ft) \tag{5-65c}$$

$$w = 0$$

where $\ell' = kf$ and $k > 0$ are arbitrary. The equations represent a horizontally polarized wave with no disturbance at the free surface. Proceeding to determine the Lagrangian description of the particle motion, as was done for the plane Sverdrup wave in equations (5-39) through (5-43) and Figure 5-7, we have that

$$\frac{d\xi_1}{dt} = e^{\ell' y} \cos(kx_1 - ft)$$

$$\frac{d\zeta_1}{dt} = e^{\ell' y} \sin(kx_1 - ft)$$

where $x_1 + \xi_1$ has again been approximated by x_1. Integrating, dropping the subscripts, squaring, adding, and rearranging, we get a result identical to (2-264); that is, the flow is inertial and the orbital paths are circles. Continuing along the Proudman wave path, at point 4, $\sigma = 0$, $\ell' = (k^2 + f^2/gh)^{1/2}$, $A < 0$, and the index equals 1. In the interval between points 3 and 4, $k < \ell'$ and $A < 0$, the index is positive, and the transverse slope is once again quasigeostrophic.

To summarize, the Kelvin wave tends to be vertically polarized and is exactly so at $\sigma = k\sqrt{gh}$. From point 1 to point 2 to point 4, the particles move forward under a crest, the orbit tips up from 1 to 2, being vertical at this last point, and continues tipping in the same direction as one proceeds to point 4. In the Northern Hemisphere the Kelvin wave is left-decreasing (right-decreasing in the Southern Hemisphere). The Proudman wave, on the other hand, tends to be horizontally polarized and is exactly so at $\sigma = f$. From point 1 to point 3, the particles move forward under a crest, while from point 3 to point 4 they move backward under a crest. From point 1 to point 3 to point 4, the orbit continues to tip in the same direction as it does from point 1 to point 3. In the Northern Hemisphere the Proudman wave is right-decreasing (left-decreasing in the Southern Hemisphere).

Platzman (1971) cites several examples of attempts to relate Poincaré and Kelvin waves to the tides of the ocean, only one of which we will examine. Figure 5-21 is a tidal chart showing the M_2 tide in the English Channel (the M_2 designates the main and lunar semidiurnal tidal constituent which has a period of 12 lunar hours or 12.42 solar hours). Cotidal lines (lines passing through all points where high water occurs at the same time) indicate a wave moving up the channel from west to east in the form of a progressive wave. The corange lines (lines passing through points of equal tidal range) show a larger tidal range along the French coast than along the English coast. Platzman assumed that the channel tide could be represented by a vertically polarized Kelvin wave, the prototype Kelvin wave of equations (5-64). To test this assumption, he made a comparison between the ratio $(\eta_F)_{\max}/(\eta_E)_{\max}$ (where the subscripts F and E refer to the French and English coasts, respectively, and max to the maximum value) as obtained from the tidal chart, with the estimate of this ratio obtained for the assumed Kelvin wave. The coordinate system is placed such that the x axis is at midchannel pointing east.

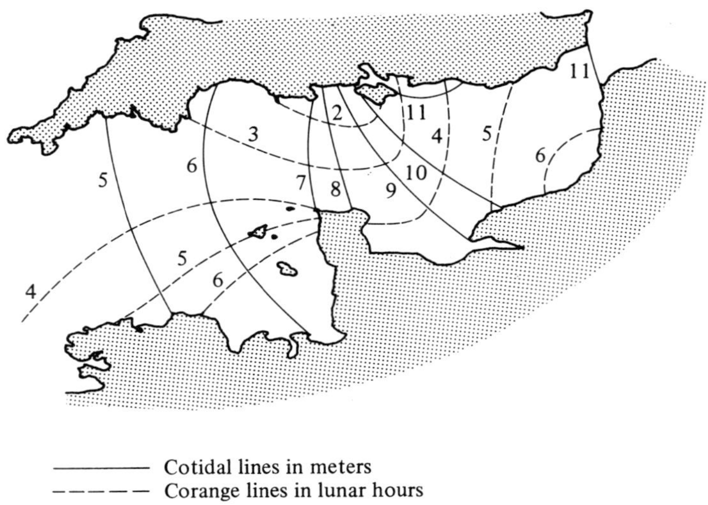

Fig. 5-21. Cotidal and corange lines for the M_2 tidal constituent in the English Channel. Proudman, *Dynamical Oceanography*, New York: Dover, 1953.

From (5-64a), we have that

$$\frac{2(\eta_F)_{\max}}{2(\eta_E)_{\max}} = \frac{e^{fY/2\sqrt{gh}}}{e^{-fY/2\sqrt{gh}}} = e^{fY/\sqrt{gh}} \tag{5-66}$$

where Y is the width of the channel. Estimating that $h = 50$ to 80 m, $Y = 100$ to 150 km, and $f = 1.12 \times 10^{-4}\ \text{sec}^{-1}$ (latitude of 50°N), $fY/\sqrt{gh}$ is about 0.5 to 0.6 and the ratio of (5-66) is about 1.6 to 1.8. This is in rough agreement with the values of Figure 5-21; for example, at the western entrance we obtain a tidal range of 6 m on the French coast and 3 m on the English coast.

Finally, the distance between the cotidal lines of the fifth and eleventh hours as scaled from Figure 5-21 is approximately 460 km. This interval of 6 hr is half the tidal period, and the tide should have a wave length of approximately 920 km. We have seen that for the prototype Kelvin wave $k = \sigma/\sqrt{gh}$. For the M_2 tide $\sigma = (2\pi/12.42)\ \text{hr}^{-1}$, and taking h as approximately 50 m we get a computed wave length $2\pi/k = 950$ km. This agreement is rather impressive considering the very complicated nature of the tidal process.

5-7 Plane Rossby Waves

Rossby (1939, 1940, 1945) laid the foundation for modern dynamic meteorology in a series of papers highlighting the importance of a simple wave equation. He believed that this equation would describe the movement of observed large-scale disturbances in the pressure field of the atmosphere. The motions of interest are slow periodic oscillations of a period of several days that can take place in a thin layer of fluid on the surface of a rotating sphere. These motions, termed *Rossby waves* or *planetary waves*, illustrate some unusual features of wave motion and are of interest in oceanography, since "the response of the oceans to a variable wind stress applied at the surface may largely appear in this type of motion" (Longuet-Higgins, 1964).

Previously, we discussed various kinds of long gravity waves that have been modified to a certain extent by the rotation of the earth and all of which are first-class waves. Rossby waves, on the other hand, constitute motions of the second class in that if the rate of rotation vanished, they would cease to be periodic in time and would instead become steady currents. In these oscillations the surface elevation is in fact negligible. Rossby (1939) first pointed out that the northward or southward displacement of fluid over the surface of the earth results in a net restoring force due to the change in the vertical component of angular velocity. In our introductory examination we follow Rossby in using the β-plane approximation, wherein it is assumed that the spherical surface of the earth can be locally approximated by a plane surface and that the local Coriolis parameter varies with latitude such that its northward derivative is a constant. The validity of this approximation has been examined by other investigators (e.g., Longuet-Higgins, 1964; Veronis, 1963; Phillips, 1963) and will not here concern us.

Consider, then, that the motion is governed by the equations of motion in the form of (5-1) and (5-2),

$$\frac{\partial u}{\partial t} - fv = -\frac{1}{\rho}\frac{\partial p}{\partial x} \tag{5-1}$$

$$\frac{\partial v}{\partial t} + fu = -\frac{1}{\rho}\frac{\partial p}{\partial y} \tag{5-2}$$

and that it is horizontally nondivergent, so that

$$\frac{\partial u}{\partial x} + \frac{\partial v}{\partial y} = 0$$

Cross-differentiation of the equations of motion and subtraction of the two resulting equations eliminates the pressure terms to give

$$\frac{\partial}{\partial t}\left(\frac{\partial v}{\partial x} - \frac{\partial u}{\partial y}\right) + \beta v = 0 \tag{5-67}$$

Introducing the stream function ψ by means of the definitions

$$u \equiv -\frac{\partial \psi}{\partial y} \qquad v \equiv \frac{\partial \psi}{\partial x}$$

yields a vorticity equation

$$\nabla_H^2 \frac{\partial \psi}{\partial t} + \beta \frac{\partial \psi}{\partial x} = 0 \tag{5-68}$$

where ∇_H^2 is the horizontal Laplacian operator. Let us try the solution

$$\psi = e^{i(kx + \ell y - \sigma t)} \tag{5-69}$$

where, as before, k and ℓ are wave numbers in the x and y directions, the wave length of the oscillation being $2\pi/K$ with $K^2 \equiv k^2 + \ell^2$. If we denote α as the angle between the x axis and the wave number vector $\mathbf{K}$, then from Figure 5-22,

$$k = K\cos\alpha \qquad \ell = K\sin\alpha \tag{5-70}$$

Substitution of (5-69) into (5-68) immediately gives us the dispersion relation

$$\sigma = -\frac{\beta k}{K^2} \tag{5-71}$$

or since σ/k represents the speed of a crest along the x axis,

$$\frac{\sigma}{k} = -\frac{\beta}{K^2} \tag{5-72}$$

This equation states that the entire wave pattern moves westward with a speed σ/k, depending only on the wave number K and not in any way on the wave crest orientation α.

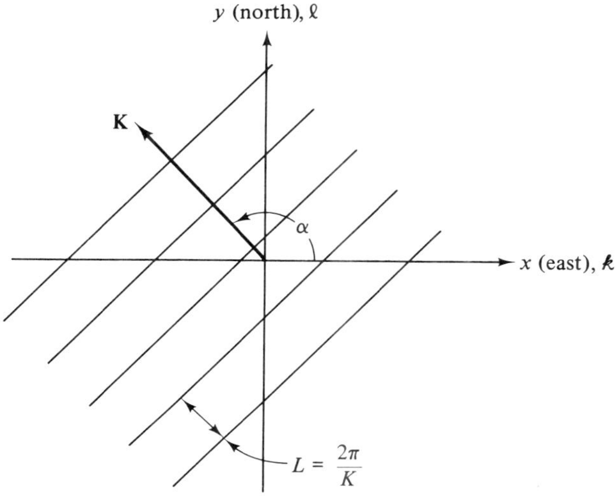

Fig. 5-22. Definition sketch for the wave number of a plane Rossby wave.

Equation (5-71) may be written as

$$k^2 + \ell^2 + \frac{\beta}{\sigma} k = 0$$

and upon adding $(\beta/2\sigma)^2$ to each side factored to

$$\left(k + \frac{\beta}{2\sigma}\right)^2 + \ell^2 = \left(\frac{\beta}{2\sigma}\right)^2 \tag{5-73}$$

which for fixed values of the frequency σ is the equation of a circle of radius $\beta/2\sigma$ in wave number space which passes through the origin and has its center $k = -\beta/2\sigma$, $\ell = 0$ (Figure 5-23). Notice on the diagram that k is always less than or equal to zero and that the phase velocity **c** has always a westward component. We can show this analytically from the definition of the phase velocity,

$$\mathbf{c} \equiv c_x \mathbf{i} + c_y \mathbf{j} = \frac{\sigma \mathbf{K}}{K^2} \tag{5-74}$$

which leads to

$$c_x = \frac{\sigma k}{k^2 + \ell^2} \qquad c_y = \frac{\sigma \ell}{k^2 + \ell^2} \tag{5-75}$$

Substituting from (5-72) yields

$$c_x = -\frac{\beta k^2}{(k^2 + \ell^2)^2} \leq 0 \tag{5-76a}$$

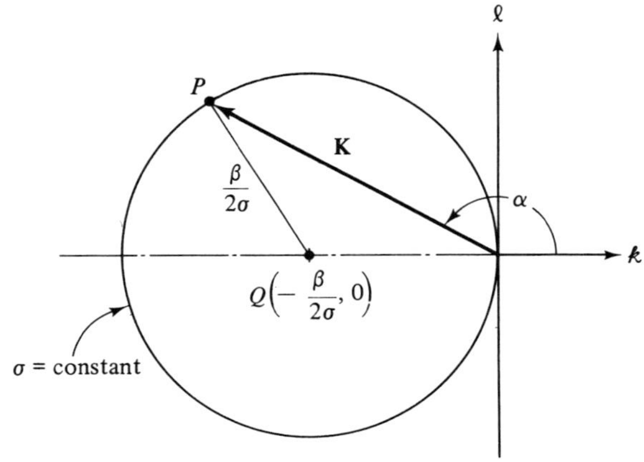

Fig. 5-23. Locus of (5-73) for σ fixed.

since $\beta \geq 0$ and

$$c_y = -\frac{\beta k \ell}{(k^2 + \ell^2)^2} \tag{5-76b}$$

from which we see in view of (5-76a) that c_y always has the same sign as ℓ.

Consider now the group velocity defined as

$$\mathbf{c}_g \equiv c_{g_x}\mathbf{i} + c_{g_y}\mathbf{j} \equiv \frac{\partial \sigma}{\partial k}\mathbf{i} + \frac{\partial \sigma}{\partial \ell}\mathbf{j} \tag{5-77}$$

From (5-71),

$$\begin{aligned} c_{g_x} \equiv \frac{\partial \sigma}{\partial k} &= -\frac{\beta}{K^2} + \frac{2\beta k^2}{K^4} = \frac{\beta(k^2 - \ell^2)}{(k^2 + \ell^2)^2} \\ &= \frac{\beta(k^2 + \ell^2)(\cos^2 \alpha - \sin^2 \alpha)}{(k^2 + \ell^2)^2} = \frac{\beta}{K^2}\cos 2\alpha \end{aligned} \tag{5-78a}$$

and

$$c_{g_y} \equiv \frac{\partial \sigma}{\partial \ell} = \frac{2\beta k \ell}{K^4} = \frac{\beta}{K^2}\sin 2\alpha \tag{5-78b}$$

It is apparent that the group velocity vector makes an angle 2α with the x axis and has a magnitude of β/K^2, Further, in wave number space the group velocity must in general be normal to the circular locus, or parallel to the radius PQ of Figure 5-22. To show this, consider a change in the vector wave number from $\mathbf{K}$ to $\mathbf{K} + d\mathbf{K}$ while keeping the frequency σ constant (Figure 5-24). Since $\sigma = \sigma(k, \ell)$, along the locus of (5-73)

$$d\sigma = \frac{\partial \sigma}{\partial k}\,dk + \frac{\partial \sigma}{\partial \ell}\,d\ell = 0 \tag{5-79}$$

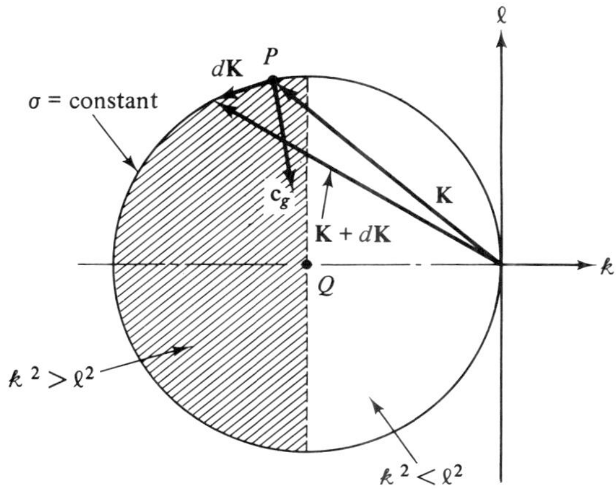

Fig. 5-24. Group velocity vector in wave number space.

But

$$d\mathbf{K} = (dk)\mathbf{i} + (d\ell)\mathbf{j}$$

and thus, from (5-77) and (5-79),

$$d\mathbf{K} \cdot \mathbf{c}_g = \frac{\partial \sigma}{\partial k} dk + \frac{\partial \sigma}{d\ell} d\ell = 0$$

which tells us that $d\mathbf{K}$ and $\mathbf{c}_g$ are normal to each other and that $\mathbf{c}_g$ is therefore parallel to the radius of the circular locus. Also, if $k^2 > \ell^2$, the group velocity, from (5-78a), always has an eastward component ($c_{g_x} > 0$), while for $k^2 < \ell^2$ it will have a westward component ($c_{g_x} < 0$). Thus, the group velocity is always directed radially inward toward the center of the circle, as shown in the figure.

The average energy flux **P** defined as

$$\mathbf{P} \equiv \overline{P_x}\mathbf{i} + \overline{P_y}\mathbf{j} \tag{5-80}$$

where $\overline{P_x}$ and $\overline{P_y}$ are given by (5-20) and (5-21), is easily found using the group velocity and the average mechanical energy density. This last quantity is the sum of the average potential energy density, $\overline{E_P}$, as given by (5-18), and the average kinetic energy density, $\overline{E_K}$, as given by (5-19). Since the free surface elevation is negligible for Rossby waves, $\overline{E_P} = 0$ and the average mechanical energy density is given in terms of the stream function ψ as

$$\bar{E} = \overline{E_K} = \tfrac{1}{2}\rho h\overline{\left[\left(\frac{\partial \psi}{\partial x}\right)^2 + \left(\frac{\partial \psi}{\partial y}\right)^2\right]} \tag{5-81}$$

Substituting from (5-69) after taking the real part,

$$\bar{E} = \tfrac{1}{2}\,\rho h(k^2 + \ell^2)\,\overline{\cos^2 (kx + \ell y - \sigma t)}$$

or, finally,

$$\bar{E} = \tfrac{1}{4}\rho h(k^2 + \ell^2) \tag{5-82}$$

$$\bar{E} = \tfrac{1}{4}\rho h K^2 \tag{5-83}$$

Applying the concept, as in the case of Sverdrup waves, that the ratio of the energy flux to the mechanical energy density gives the group velocity,

$$\begin{aligned}\mathbf{P} &\equiv \bar{E}\mathbf{c}_g \\ &= \left(\frac{1}{4}\rho h K^2\right)\left[\frac{\beta(k^2 - \ell^2)}{K^4}\mathbf{i} + \frac{2\beta k\ell}{K^4}\mathbf{j}\right] \\ &= \frac{1}{4}\frac{\rho\beta h}{K^2}[(k^2 - \ell^2)\mathbf{i} + 2k\ell\mathbf{j}] \\ &= -\frac{1}{4}\frac{\rho\sigma h}{k}[(k^2 - \ell^2)\mathbf{i} + 2k\ell\mathbf{j}]\end{aligned} \tag{5-84}$$

having used (5-71) and (5-78a, b). The magnitude of the average energy flux,

$$\begin{aligned}|\mathbf{P}| &= \bar{E}|\mathbf{c}_g| \\ &= (\tfrac{1}{4}\rho h K^2)\left(\frac{\beta}{K^2}\right) \\ &= \tfrac{1}{4}\rho\beta h\end{aligned} \tag{5-85}$$

is seen to be independent of the wave numbers k and ℓ.

As pointed out by Longuet-Higgins (1964), the group velocity may also be considered as the sum of the two velocities,

$$\mathbf{c}_g \equiv \frac{\sigma}{k}\mathbf{i} + \mathbf{c}'_g = -\frac{\beta}{K^2}\mathbf{i} + \mathbf{c}'_g$$

where σ/k is the speed of a crest along the x axis as in (5-72). Solving for $\mathbf{c}'_g$,

$$\begin{aligned}\mathbf{c}'_g &= \mathbf{c}_g + \frac{\beta}{K^2}\mathbf{i} \\ &= \frac{2\beta k^2}{K^4}\mathbf{i} + \frac{2\beta k\ell}{K^4}\mathbf{j} \\ &= \frac{2\beta k}{K^4}(k\mathbf{i} + \ell\mathbf{j}) \\ &= \frac{2\beta k}{K^4}\mathbf{K} \\ &= -\frac{2\beta|k|}{K^4}\mathbf{K}\end{aligned}$$

which is a vector parallel to the wave number vector $\mathbf{K}$ but opposite in direction. Therefore, to an observer moving westward with the characteristic velocity $(\sigma/k)\mathbf{i}$, the wave crests would appear to be stationary and the group velocity would appear to be perpendicular to the wave crests and directed so as to have an easterly component.

Bibliography

BAKER, JR., B. B., W. R. DEEBEL, and R. D. GEISENDERFER, ed. (1966): *Glossary of Oceanographic Terms*, 2nd ed. U.S. Naval Oceanographic Office (U.S. Government Printing Office), Washington, D.C., 204 pp.

COX, R. A., and N. D. SMITH (1959): The specific heat of sea water. *Proc. Roy. Soc.*, A, Vol. 252, pp. 51–62 (London).

DEFANT, A. (1929): Stabile Lagerung ozeanischer Wasser Körper und dazu gehörige Stromsysteme. *Veroff. Inst. f. Meereskunde*, Univ. Berlin, N. F., Reihe A, No. 19 (Berlin).

——— (1936): Ausbreitungs und Vermischungsvoränge im Antarktischen Bodenstrom und in Subantarktischen Zwischenwasser. *Deutsche Atl. Exped. "Meteor" 1925–27, Wiss. Erg.*, Vol. 6, Teil 2, Liefg. 2, pp. 55–96 (Berlin).

——— (1961): *Physical Oceanography*, Vol. I, Pergamon Press, London, 729 pp.

DURST, C. S. (1924): The relationship between current and wind. *Quart. J.R. Met. Soc.*, Vol. 50, p. 113 (London).

ECKART, C. (1947): Excerpts of notes on Scripps Institute of Oceanography Course 217. Unpublished, 72 pp. (La Jolla).

FOFONOFF, N. P. (1962): Physical properties of sea water. *The Sea*, Vol. 1, pp. 3–30. Edited by M. N. Hill. Interscience, New York.

FOFONOFF, N. P., and C. FROESE (1958): Program for oceanographic computations and data processing on the electronic digital computer ALWAC III-E. PSW-I Programs for properties of sea water. *Manuscript Rep. Ser. No. 27, Fish. Res. Bd. Can.*, Pac. Oceanogr. Group. Unpublished (Nanaimo, B.C.).

HOUGH, S. S. (1897): On the application of harmonic analysis to the dynamical theory of tides. Part I. On Laplace's "oscillations of the first species," and on the dynamics of ocean currents. *Phil. Trans. Roy. Soc.*, A, Vol. 189, pp. 201–257 (London).

——— (1898): On the application of harmonic analysis to the dynamical theory of the tides. Part II. On the general integration of Laplace's dynamical equations. *Phil. Trans. Roy. Soc.*, A, Vol. 191, pp. 139–185 (London).

ISELIN, C. O'D. (1936): A study of the circulation of the western North Atlantic. *Papers Phys. Oceanogr. and Meteorol.*, Vol. 4, No. 4, 101 pp. (Woods Hole and Cambridge, Mass.).

JACOBSEN, J. P. (1913): Beitrag zur Hydrographie der Dänischen Gewässer. *Medd. Komm. Havunders.*, Ser. Hydr., Vol. 2, No. 2, p. 94 (Copenhagen).

——— (1927): Eine graphische Methode zur Bestimmung des Vermischungskoeffizienten im Meer. *Gerl. Beitr. z. Geophysik*, Vol. 16, p. 404 (Leipzig).

KAMENKOVICH, V. M. (1977): *Fundamentals of Ocean Dynamics*. Elsevier, Amsterdam, 249 pp.

KESTIN, J. (1966): *A Course in Thermodynamics*. Ginn-Blaisdell, Waltham, Mass., 615 pp.

KINSMAN, B. (1965): *Wind Waves: Their Generation and Propagation on the Ocean Surface*. Prentice-Hall, Englewood Cliffs, N.J., 676 pp.

LAPLACE, P. S. (1755): Recherches sur plusieurs points du systeme du monde. *Mem. Acad. Roy. Soc.*, Vol. 88, pp. 75–182. See also (1776): *ibid.*, Vol. 89, pp. 177–267.

LONGUET-HIGGINS, M. S. (1964): Planetary waves on a rotating sphere. *Proc. Roy. Soc.*, A, Vol. 279, pp. 446–473 (London).

MONTGOMERY, R. B. (1939): Ein Versuch, den vertikalen und seitlichen Austausch in der Tiefe der Sprungschicht im äquatorialen Atlantischen Ozean zu bestimmen. *Ann. d. Hydr. u. Marit. Meteorol.*, Vol. 67, p. 242 (Berlin).

MONTGOMERY, R. B., and E. PALMÉN (1940): Contribution to the question of the Equatorial Counter Current. *J. Mar. Res.*, Vol. 3, pp. 112–133.

MUNK, W. H. (1950): On the wind-driven ocean circulation. *J. Meteor.*, Vol. 7, No. 2, pp. 79–93.

NEUMANN, G., and W. J. PIERSON, JR. (1966): *Principles of Physical Oceanography*. Prentice-Hall, Englewood Cliffs, N.J., 545 pp.

PHILLIPS, O. M. (1963): Energy transfer in rotating fluids by reflection of internal waves. *Phys. Fluids*, Vol. 6, pp. 513–520.

PLATZMAN, G. W. (1971): Ocean tides and related waves. *Mathematical Problems in the Geophysical Sciences*, 2: *Inverse Problems, Dynamo Theory, and Tides*, Lectures in Applied Mathematics, Vol. 14, pp. 239–291. Edited by W. H. Reid. Amer. Math. Soc., Providence, R.I.

PROUDMAN, J. (1953): *Dynamical Oceanography*. Metheun, London. Dover, New York, 409 pp.

REICHARDT, H. (1938): Messungen turbulenter Schwankungen. *Naturwissenschaften*, 404.

REID, R. O. (1948): The equatorial currents of the Pacific as maintained by the stress of the wind. *J. Mar. Res.*, Vol. 7, No. 2, pp. 74–99.

——— (1954): Part one, fundamental principles of hydro-thermo-dynamics. Unpublished, 255 pp. (College Station, Texas).

REYNOLDS, O. (1895): On the dynamical theory of incompressible viscous fluids and the determination of the criterion. *Phil. Trans. Roy. Soc.*, A, Vol. 186, pp. 123–164 (London).

ROSSBY, C.-G. (1939): Relation between variations in the intensity of the zonal circulation of the atmosphere and the displacement of the semipermanent centers of action. *J. Mar. Res.*, Vol. 2, pp. 38–55.

——— (1940): Planetary flow patterns in the atmosphere. *Quar. J. Roy. Meteor. Soc.*, Suppl. 66, pp. 68–87.

——— (1945): On propagation of frequencies and energy in certain types of oceanic and atmospheric waves. *J. Meteor.*, Vol. 2, pp. 187–204.

SCHMIDT, W. (1917): Wirkungen der ungeordneten Bewegungen im Wasser der Meere und Seen. *Ann. d. Hydr. u. Marit. Meteorol.*, Vol. 45, pp. 367–381, 431–445 (Berlin).

STOCKMAN, W. B. (1945): Horizontal circulation of water in the sea, as caused by non-uniform velocity distribution in a shoreward wind across its movement. *C. R. Acad. Sci., U.S.S.R.*, Vol. 49, No. 2, pp. 102–106 (Doklady).

STOMMEL, H. (1948): The westward intensification of wind-driven ocean currents. *Trans. Amer. Geophys. Un.*, Vol. 29, No. 2, pp. 202–206.

——— (1957): A survey of ocean current theory. *Deep Sea Res.*, Vol. 4, No. 3, pp. 149–184.

——— (1965): *The Gulf Stream*, 2nd ed. Univ. of Calif. Press, Berkeley, Calif., 248 pp.

SVERDRUP, H. U. (1939): Lateral mixing in the deep water of the South Atlantic Ocean. *J. Mar. Res.*, Vol. 2, No. 3, pp. 195–207.

——— (1947): Wind-driven currents in a baroclinic ocean; with application to the equatorial currents of the eastern Pacific. *Proc. Nat. Acad. Sci.*, Vol. 33, No. 11, pp. 318–326 (Washington, D.C.).

SVERDRUP, H. U., M. W. JOHNSON, and R. H. FLEMING (1942): *The Oceans, Their Physics, Chemistry, and General Biology*. Prentice-Hall, Englewood Cliffs, N.J., 1087 pp.

TENNEKES, H., and J. L. LUMLEY (1972): *A First Course in Turbulence*. The MIT Press, Cambridge, Mass., 300 pp.

THOMSON, W. (LORD KELVIN) (1857): On the alterations of temperature accompanying changes of pressure in fluids. *Proc. Roy. Soc.*, Vol. 8, pp. 564–569 (London).

——— (1879): On gravitational oscillations of rotating water. *Proc. R. Soc. Edinburgh*, Vol. 10, pp. 92–100.

THORADE, H. (1914): Die Geschwindigkeit von Triftströmungen und die Ekman'sche Theorie. *Ann. d. Hydr. u. Marit. Meteorol.*, Vol. 42, p. 379 (Berlin).

——— (1931): Strömung und zungenförmige Ausbreitung des Wassers. *Gerl. Beitr. z. Geophysik*, Vol. 34, pp. 57–76 (Leipzig).

VERONIS, G. (1963): On the approximations involved in transforming the equations of motion from a spherical surface to the β-plane. I. Barotropic systems. *J. Mar. Res.*, Vol. 21, No. 2, pp. 110–124. II. Baroclinic systems, *ibid.*, Vol. 21, No. 3, pp. 199–204.

——— (1966): Wind-driven ocean circulation—Part I. Linear theory and perturbation analysis. Part 2. Numerical solutions of the non-linear problem. *Deep-Sea Res.*, Vol. 13, pp. 17–29, 31–55.

——— (1973): Large scale ocean circulation. *Advances in Applied Mechanics*, Vol. 13, pp. 1–92. Edited by Chia-Shun Yih. Academic Press, New York.

WILSON, W. D., and D. BRADLEY (1968): Specific volume of sea water as a function of temperature, pressure and salinity. *Deep Sea Res.*, Vol. 15, No. 3, pp. 355–363.

WÜST, G. (1936): Kuroshio und Golfstrom. *Veröff. Inst. f. Meereskunde*, Univ. Berlin, N. F., Reihe A, No. 29.

Books for Supplementary Reading

Allen, Theodore, Jr., and Richard L. Ditsworth (1972): *Fluid Mechanics.* McGraw-Hill, New York, 415 pp.

Aris, Rutherford (1962): *Vectors, Tensors, and the Basic Equations of Fluid Mechanics.* Prentice-Hall, Englewood Cliffs, N.J., 286 pp.

Batchelor, G. K. (1967): *An Introduction to Fluid Dynamics.* Cambridge University Press, New York, 615 pp.

Bird, R. Byron, Warren E. Stewart, and Edwin N. Lightfoot (1960): *Transport Phenomena.* Wiley, New York, 780 pp.

Brodkey, Robert S. (1967): *The Phenomena of Fluid Motions.* Addison-Wesley, Reading, Mass., 737 pp.

Defant, Albert (1961): *Physical Oceanography,* Vols. I and II. Macmillan, New York, 729 and 598 pp.

Dutton, John A. (1976): *The Ceaseless Wind; An Introduction to the Theory of Atmospheric Motion.* McGraw-Hill, New York, 579 pp.

Fitts, Donald D. (1962): *Nonequilibrium Thermodynamics; A Phenomenological Theory of Irreversible Processes in Fluid Systems.* McGraw-Hill, New York, 173 pp.

Flugge, Wilhelm (1972): *Tensor Analysis and Continuum Mechanics.* Springer-Verlag, New York, 207 pp.

Frederick, Daniel, and Tien Sun Chang (1969): *Continuum Mechanics.* Allyn and Bacon, Boston, 218 pp.

Greenspan, H. P. (1968): *The Theory of Rotating Fluids.* Cambridge University Press, New York, 327 pp.

Haskell, Richard E. (1972): *Introduction to Vectors and Cartesian Tensors; A Programmed Text for Students of Engineering and Science.* Prentice-Hall, Englewood Cliffs, N.J., 365 pp.

Hinze, J. O. (1975): *Turbulence*, 2nd ed. McGraw-Hill, New York, 790 pp.

Holton, James R. (1972): *An Introduction to Dynamic Meteorology*. Academic Press, New York, 319 pp.

Kamenkovich, V. M. (1977): *Fundamentals of Ocean Dynamics*. Elsevier, Amsterdam, 249 pp.

Kestin, Joseph (1966): *A Course in Thermodynamics*. Ginn-Blaisdell, Waltham, Mass., 615 pp.

Kinsman, Blair (1965): *Wind Waves; Their Generation and Propagation on the Ocean Surface*. Prentice-Hall, Englewood Cliffs, N.J., 676 pp.

Lamb, Horace (1945): *Hydrodynamics*. Dover, New York, 738 pp.

Landau, L. D., and E. M. Lifshitz (1959): *Fluid Mechanics*, Vol. 6 of Course of Theoretical Physics. Addison-Wesley, Reading, Mass., 536 pp.

Long, Robert R. (1961): *Mechanics of Solids and Fluids*. Prentice-Hall, Englewood Cliffs, N.J., 156 pp.

Malvern, Lawrence E. (1969): *Introduction to the Mechanics of a Continuous Medium*. Prentice-Hall, Englewood Cliffs, N.J., 713 pp.

Milne-Thomson, L. M. (1960): *Theoretical Hydrodynamics*, 4th ed. Macmillan, New York, 660 pp.

Neumann, Gerhard. (1968): *Ocean Currents*. Elsevier, Amsterdam, 352 pp.

Neumann, Gerhard, and Willard J. Pierson, Jr. (1966): *Principles of Physical Oceanography*. Prentice-Hall, Englewood Cliffs, N.J., 545 pp.

Pedlosky, Joseph (1979): *Geophysical Fluid Dynamics*. Springer-Verlag, New York, 624 pp.

Proudman, J. (1952): *Dynamical Oceanography*. Methuen, London; Dover, New York, 409 pp.

Robertson, James M. (1965): *Hydrodynamics in Theory and Application*. Prentice-Hall, Englewood Cliffs, N.J., 652 pp.

Schlichting, Hermann (1968): *Boundary-Layer Theory*, 6th ed. McGraw-Hill, New York, 747 pp.

Sokolnikoff, I. S. (1956): *Mathematical Theory of Elasticity*, 2nd ed. McGraw-Hill, New York, 476 pp.

Stommel, Henry (1965): *The Gulf Stream; A Physical and Dynamical Description*, 2nd ed. University of California Press, Berkeley, Calif., 248 pp.

Sverdrup, H. U., Martin W. Johnson, and Richard H. Fleming (1942): *The Oceans; Their Physics, Chemistry, and General Biology*. Prentice-Hall, Englewood Cliffs, N.J., 1087 pp.

Tennekes, H., and J. L. Lumley (1972): *A First Course in Turbulence*. MIT Press, Cambridge, Mass., 300 pp.

Author Index

Subject Index